HALF A CENTURY IN THE MILE-HIGH CITY

TIMBER LINE is the story of the Denver POST, and the men who made it.

F. G. BONFILS

was a graduate of West Point, a descendant of Napoleon Bonaparte—and a "speculator" who had to move fast to keep ahead of crooked lotteries that had earned him the nickname "Napoleon of the Corn Fields."

H. H. TAMMEN

was a bartender who showed his own brand of ingenuity when he made a fortune on worthless "peacock ore," selling the chips in what may well have been the first western souvenir shop.

****

"Gene Fowler loves people, particularly the unrighteous, the salty, and the doers of grandly macabre deeds. In this book he does a job of reporting and writing which far surpasses anything he has ever done before." —BOOKS

A COMSTOCK EDITION

TIMBER LINE

Gene Fowler

A Story of Bonfíls and Tammen

A COMSTOCK EDITION

SAUSALITO • CALIFORNIA

SBN 0-89174-007-4-275

First Printing: July 1974
Second Printing: May 1977

Printed in the United States of America

Cover photo: Larimer Street, looking east from 16th Street in 1884.
Courtesy of Library, The State Historical Society of Colorado

COMSTOCK EDITIONS, INC.
3030 Bridgeway, Sausalito, CA 94965

TO
THOMAS MEIGHAN

They are not long, the days of wine and roses:
  Out of a misty dream
Our path emerges for a while, then closes
  Within a dream.

—ERNEST DOWSON

Contents

TIMBER LINE

1

The Amorous Senator

THE SENATOR was the finest animal I ever owned. He was sturdy of back and strong of will. Occasionally he seemed sullen with captivity—I think an enforced celibacy weighed on his mind—but withal, the Senator was an effervescent fellow, unbroken in spirit and possessed of a flair for life.

The Senator was a mouse-colored burro, with one lop-ear which lent his port side a docile quality, entirely misleading, if not actually libelous. The other great ear stood like a member of the Coldstream Guards, forever hearkening to some oracular summons inaudible to humans. I am sure he was clairvoyant, for he could foretell a rattlesnake half a mile away. No doubt he had stemmed from Balaam's beast.

At sunrise, the Senator was as loquacious as any chanticleer, and he sang the dawn in terms of the soul undaunted. Most critics eschew the brayings of a burro, holding that the bestial *bel canto* is unmusical, quite, as the cascading hee-haws roll through glens where the quivering aspens dance in a west wind. But when the Senator cut loose with stentorian treble, I swear there was launched a commanding voice that would have won him top billing at the Metropolitan—at La Scala of Milan, even—had he deigned to become civilized, to bow, to scrape, to barter the brave pleasures of the wilderness for the perfumed ateliers of Park Avenue, his high mountain trail for a pent-house and a salon of dyspeptic fawners. He performed vocally in the Chinese scale of quarter notes, this Bing Crosby of the crags.

Somehow, it seemed the Senator was re-born each morning. By noon he had attained a sort of middle-age, and forthwith began to act rather foolishly and sophomoric. It was hard to curb him from a beckoning mesa, where trollop-eyed Jennies were waiting like spiders for just such a blade as the Senator.

Then, at nightfall—his one upright ear giving him a unicorn profile against the moon—the Senator seemed to be smelling Death. But he did not cringe. He stood statuesquely and would have presented a kingly mien, were it not for the pendulous ear. This blemish robbed him of majesty, and perhaps imbued a twilight tinge of inferiority. Artists unwittingly have revealed this quality in crippled rulers by their very effort to gloss over deformity, emphasizing, rather than concealing in stone or canvas, the regal funks.

Thus, the boomerangs of flattery so frequently have resulted in non-payment of royal bills, imprisonment in dungeons, market-place flagellations, disgrace and banishment. All of which does not prove that monarchs are stupidly ungrateful; rather, it defines the artist as a traitor to his craft whenever he counterfeits a rose where the wart or wen has priority.

I, too, endeavored to correct a noble infirmity, the Senator's auricular lassitude, by application of birch-splints. He roundly booted me into a cold and shallow creek for my pains. The Senator wanted no pity, and of pity he gave none. A gentleman, and my first tutor.

He stood still enough—outwardly—as the night stars came, but his hackamore was askew with straining, his tether rope taut with yearning.

The Senator had legs of the Queen Anne period of furniture, and his feet were as dainty as Cinderella's—albeit more lethal. He had porphyry eyes that looked wistfully toward the far mesa, where the wild burros herded. When distant throatings of the wanton Jennies rode the wind with nymph-like invitations, the Senator would mumble love ditties. His body was Prometheus; his soul, Orpheus.

I did not quite understand him . . . then.

I was nine years old when I assumed ownership of the

Senator. I believe he was six or upwards; but no one knew for certain—nor cared, either, since the day the Senator almost amputated the exploring hand of Tom Aldrich, who had said such matters could be determined beyond dispute by an examination of equine teeth.

Aside from a dose of the Senator's heels the time I played plastic surgeon on his flaccid ear, he and I got along together. The Senator had served in Axel Carr's small pack-train, carrying crates of dynamite from Empire, thirty miles away, to Red Mountain. He was named in honor of Senator ("Silver Dollar") Tabor, who had grubstaked Carr, as well as my grandpa and a hundred other prospectors in Leadville days. During a laborious era of powder-toting, the Senator had *two* upright ears, and had sampled the joys of fatherhood.

One autumn evening Carr's cabin went to Heaven with a grand roar, and Carr with it. Fragments of Carr were found every few minutes after the mysterious detonation—a leg, as I recall it, being the largest recoverable part of the confetti-like decedent. It was hanging, in a weird yuletide fashion, from the top branch of a Douglas fir tree.

The Senator had escaped whole—but his left ear had been slapped down by the blast—and I took him over in fee simple.

How glad the days! How sweet the smell of the pines, and how caressing the wind against the smooth cheeks of nine! We would ride along the old Mormon Road, where Brigham Young, Joseph Smith and other whiskered saints had trod during the long-ago pilgrimage to their Utah Canaan. I would sit astride the Senator—well back on his maltese rump—a Sir Launcelot, listening to the Merlin-tones of the great winds. There were voices, prophecies, soothsayings that only a child might understand—men have forfeited the right to dwell in enchanted places.

And often I looked at timber line, some three thousand feet above our remote valley. Timber line, above which no trees grow. From a distance, timber line is a strangely level hedge. The peaks rise baldly, a congregation of tonsured monks. Red Mountain was a fat, jovial fellow with a pink sandstone pate—a Franz Hals' portrait, a Falstaff of the range. To the east of Red Mountain lay the mesa, where the bawdy Jennies kicked up their heels at convention and domesticity.

At timber line, once you are there, a measureless rug of white sheep's wool quite frequently is spread beneath your feet. There you will find your hedge greatly changed. The trees are gnarled, stunted, rheumatic. There the whirlwinds have their home. And you come to feel that timber line is symbolic of something, a barrier perhaps—you never quite know what it means, but you sense that timber line is a frontier, where the work of man leaves off and that of God begins.

Seldom is there foliage on the storm-facing bosoms of the strained, tenacious old trees that have breasted a century or more of prevailing winds. The mountaintops have a climate entirely their own, seasons exclusively their own, and weather as fickle as the vows of diplomats. If you scale a peak crowned by an acre or so of tableland, you will encounter amazing paradoxes. You have climbed through a violence of upthrusting winds—mostly from the east—which have endeavored to hurl you from crag to ravine; then suddenly you win the flat summit and stand in an area so calm as to be uncanny. The nearby gale, mounting a corkscrew to heaven, rockets past your peak to wrestle with the stars. And it is a vastly impressive experience to stand in the calm zone, hearing the unruly cry of the great wind, and with never a whisper of it straying through your young hair. At timber line you become Olympian.

There are times, too, above timber line, when an electric storm actually touches you, and it is not at all dangerous—I have no scientific basis for saying so—even though your young hair stands on end, and sparks pop from your finger-tips when you flex your knuckles. You are shaking hands with Jove himself, and a very merry fellow he is up there.

And again, to be above timber line, when tarnished clouds are spilling snow on cabins thousands of feet below—but above and about you there is no snow! Only the sunlight, and here and there a peak, rising island-like, from the cloud-sea. All these are wonderful things. But they belong to a past, to a boyhood, and I believe I was talking about the Senator.

Where are his grandly lecherous bones today?

The Senator waited one noon for Grandma to hang out the sugar-sack dishcloths; and he promptly ate them, gobbling every one of them with gargantuan optimism. This gluttony was an overt act. It was my first intimation that Fate does not travel a direct and simple route, but moves along oblique lines. An assassin kills a prince in a provincial town, and a world war is born. A burro eats Grandma's dish-clouts, and . . . well, the old lady was a person of dominance and occasional choler.

"The burro must go," she said; "he's destructive."

The fabric-sated Senator now was braying in delirious crescendo, his great trowel of an ear cocked toward the mesa.

"I never liked him anyway, Grandma said; "he's shameless."

Grandpa had had another futile day of blasting the granite hill for gold. (He had had twelve years of futile days—but was there not always a Tomorrow?)

Grandpa was busy, putting half a dozen yellow sticks of dynamite into the oven, to dry the powder for next day's salvos. "You might of known better'n hang anything near that jackass," Grandpa said. "How was he to know?"

"*Must* you put that dynamite in the oven?" Grandma asked. "It drives me crazy."

"Keep your shirt on," Grandpa said. "I know my business."

These good folk were a bit critical of each other. It seems that General Lew Wallace, when Governor of New Mexico, had consulted Grandma in a literary way. He had given her a warmly-autographed copy of *Ben Hur*. Since that time, Grandma and Grandpa had been a bit critical of each other. Heigho!

Grandpa was handling the sticks like a farmer placing eggs in an incubator. "There's not the slightest danger. We die when our time comes."

"That's what Axel Carr thought," Grandma said. "It's tempting Providence. Anyway, get rid of that burro. What good is he?"

I lay in my bunk, listening. I heard the Senator braying right lustily. I heard Grandma pronounce his doom. I heard Grandpa grunt and go out to drive the Senator away, like a grim father giving the gate to an erring son. I did not cry when he turned the Senator loose upon a wide world. But I did hope the log cabin would blow up, like Carr's, myself with it. I planned to slip from my bunk and set off the powder in the oven. I waited a while for it to go off of its own accord. Then it was too late, for Grandpa returned, grumbling, took the powder from the oven, and laid it aside for the next of his countless Tomorrows.

"Women are a problem," he was saying to himself.

For a week thereafter I wandered the Morman Road afoot. It was small fun. I fashioned spears from lodge-pole pine. I roamed the groves, hurling javelins at chipmunks, pretending they were Saracens. But the joys of crusading were ephemeral. I kept asking Grandpa if he had run across the Senator, during the trips up and down the trail.

One evening in late spring, Grandpa said: "I seen the Senator today. He was on the mesa with the other Jacks and Jennies."

"How did he look? Homesick?"

Grandpa stuck a stout hand into his beard. An hour-glass stream of ore dust spilled from his pancake hat. "Depends on what you call homesick. I dunno. Come with me tomorrow. I think I'm nearing a lode."

I went to bed, but couldn't stop. Just to see the Senator would be something. Maybe . . . but it's hard to dream when an argument is in progress near one's bed. Near one's soul. Grandma and Grandpa were having a few words in regard to a pan of Grandpa's personal sourdough, from which he made his own biscuits. Somebody had thrown it away. It was a breed of pastry second to none, in the opinion of this gritty old citizen of the hills. I fell asleep during these log-cabin recriminations. I remember vaguely having heard a reference to General Wallace's amateur standing as a biscuit connoisseur.

"Just leave his name out of it," Grandma was saying. "The General had background."

Next morning we went toward the claim. "I'm likely to get into the vein today," Grandpa said. "We'll be richer'n all hell, and you can have a pony. A dozen ponies."

We finally reached the mesa. In the distance were about forty burros of various conformations. Some were grazing, others immobile, their ears vaned toward us querulously. Standing guard, a Lucifer of pride and apparently monarch of the whole herd, was the Senator. Beside him was the most dissipated, flea-bitten Jenny I ever had seen, frisking and cavorting like any old scut.

I ran with all speed, stumbling over boulders, but never slackening. Grandpa was howling: "You got no rope! You got to have a rope."

The Senator appraised me rather coldly, in the manner of a friend to whom you have lent money. "Senator!" I called. "Senator!"

I don't think titles meant much to him. He turned philosophically; then, with a most obscene gesture, lifted his heels, snorted and galloped swiftly off, the herd following him like sycophants. I felt whittled-down. I could sense, even at nine, that the Senator's affections were not for me, but with that unwholesome Jenny—that Moll Flanders of the mesa—whose sly countenance haunts me even to my forty-third year. God, what a strumpet!

I limped back to Grandpa. "Won't he ever come home?"

The old man lifted his pack. "I guess not, son. And I guess you're too young to take telling. But when you grow up, remember what I say: don't ever expect anything that's in love to listen to reason."

To comfort me, Grandpa told the story of Alfred Packer, the man-eater. It was a grand story. I wish I could tell it in the first person, as Grandpa told it. The story of Packer—with Fate moving along oblique lines—leads into the saga of Bonfils and Tammen. And it all seems connected with timber line.

I've often wondered if Love killed the Senator, or if the Senator killed Love.

2

Coals In A Coffee-Pot

THE ONE-ARMED pawnbroker at Provo finally allowed Frank Swan, the faro-bank dealer, four hundred dollars on his diamond horseshoe pin.

"I can't help what it stood you," the broker said; "I ain't in business for my health."

The wind was bouncing the pawnbroker's shop-sign, three huge brass grapes that hung above the old door. "It's highway robbery," Swan said, "but I got to take it."

"What time you boys leaving town?"

"Tomorrow at sun-up."

"Colorado, eh? My sister died of consumption there. Denver."

"We aim to explore the San Juan. Well, gimme the four hundred."

The pawnbroker was slow, with his one hand, in counting the bills. "Packer was here looking for you."

Swan tucked the money and the pawn ticket in his wallet. "We're taking Packer along as camp-helper."

"I'd like to go," the pawnbroker said, tapping his stump with his lone hand, "only I'm a wingless. A wingless is only fit to stay home with skirts on."

"You got no kick," Swan said. "You do pretty good burglary with one mitt. You should ought to wear a mask."

"I pay top prices," the pawnbroker said. "Can I help it if people get hard up? They say the San Juan is a bonanza."

"We can't miss."

"You got to watch out this time of year, though. The blizzards."

Swan was on his way out. "We'll handle the blizzards. Which way did Packer go? 'The Black Cat'?"

The pawnbroker was painfully cleaning his spectacles with a bandana. "I didn't notice. Me and him ain't intimate."

Swan turned at the door. "The counterfeit bill, eh? You got him all wrong. He had nothin' to do with it. Absolutely nothin'."

"Maybe not," the pawnbroker said. "Maybe not."

It was autumn of 1873. Men were speaking of the San Juan assays, their nostrils dilating, as though they actually smelled a boom. The mountains of Southwestern Colorado, among the highest of the Territory (it was almost three years later that Colorado became a State), promised much gold and silver for the adventurer. The Spaniards had penetrated the San Juan uplands more than a century before, but had failed to conquer the peaks or to unlock the great treasure chests there.

The valleys of that region now were in possession of the gallant Indian Chief, Ouray. He was a friend of the white man—this despite his fiery lineage—his father a Ute, his mother an Apache. Since boyhood in his native Taos, Ouray had demonstrated both bravery and wisdom. He had met and vanquished the Arapahoe, Cheyenne and whirlwind Sioux. In 1858, when he was but nineteen, Ouray, with thirty Utes at his side, had repulsed some seven hundred Arapahoes near the newly born city of Denver—this after a combat of fourteen hours.

In his fighting prime, Ouray was a magnificent figure, the beau ideal of a leader, mounted on a buck-skin horse, his war bonnet of eagle feathers giving him added height, fierce stripes of ochre and black across his bronze face, his body naked except for a scarlet sash. But an inner light, a knowledge that nothing could check for long the march of the white man, caused Ouray to subdue in himself the violent, barbaric flame, to set up in its stead the tablets of wisdom. He married Chipeta, dark, young and as graceful as a doe, and their romance became a legend. She, of all

Indian women, was more wife and less squaw than were her sisters.

Ouray was to die in his forties, racked by fever, but still endeavoring to salvage the remnant of his race by teaching them to gain treaties instead of scalps. How well the white man kept these covenants is a subject not taught in public schools—this and kindred topics might tend to make cynics of our children before their time.

Elsewhere in Colorado, Indian troubles were to be met then and thereafter. There was a Ute uprising in September of 1879 at the White River Agency—the Meeker massacre. Trappers occasionally shed their own skull-fur while seeking pelts of lesser animals, and there were reports of copper-skinned marauders from time to time. But white men faced no opposition in coming peaceably to Ouray's domain—their enemy was the swift-striking storm, the wilderness, the hardships of timber line.

Twenty men of varying ages pooled their interests at Provo, Utah, in the autumn of 1873. They held meetings, at which Mexican whiskey—"Taos Lightning"—was served, and stories of the San Juan's waiting gold and silver were told. These men outfitted a pack-train, to push southeast to the lofty San Juan country.

Alfred Packer, a newcomer at Provo, was the twenty-first man to be accepted by the fortune hunters. He was twenty-four years old, tall, gaunt, illiterate, taciturn. He was dark and wore a frowsy black beard. He was not well known to many of the adventurers; to some he was an unwelcome addition. He had applied for a position as guide, but a majority of the party demurred.

Packer was presumed to have been a trapper; and, like many of that sort—grandiose and colorful tales to the contrary notwithstanding—was not a fellow one chose for a bosom friend. It was thought he had had something to do with the recent circulation of spurious banknotes in a country where metal was the usual legal tender, and where almost anyone might be deceived by counterfeit bills.

The party of prospectors kept late hours, drinking, singing and boisterously counting their chickens before they were hatched. There was a glassy, far-away look in their eyes. A fever possessed their brains.

Swan went to "The Black Cat" saloon and gambling house to tender his resignation as faro-dealer. That mission accomplished, he would go at once to the boarding house to inform Nona, his girl, that he was leaving on the morrow for the hills. What if she did object?

Swan refused a drink and said good-bye to his tavern intimates. "Packer was asking for you a while back," the bartender told Swan.

"He knows where to find me," Swan said. "Set aside a gallon for the boys with my compliments."

Swan found Nona embroidering a motto on a pillow-top. She was putting stiff-stemmed violets among the words: "No Knife Can Cut Our Love In Two."

"The Packer fellow just left," Nona said.

Swan tossed his floppy black hat to the bed. "Why didn't you ask him to hang around?

"You know why. Take that hat off'n the bed."

"No, I don't know why, either. We got business." He lifted his hat from the bed and flung it to a chair. "What difference is it where you put a hat?"

"It's unlucky. My uncle died of spasm right after he slung his hat onto a bed. I've knowed plenty of cases."

"Did Packer say he'd be back?"

She was watching him curiously as he removed his frock coat, his black scarf and his stiff-bosomed shirt, to wash at a big china bowl on a stand. She was looking at the scarf particularly, as though wondering where the horseshoe pin had gone.

Nona put down her fancy-work. "He said he'd be back."

Swan was pouring water from a tall white pitcher. "You don't like Packer. Why?"

She was still looking at the scarf. "He's bad medicine."

He was making canoe-paddle sounds with his hands in the water. "The same old song, eh? Counterfeit bills." He was cleansing his face, and there were bubbles of soap in the corners of his shut eyes as he groped for the towel. "For Gods sake, why do you keep changin' that towel around?" She rose and handed him the towel, wordlessly. He went on: "Well, they talk a lot, these tin-horns and their gals. But Packer never shoved the queer. . . . Where'd you hide the comb? Is this a game?"

She gave him the comb. "Just keep an eye on Packer, that's all."

"You and your spirit dreams!" he said as he combed his graying hair. He smoothed down the cow-licks with small, nimble hands. "They didn't find anything when they frisked him, did they? Answer me! I'm not talkin' to myself."

"It's your do." She again was looking at the scarf, and he seemed uneasy. "Frank," she said, "somethin's eatin' on you."

"You're crazy."

"No," she said, "I can read you like a book. What is it?"

He slipped into his shirt and snapped on his starched cuffs. "I'm goin' with the prospectin' party," he said. "We leave at sun-up."

She picked up the scarf and held it toward him. "Oh," she said. "Oh!"

He snatched the scarf savagely from her hand. "Yes," he said. "Oh!"

He turned away guiltily. "Let me tie it for you," she said.

"No thanks. I'd of told you sooner, only . . . well, you heard me. I'm goin'. Sun-up. And no squawkin'."

"What's the matter with the job here? You're pop'lar, Frank. The house likes you.'

He was having trouble with the scarf, despite the skill of his gambler's hands. "I'm tired of settin' down, mainly. I got my belly full of aces. A man goes to seed."

"Is that the only reason?"

"What in hell else? I'm tired."

Nona held his coat. "Well, then, I'm goin' to the San Juan with you."

"Like hell you are! You'll stay right here and like it."

Nona was choking up. "Is it . . . is it some other woman, Frank?"

He caught his hand in the coat-sleeve. "Hold it up higher," he said. "I ain't a dwarf." A pleading note entered his voice. "Now looka here. I'm tired of settin' down. I've set and set. For what? It's a wonder I got a behind left, I've set so much. Let's go eat."

"Ain't you got to wait for Packer?" She began to snivel. "Frank, I been havin' bad dreams all summer. Has Packer got somethin' on you?"

"Didn't I just finish tellin' you I was tired?"

"I can read you like a book," she said. "You're tired of me. It's *me* you're tired of."

"Quit blubberin'," Frank said, "and let's put a steak into us."

She clutched his arm. She began to fumble at his sleeve, fingering it as though it were an oboe. "Frank, it's some chippy. People been sayin' that you and Packer and . . ."

Frank shook himself free. "The hell with what they say! Let's go eat."

She took hold of his sleeve a second time. "Frank, look me in the eye."

Again he pulled away. "*Look* you in the eye! I'll *punch* you in the eye! Damn it! Can't a fellow get tired? I'm so damned tired I could faint, if I was a sissy."

Nona's hospitable bosom behaved like a tide. "Tell me you ain't tired, honey."

Swan began to chortle and scream. "But I *am* tired! What do you s'pose gives me them circles under the eye?"

"I mean tired of me, honey."

"Oh, for God's sake, lemme alone! Will yah?"

She stepped back. "The spirit-dreams was right. You've outgrowed me. Yes, you have. And me givin' you my youth, puttin' up with what no other woman in God's world would put up with. Ever since you been runnin' with that Packer, bein' pop'lar's gone to your head. You and that sneak, Packer!"

"Shut up!"

"I won't shut up. Don't you tell me to shut up."

"I'll make you shut up."

"I ain't one of your dance-hall gals. Where's the diamond horseshoe?"

He made a last effort to throttle his emotions. "Now never mind the pin. Let's go eat."

"No? Well I *will* mind the pin. Who gave it to you? Me. And you go and give it to that chippy."

"What chippy?"

"Never mind what chippy. I can read you like a book. Chasin' off with a chippy. You and that sneak, Packer! You spineless, weak . . ."

She got no further. A roundhouse right landed between her first and third chins. She went down like a shot duck. Frank stood over her, looking blankly at the pile of flesh. He was sucking at his knuckles and breathing hard. "You brung it on yourself," he said, "spirits or no spirits."

The door opened and Packer came in without knocking. He walked with a lazy, silent stride. He paid scant attention to Nona, who was coming to with wheezy moans, like a surgical patient emerging from an anaesthetic.

Swan jerked his thumb, in an embarrassed way, toward Nona. She now was on her knees, her eyes dull, like those of a half-slaughtered ox. She shook her head instinctively, to clear it.

"I hear you, Red Feather," she said confusedly.

"She allus keeps fainting," Swan said. "Bilious." He assisted Nona ceremoniously to her feet. "Feel better, honey?"

She didn't reply. She went slowly to a rocking-chair, and sat down heavily. Packer got a wet towel, but she brushed it aside. She picked up her embroidery. She stared at the motto as the two men began to talk things over.

"I only raised four hundred," Swan told Packer. "It'll have to do."

"What about the job?"

"Well, some of the boys voted for you, but the most of 'em said you shouldn't guide the party."

"Why not?"

"I stuck out for you, but they claimed you mightn't know the San Juan country like you say you do."

Packer snorted. "I've hunted and trapped there the last five years."

"I brung all that up at the meetin', but the best they can do is give you a job as camp-helper."

Packer glared. "Theyll find out when we get there."

"Would you have a bite with me and Nona?"

Packer was on his way out. "No. I just et."

"See you at sun-up then. So long."

After Packer had gone, Frank stood uneasily beside a window. It was growing dark. The first night-notes of a honkey-tonk piano came on the autumn wind. Nona had not moved. She sat in a trance-like posture.

"It's gettin' colder," Swan said. Nona did not answer him. She was staring unblinkingly at the embroidery. "You better light a lamp," he said; "that is, if you ain't goin' to commune." Still she did not reply. He leaned over her. "Sorry I lost my temper," he said. "I'm sure sorry. Hungry?" He watched her for a long time. Suddenly he blurted: "Then to hell with you! Sulkin' all the time! You make me sick, you and your fortune tellin'! Honest to God, you make me sick." He picked up his big black hat, slapped it on his head and started out.

"Where are you goin', Frank?"

He turned at the door. "Oh, so you can talk, eh? Well, don't let me bust up your trance. Goin'? Why, I'm just goin', that's all. And what's more, I ain't never comin' back."

"No," she said quietly, "you won't never come back."

He seemed deflated for the moment; he actually had great respect for her psychic utterances. "What's that you say?"

She was not looking at him, but at the motto, which could not be read in the dim light. "I said you ain't never comin' back. You're goin' with Packer. You won't never come back."

He fumed for a moment. Then he gritted his teeth, slammed the door and was gone.

The party of twenty-one gold-hunters from Provo reached the Colorado boundary late that November. Although the great range now could be seen clearly, it seemed the prospectors made little progress. The thin atmosphere of high altitude had a magnifying quality. A peak apparently fifteen miles distant today, on the morrow would still have a mirage-like illusion of being fifteen miles distant. Lieutenant Zebulon M. Pike had suffered a similar optical experience November 15, 1806, on first view of the peak which now bears his name. The mountain "appeared like a small blue cloud" as Pike squinted through his spy-glass that November 15. It was not until November 23 that he and his party reached its sprawling base, there to build a log breastwork against possible attack—the first structure raised by Americans in Colorado.

The Provo men were grumbling. They had not seen a roof of civilization for a full month. Their spirits were low; their supplies lower. They were footsore, their clothing frayed, their belts notched-in to bolster wasting bodies.

The picture of oncoming winter was apparent even to an inexperienced eye. There are frosts nearly every month in this high land, and by late autumn the night air is brutal. Winds were ripping through the cañons, wild stallion winds stampeding toward the Pacific slope. The ravines were hell-gates of wind.

The less hardy ones of the unkempt party began coughing—a bronchial Wagnerian troupe. Some of the men bled at the nose from exertion in an altitude of two miles above sea-level. Game was scarce, for a hibernating slumber was upon the land. The beaver had finished harvesting, their colonies seemingly lifeless and forsaken on the great moraines. The aspen stumps, from which these amazing engineers had gnawed logs for their dams, protruded from the frozen earth like fat pencils. A landslide occasionally out-roared the gale with a cannonade of plunging rock and sand.

The high world was awaiting its first general snow—often the heaviest of the year. It would swirl from heaven in a chokingly fine, directionless blast—snow as fine as confectioner's sugar. It would blind the straggler, choke him, lulling him with treacherous coma, finally burying him beneath a thick robe of ermine. Tragedy is not always veiled in black.

The party reached a meadow, formed after a forest fire of two centuries ago. They checked landmarks with their crude maps, the metallic content of the terrains having put their compasses at fault. They resumed the march, their backs to the descending sun, and at nightfall came to a river. It was frozen over from bank to bank, a serpentine of gray silver. It was the Uncompahgre, asleep until spring freshets would flood its forks from the south and southeast. Then it would awaken with torrential rage, surging to the Gunnison, which in turn poured into the storied Colorado.

How rich in history is each wave of the great salt sea, and how far its waters have come!

The vanguard set up a shout. They had sighted an Indian camp. The chieftain of the village had been apprised of the advancing party. He was on his way now to meet them, accompanied by an escort of tribesmen. He was holding his right hand high. There was a Roman quality about him that made of his loose blanket a senatorial toga. The greatest chief of his generation, Ouray, the peacemaker.

"You are welcome," Ouray said in English. "These are my people."

"We're lost," Swan said, "and we have no meat."

Ouray led the party to his encampment. "My fires are warm. You are welcome."

At the camp the men met Ouray's wife. "Chipeta is a Christian," Ouray said with quiet pride. "She welcomes you."

After they had eaten, the men sat about a great fire. Ouray gave them tobacco. He told stories of mighty storms, of blizzards in winter, of avalanches in spring. Then he rose.

"Good night, my brothers," he said.

"We'd like to move in a few days," Swan said. "But we're short of grub."

Ouray studied Swan's garments. "My white friend's clothes are big with fasting. His boots are thin. My meat is your meat; but you are the storm's meat. Stay."

The party spent several days at the chief's encampment. The soreness left their hamstrings; the wrinkles vanished from their bellies. But once those bellies were full and warm, they began anew to twitch with the narcotic craving for gold.

Ouray counseled the men against hurling themselves on the horns of winter. A majority voted to lie low until spring. There was a schism, however, gold-hunters being like over-pious citizens, ready on small provocation to begin a war to prove a point. Six of the twenty-one elected to continue toward the promised Eldorado.

These six were Swan, Packer, Miller, Noon, Humphreys and Bell. The last-named was a lad of trusting eye. He had promised to bring his sweetheart a mountain of solid gold. The six dissenters chose Packer as their guide. Packer then applied to Ouray for provisions.

"You do not know this country," Ouray said. "Why lead your companions to their death?"

"I know the country," Packer said. "How about some grub?"

"Will you not listen to wise counsel?" Ouray asked.

"When I want advice I'll let you know," said Packer.

"You are a man," Ouray said. "Do not behave like a child. My fathers were wise. They taught me the ways of the storm. They taught me to meet death, not to seek it. Gold is in the mountains, yes. But gold will be there when the snows melt again. It will be there long, long after we are gone. There are more dead men than there are men alive. What good is gold to the dead? Can the fire-killed pine put back its green boughs? Can the bones of the buffalo again wear skin?"

"How much grub you givin' us?" Packer asked.

"If you were on the way to rescue some white brother," Ouray said, "my people would not count their own mouths. But you are not on a good journey. We will give you food, yes, but not all you can carry. An evil spirit is with you."

"You know a heap, don't you?" said Packer.

"I know nothing beyond the words of wise men, and what I have seen. But I know you are a fool."

Ouray turned away. He directed his braves to give Packer's detachment one week's rations, no more. The six adventurers strapped on their knapsacks.

Ouray spoke to the six men in general. "Follow the Uncompahgre northward on the ice," he said. "On reaching the mouth of the Lake Fork of the Gunnison, take the left fork. You will come to the Los Pinos Agency, where, if you listen to good counsel, you will remain till spring."

"Let's get going," Packer said.

Ouray shook hands with Swan. "You are too wise to die," he said. Then he shook hands with the lad, Bell. "And you are too young to die." He raised his hand in slow salute. "But all must die sometime."

Packer followed the river ice. However, when the party reached the mouth of the Lake Fork, Packer insisted on pursuing the *right* fork.

"Just leave it to me," Packer said, when Swan recalled Ouray's directions. "I know this country."

The wilderness each day became higher, wilder, more forbidding. If there had been hardships before, now there was a terror of the almost Arctic cold, and shortage of food. The men began quarreling and bickering. Packer was lamed in falling from a spur while reconnoitering. The party delayed two days, waiting for his swollen ankle to reduce. Packer became more silent than ever. He kept his Winchester rifle constantly in his grasp. There was but one cartridge left of all the party's ammunition. That cartridge was in Packer's rifle-breech—the lone bullet his only safeguard against mutiny.

The last box of matches was lost. Packer took coals from the campfire, carrying them in a huge coffee-pot each time he broke camp, to light a fresh fire without resorting to even more primitive methods.

Two weeks of wandering and the party was in Hinsdale County. Swan implored Packer to turn back.

"Who's guiding this party?" Packer said.

"You don't know the country," Swan said. "All we've had for three days is roots and berries."

"You ain't heard me kick," Packer said.

"We've decided to go back, whether you do or not."

"Remember me to the folks," Packer said.

Swan was faint. He leaned on his axe. "You said you knew the way to the Los Pinos Indian Agency. You promised to have us on the Cochetopa, or at Saguache, last Wednesday. That's a week ago. Now look at us."

The first flurry of snow-flakes set in. The sky became white, gray; then a whine, like that of some great wolf, came warningly, forebodingly, up the ravine. The mountain buttresses grew indistinct beneath the gray banner of the snow, the hills standing like ships anchored in a mist, the fog-horn sounds of the wind rising swiftly in sinister volume.

"Good God!" Swan said, "and we didn't bring snow-shoes!"

"I'll climb higher," Packer said. "Maybe I can get above the storm to see just where we're at."

Swan's lips were blue. "I knew it! We been lost all the time. God damn you!"

"Shet up, and get everybody under that ledge."

Packer limped into the growing storm. Swan tremblingly replenished the fire. He assigned the men to guard duty, each to stand his turn in rotation while the others slept. The storm became a fury. Drifts were built high, only to be torn down again, like shifting dunes. The men kept from freezing by burying their blanketed bodies beneath the snow.

The blizzard continued till dawn. Packer had not returned. At sunset the wind modified, Packer, his eyes almost blinded, his hair and beard matted with ice, stumbled to the ledge. Young Bell was on guard. The others were asleep, numbed by hunger and exposure.

Packer found Bell wide-eyed, almost hysterical beside the fire. Packer later claimed that Bell attacked him, knocking out two of his front teeth with the butt of a gun. Circumstantial evidence, however, indicated that Packer, half-demented, tried to induce Bell to join him in slaying and robbing the four sleepers. Slay them, he certainly did—as they slept—using Swan's axe on their skulls. He then took his bowie and stripped from Swan's thigh a slice of flesh. He cooked it over the small fire, and it is probable—as Packer claimed—that the famished Bell partook of this macabre meal.

The man-eater declared that Swan had died from hunger, that the survivors practised cannibalism upon the cadaver. He said that another, then another died of starvation, and that finally he and Bell were the two remaining. He averred that he and Bell entered a solemn pact not to kill each other, but to chance fighting it through together. Who broke the vow, if vow there was, between these two starkly marooned men?

All we know is there had been no struggle where the four sleepers were killed, their skulls bashed in; that there *was* a terrific battle where Bell finally sank to stain the white snow—the one remaining bullet, Packer's, drilling him through the heart.

"It was Bell who was crazy," Packer claimed. "Why, when I shot him in self-defense, he was howling out: 'You took away my mountain of solid gold.'"

Packer now swung the death-axe on some small trees. He fashioned a rude hut. He wore a trail between the hut and the place of massacre. He lived there for four weeks until he could travel, waist-deep, through the snow. He lost his lean look, but was beset by lameness and sustained nausea.

Six weeks after the crime, he walked slowly into the Los Pinos Agency, seventy-five miles from Lake City. He told a bizarre tale of a quarrel, of his five companions having gone on without him, leaving him to travel alone.

General Adams, the agent, was on business in Denver. The General's secretary, Stephen A. Dole, heard Packer's story and believed it. He thought it odd, however, that a man who had subsisted on roots and berries should bear such a sleek look; that a man who had hungered much should call for whiskey instead of food on emerging from the wilderness.

Packer refused to stay for long at the Agency. "I'm anxious to get back to my folks in Pennsylvania," he said. He left for Saguache.

General Adams, returning to the Agency from Denver, stopped over at Saguache. He asked concerning Packer, who was displaying large sums of money, carousing and drinking. The General heard disquieting stories from members of the original party, wintering at Saguache.

One of the prospectors, a man named "Frenchy," said: "Packer was arrested for counterfeiting in Salt Lake. He once killed a Mexican sheep herder, and he's no good. I think the men who went with him are dead."

Packer, when drunk, showed a pipe and some trinkets known to have belonged to the five adventurers. He bought a horse from Otto Mears, the sutler, and drunkenly tendered a counterfeit bill. He snatched it back and produced a good bill from a second pocketbook.

General Adams inveigled Packer into going back to the Agency, ostensibly to guide a searching party for the five men. After numerous excuses, Packer, Mears and three of the Provo men set out for the Agency. Under threats of lynching, Packer fabricated a tale of starvation, survival of the fittest, enforced cannibalism and self-defense.

On the way to the Agency, and while crossing Cochetopa Creek, Packer flung two objects into the water. "Just some trash," he explained.

He had dispensed with his pocketbooks.

While guiding a party of searchers under the supervision of Constable H. F. Lauter, Packer deliberately misled them. Then he tried to slay Lauter with a knife. The search was abandoned, for the snows still were deep. Lauter turned Packer over to the Saguache authorities.

The thaws of 1874 came to the highlands. Three prospectors—Nichols, Wells and Graham—were camping near a pine grove below Lake Cristobal. Their dog brought in a bone, a human ulna. They investigated.

They came upon five bodies. Four were lying in a row. A fifth lay some distance from the others, headless.

Faced by sure evidence, Packer escaped from the sheriff's home, where for some months he had been "tending" that officer's children—a grisly sort of nursemaid.

A man-hunt of nine years failed to flush Packer. In March of 1883, nine and a half years after the crime, Prospector Frenchy was lying on his bed in a Fort Fetterman, Wyoming Territory, rooming house, thinking of women. Voices drifted through the thin partition of Frenchy's room. He recognized one voice as that of the man-eater. He got in touch with General Adams, now a United States Post-office Inspector.

"He calls hisself John Schwartze," Frenchy informed.

Packer was trapped. On April 13, 1883, in Lake City, a jury convicted him of premeditated murder. Judge M. B. Gerry, a Southern gentleman of the old school, a fiery Democrat, sentenced Packer to be hanged on May 19, following. Although Judge Gerry delivered what was considered the most eloquent hanging speech in Western court history, an apocryphal sentence is the one that persists, and by which this scholarly gentleman's name still lives.

It happened this way: Larry Dolan, who had a grudge against Packer, attended every session of the trial. Between times, Larry filled himself to the larynx with "Taos Lightning." Sometimes he was so beautifully boiled that he had to be restrained by the court bailiff. In fact, just as the fluent Judge Gerry began a classical pronouncement of doom, Larry let out a cheer and fell from a bench. Then he rose, bowed and ran drunkenly to his favorite saloon, bellowing like ten Apis bulls of Egypt:

"Well, boys, ut's all over; Packer's to hang. The Judge, God bless him! says, says he: 'Stand up, yah man-eatin' son of a bitch, and receive your sintince!' Then, p'intin' his tremblin' finger at Packer, so ragin' mad he was, he said: 'They was siven Dimmycrats in Hinsdale County, but you, yah voracious, man-eatin' son of a bitch, yah eat five of thim! I sintince ye t' be hanged by th' neck ontil y're dead, dead, dead; as a warnin' ag'in' reducin' the Dimmycratic popalashun of th' State."

Because of a technicality (that the law under which Packer was tried and convicted, and which went into effect March 1, 1881, was *ex post facto*, etc.) Packer gained a new trial and a change of venue to Gunnison County. There he was convicted and sentenced to a lifetime of hard labor at the State Penitentiary at Canon City.

He entered those walls, vowing never to speak again. He worked wordlessly, fashioning hair bridles, hair rope and other cowboy paraphernalia. It was not until the close of 1900 that Packer broke his silence of eighteen years.

A beautiful and gifted woman journalist, Polly Pry, came upon him at the prison. She was working on a paper, the *Denver Post*, owned by the sensationally colorful partners, Frederick G. Bonfils and Harry H. Tammen. That paper began a crusade for the man-eater's pardon. And that campaign's aftermath brought to them an amazing interlude of tribulation, physical injury and—nationwide advertising.

Of Bonfils and Tammen, I sing (albeit in quarter notes, after the manner of my beloved Senator).

3

Argonauts At Play

When I was a young and virginal news reportor, Artie MacLennan, of the *Republican*, used to send me downtown to "cover the Lowers." This humble beat took one among the more ancient hotels of Denver; once-gay caravansaries, now a forlorn group of exiles.

It was a land of last chapters, this zone of decaying inns. To ambitious journalists, the assignment was a stale loaf of beginner's bread, unimportantly routine—almost a degradation. To myself, forever cavorting in a suit of sardine-can armor, it was a grand foray.

There were ruins of lobbies where a vital company had partaken of Life in roistering years of gold and silver. The sway-back furniture served to accentuate a feeling for the past. The surviving crystals of aged chandeliers seemed the crown jewels of a ghost.

The people of these hotels now had a Rip Van Winkle quality. The senescent taverns had become sanctuaries for tattered men—their lives behind them. An unshaven wreck slept off his rot-gut whiskey in a spavined chair, the selfsame leather which once supported the posterior of Ulysses S. Grant, not to mention the aristocratic rear of Lord Dunraven, the thespic nates of Richard Mansfield, John McCulloch and Lawrence Barrett. One might see a deflated gentleman, shuffling with the paretic badge of a long-ago indiscretion, his choo-choo feet sending him crabwise across rutted tiles. In an earlier day, these flags had known the boots of Adelina Patti, Helena Modjeska, the gifted Pole, and the compelling Lotta Crabtree.

There is an irony implied in every man-made structure,

whether a home or a civilization. A child's blocks find the ash-can. The Roman Forum is a Mills' Hotel for stray cats. Mattie Silks' place, a giddy *maison d'amour* of Denver's former tenderloin, today is a Buddhist church.

The story of Greece is written in its monuments; that of America in its hotels. Of the West's fabled hostelries, the Windsor of Denver had the noblest tradition. It still stands, this five-storied edifice of Victorian stone, at Eighteenth and Larimer Streets, a cathedral of Poverty Parish. It is hub to a wheel of Dickensian pawn-shops, rescue missions, patent-medicine sanctums, and small, hodgepodge businesses. Nightly the drums of the Salvation Army bombard its venerable walls, and there come thin, carrying voices: "Bringing in the Sheaves." Patient Joshuas encircling a down-at-the-heel Jericho.

The Windsor's huge inside doors of black walnut are ignorantly painted over, but the engraved hinges, hand-tooled and of solid brass, testify to the vigor of a departed order. The ceilings—in the bedroom suites even—spread twenty feet above hard-wood floors. The burly mouldings have a personality undisturbed by decades of neglect.

The Windsor's grand ball-room, where Empire Builders led their glittering ladies in square dance and waltz, shelters Billy Papke's Gymnasium. The afternoon sun, shining through tall, plate-glass windows, finds a troupe of pork-and-beans pugilists working with the solemnity of blacksmiths on one another's bodies. A training-ring is pitched beneath dim frescoes of cupids, now an array of unsanitary foundlings.

Off the ball-room, with its tobacco fumes and sour, athletic odors, is the banquet kitchen, where French chefs formerly labored like evangelists at a revival altar. One has to squeeze past a rubbing-table on entering the kitchen, now a store-room, the plaster falling; a corroded sprinkling can ashore on a rat-eaten mattress; a ruptured punching-bag in a corner, limp on a mound of bricks. The only thing lacking in this kitchen-midden is a mild archeologist.

There were two bars on the first floor—one off the lobby, and a back-bar off the billiard room. These bars were world-famous, beginning with the year 1880, when an English land company built the Windsor and put Bill Bush in charge. There usually were six bartenders on duty. Of course, when General Grant, Buffalo Bill Cody, or the heavyweight champion, John L. Sullivan, visited town, a shift of twelve artists was imperative. A China boy in native costume kept the marble floor free of cigar-butts and quids; the social marksmanship of pioneers being superb in fiction but mediocre in fact.

Eugene Field, his great talent as yet not fully appreciated, because he chose to play the clown, once wrote verses here with microscopic penmanship. Not to mention such *Tribune* squibs as his thrust at the grim Colonel Cooper.

"Colonel G. K. Cooper went swimming in the hot-water pool at Manitou last Sunday afternoon, and the place was used as a skating rink in the evening."

Also certain of his drama criticisms, including the one about McCulloch:

"He played the king as if he was afraid somebody else would play the ace."

Of the Windsor's bar-crew, one was destined to grow to a full, swashbuckling manhood. His name was Harry Heye Tammen, and he was born with his tongue in his cheek.

Tammen had a genius for lasting friendships, but he enjoyed nothing quite so much as seeing a sucker squirm. He shunned society, as such, made sport of his work, and regarded his material possessions as so many toys. He had a clear disdain for criticism of any sort, good or bad. He laughed in the faces of detractors, when they sought to compel him to tryst with the pious mob—to marry his talents to mediocrity in a shotgun wedding of the soul. He surreptitiously shared his bread, his wealth and his heart with many. He belonged neither on a pedestal nor a perch. Harry Tammen was a free soul—perhaps the last one of our time.

Tammen had an elfin habit of publicizing his sins, real

or imaginary, thus leaving potential quibblers without ammunition. Only his intimates saw the man behind the postures. Elbert Hubbard, J. Ogden Armour; his partner, Bonfils; the beloved priest, Father Thomas Malone, and Agnes Tammen, his wife—these knew his great and mischievous heart. To the world his posed as Robin Hood with a megaphone, Taras Bulba with a hurdy-gurdy.

If I write freely concerning Harry Tammen's antics, his chicaneries, his seeming brigandage, I do so with the utmost love and good cheer—for he brought me up, counseling me never to lose my zest for life, nor for live things. And I never have.

I had come from the Union Depot, and from an interview with Buffalo Bill Cody. He had seemed a lone, sad figure in the cold night. There were no gay bands to meet him. He had lost his fortune and his Wild West Show. His day was done, and he was getting ready to die.

I remembered tales told by Major Burke concerning the long-haired Colonel, brave stories of circus campaigns abroad—particularly his presumed liaison with the Duchess of B——.

"Colonel," I asked, with all the curiosity of journalistic youth, "what did the Duke really say when he caught you in the lady's bathroom?"

The Colonel—yclept "Pahaska," the Long-haired, by the Indians—began blowing like a bull elephant. "Young man," he snorted, "my hair is hoary . . ."

"Yes," I interrupted, "but not with years."

When I returned to the *Denver Post,* to write a piece for the Sunday morning edition, Tammen called me into his office. Cody quite justly had demanded that I be fired.

"Were you always impudent?" Tammen asked.

"Yes," I admitted. "I always was."

He put his hand on my shoulder. "Keep it up, son," he said. "It's something you can't buy."

I attended the burial of the lusty old Colonel on Lookout Mountain, less than a year after that interview. He had lain for some months in a mortuary vault. It was a remarkable funeral, with thousands traveling the long trail to the summit of the hill. There was a circus atmosphere about the whole thing. A lot of us drank straight rye from bottles while speeches were being made by expert liars. Six of the Colonel's surviving sweethearts—now obese and sagging with memories—sat on camp chairs beside the grave of hewn-out granite. The bronze casket lay in the bright western sun. The glass over the Colonel's amazingly handsome face began to steam on the inside. You could not see the face after a while, on account of the frosted pane.

One of the old Camilles rose from her camp chair, with a manner so gracious as to command respect. Then, as though she were utterly alone with *her* dead, and while thousands looked on, this grand old lady walked to the casket and held her antique but dainty black parasol over the glass. She stood there throughout the service, a fantastic, superb figure. It was the gesture of a queen.

Harry Tammen was a roly-poly fellow, clear-skinned and blond. He had large blue eyes that never grew dull. He was born in Baltimore, Maryland, in March of 1856. His mother, Caroline, was a German woman of fine bearing. His father was Henry Heye Tammen, a Holland Dutchman. The elder Tammen died in the '60's, and while employed at Philadelphia in the consular service of the Netherlands.

Mrs. Tammen was left penniless. She was unable to support her boy. He was not yet eight years old, but he had a strong will.

"I won't go to an orphans' home," he said.

Mrs. Tammen dressed the lad in a Dutch homespun suit, much too large for him. As she hugged and kissed him good-bye, she said:

"My little son, may love and good cheer always go with you."

Forever after, Tammen closed each of his letters with the phrase: "With love and good cheer."

When little Harry set out, he could not speak English—only German. The owner of a beer garden was looking for a boy to help the porter. Tammen got the job.

From that time on he worked in many saloons, but never became a hard drinker.

Harry had practically no schooling. He learned to speak English from listening to the talk of bar patrons. He learned to read from newspapers left on saloon tables.

"The principal thing I found out in saloons," he would say, "was *what not to do* with my life."

He grew up behind a bar. Before he was twenty-one, Tammen became head bartender at the Palmer House, Chicago.

"Were you *actually* a bartender?" a lorgnetted dowager once asked Tammen in J. Ogden Armour's drawing-room.

"My sweet lady," he said, "I was the best booze-juggler in the world."

"You know," Armour once told Tammen, "I used to be shocked by those frank stories of your past. But I am thinking of rewriting my own history, and in it I shall claim I pushed a wheelbarrow in a slaughter-house."

Bill Bush saw Tammen performing behind the Palmer bar in 1880. He drafted the little Dutchman for the Windsor. Tammen took to the West, and the West to Tammen. The Rocky Mountain region's most important bibbers regarded him highly. He was a Napoleon of the mahogany. If he didn't like a man's looks, his behavior, his conversation, Tammen would not permit him to be served. He became well acquainted with Pullman, who got inspiration for sleeping-car berths from the double-deck bunks of Colorado log-cabins. (Personally, I think miners' bunks the more comfortable, but that is only one man's opinion.)

"Silver Dollar" Tabor hoisted many a powder at Tammen's bar—later he built his famous Tabor Opera House and installed a saloon more costly than that of the Windsor. Tabor usually had a splashed look, no matter how finely dressed, which caused Tammen to enquire. He learned that the Senator had a peculiar ritual while shaving. It seems he would rise, dress fully, *then* shave. Consequently he spattered lather all over his Prince Albert.

Tammen had an eye for opportunity. "I got my first real capital at the Windsor," he would say, originating an American classic in economic opportunism; "you see we had no cash registers. I used to toss a dollar to the ceiling.

If it stuck there, it belonged to the boss. If it fell, it was mine."

Whether or not this boast were true, Bill Bush failed to reprimand Tammen. The practice of till-dipping was rather widespread in pre-cash-register days. It was much easier to get a good wife than a good bartender. Witness the case of Bald Pete, bartender for Con ("Cooney-the-Fox") Keller. Pete showed up in Denver entirely hairless; he had no brows and not a sign of an eyelash. Two years later, he began to mumble about a hair-restorer formula. Then, to everyone's amazement, he began to sprout from the scalp, brow and eyelid, with hair magnificently thick and black. He sold hair-restorer faster than he could make it—the recipe was linseed oil, camphor and quinine. That was his racket, and he moved slowly from city to city, always showing up as bald as a melon, then sprouting hair, selling his restorer and moving on. It took two years to work the game, but he judged it worth while.

In Denver, Pete was waiting for the psychological moment to make the town hair-conscious. Meanwhile he dispensed drinks for Cooney-the-Fox. The very first night at Cooney's bar, Pete dipped into the till and pocketed five dollars. Cooney was watching through a peep-hole, designed for the purpose of discouraging peculation.

"Did you take something from the till?" Cooney asked.

Bald Pete fidgeted. "Only a little carfare."

"Well," said Cooney, "just go easy."

The next night Cooney came from his peep-hole. "Did you knock down some more money?"

"Only a little carfare home," Pete said.

For a month this condition obtained, but Pete was a rattling barman. Finally Pete overstepped all decent bounds, clipping Cooney for an even hundred dollars at one swoop. Cooney galloped from his peep hole to the bar.

"How much did you knock down this time?" Con stuttered.

Bald Pete gave the same old story. "Oh, just a little carfare home."

"For God's sake!" Cooney said, "where in hell do you live? Constantinople?"

Buffalo Bill Cody once took a knotty problem to Tammen. Now the good Colonel was the handsomest fellow of his time, and no amount of hard liquor could bloat his stalwart figure, crimson his Alexander-the-Great nose, or otherwise rob him of noble charms that stirred alike the hearts of Kickapoo squaws and Mayfair duchesses.

However, his vasty tippling sometimes affected his target-shooting abilities, and befogged his oratory. His Wild West Show was making him an international character, and his manager, Nate Salisbury, wished to safeguard Cody's popularity.

"You got a duty to your public," Salisbury said.

"To hell with 'em!" said Pahaska, the Long-haired.

On one memorable occasion, and while playing a town in Wyoming, Pahaska got enchantingly stewed. He didn't know where he was, geographically—nor did he care, either. At Grand Entrance time, Cody thought he was in Dakota.

"By the way, where are we?" he asked Salisbury.

"Cheyenne," Salisbury said, guiding Pahaska's usually unerring boot into a stirrup. "Cheyenne, Wyoming."

"The hell you say!" said Pahaska. "We're in Dakota!"

Now at that time, Cheyenne, for some forgotten reason, had a municipal hatred for the town of Lead, South Dakota. Bill charged into the tented Cheyenne arena aboard his white mustang. He brandished his trusty rifle amid cheers, then launched into a loud harangue before the Wyoming audience.

"Fellow citizens of Dakota," he began. There were some warning shouts, and then a few sad noises—grandfathers to the modern raspberry. But Pahaska had the voice of ten locomotives. "Friends of *Dakota,*" he insisted roaringly, "I am happy to be in *Lead* tonight, and away from our recent dreary residence in *Cheyenne.*"

The Cheyenne citizens began to hoot, to throw things. This and similar experiences led Manager Salisbury to take steps. During one of Pahaska's mornings of spiritous remorse, Nate prevailed on Cody to sign an agreement. This document limited Pahaska to ten glasses of whiskey a day, no more. In case of violation, Cody would be compelled to pay Salisbury a forfeit of five thousand dollars.

The Colonel soon repented his signature. He came downstairs at the Windsor to consult Tammen. "Harry," he said, "I'm unhappy. Here I am, tied down to a meagre ten drinks *per diem*. And look at me."

"Spread them out," Tammen said.

"Ah," said the gallant buffalo-slayer, "that's just it. I can't. I rise at eight, and have an eye-opener. Then breakfast. And before I know it, it's ten o'clock, and I've had all the contract allows." He sighed. "Sometimes I wish I wasn't a man of honor."

Tammen thought for a while. "I have it," he said.

"I pray God," Bill said.

"Ten glasses! All right. Instead of taking 'em from a whiskey-glass, drink 'em from the chaser-glasses; they're three times as big."

Cody blushed like a schoolgirl. "God bless you," he said. "God bless you."

Salisbury accused Cody of violating the bond. Not only was Pahaska growing wobbly in his feats of marksmanship, but he insulted Oklahoma citizens with praises meant for Kansans.

"Pay me the forfeit," Salisbury said.

"Ten glasses are ten glasses," Pahaska pointed out.

"I know, but you're drinking from beer glasses! Seidels even. My God!"

Salisbury resorted to the courts. The judge pondered the contract. Then he handed down an opinion.

"Colonel Cody is acting entirely within his rights," he ruled, "when he takes his beverage from a chaser-glass, a beer glass or a seidel. There is nothing in this instrument to specify the size of the glasses of whiskey. If it were worded to read: 'whiskey glasses' instead of 'glasses of whiskey,' then the plaintiff would have had a sustainable contention. But it is evident the Colonel is living up to the agreement as worded. True, he shall not drink from a hat, or a tin bucket; but so long as he sips from a chalice made of glass, he may continue, were that glass the size of a water-tower. I rule for the defendant, the plaintiff to pay costs of this action."

"Judge," Colonel Cody said, "you are a pillar of society. Here are some passes for my show."

Denver was wide open in the '80's and '90's. There was gambling at the Interocean, the Jockey Club, Mark Watrous' place, at the Arcade, and one could bet on the ponies at Overland Park. Ed Chase, Ed Gaylord and Chuck Chucovitch were the triumvirate controlling "square" gambling in Denver. Many others conducted crooked games, with a three-shell variety of cunning.

Men hammered the hills for as long as two or three years, sometimes striking it rich, then coming to Denver to toss off their fortunes in a single night.

"A gopher should always stay in his hole," Grandpa would say.

There were many colorful confidence men. Lou Blonger, a kind-hearted fellow and a life-long friend of Tammen, was one of them. Lou got mixed up with the Maybury gang, and in his declining days was sent to Canon City's penitentiary for indulging in a racket that included a potential sale of gilt from the Capitol Building's dome.

Old Doc Baggs was another humdinger. He took bets from the unwary on a fake, wire-tapping race-track game. His huge silverplated safe was the wonder of the town. One night, the building in which Doc ran his game took fire. An assembled crowd stood popeyed. Coming down the stairs, and *carrying the enormous safe on his shoulders,* was old Doc. The safe, it seems, was made of papier-mâché, overlaid with silver paint.

"I couldn't leave it behind," he said, "for sentiment's sake."

The foremost of all the town's con men was Jefferson ("Soapy") Smith. Soapy operated in the lower part of Seventeenth Street, near the Union Depot. Newly arrived tourists were his specialty. His best-known dodge was the sale of lucky soap. He wrapped small pieces of soap—and in full view of his audience—with ten, twenty, fifty-dollar bills. Next, he put plain paper over the moneyed soap. Then he tossed the wrapped soap-cakes into a basket, allowing a customer to choose any package for a fee of one dollar.

Members of the audience bought an abundance of this soap. But the winners always were Smith's confederates.

Soapy skippered a gambling house in Larimer Street, but no one ever picked up a sizeable bet here. This genius was run out of Denver. He drifted to Creede, when that camp began to boom. He left Creede (by request) and joined the gold-hungry mob in the Yukon. There he became king of Alaskan confidence men, ruling whole towns with his wits and his pistol. He finally succumbed up North from a gun-shot wound, adroitly placed in his pancreas.

Probably no one knew Soapy Smith better than did the late Wilson Mizner.

"Talk about modern killers," said the dear old sage of Vine Street. "They are schoolboys with bean-blowers. Soapy topped them all. Why, I remember the time he was running a town in Alaska. That village consisted of one straightaway mile of gin mills, gambling halls and love stores. A preacher once wandered in with a donkey-load of Bibles. He visited Soapy right off.

" 'Brother,' he said, 'I'd like to build a house of God in your midst.'

" 'The hell you would,' said Soapy, mentally measuring the gospel giver for a casket. Soapy turned to his personal sexton. 'Size number four,' he said.

" 'God has sent me,' said the parson.

" 'And I'm returning you by a short cut,' said Soapy, reaching for his gun.

" 'Brother,' said the pulpit-pounder, 'let's do business. I know I couldn't last here a minute without your sayso.'

" 'You got some sense, anyway,' Soapy said.

" 'All I want is permission to pass a paper for signatures. If you'll sign your name at the top, and give a small sum, I know the other boys will follow.'

" 'That's reasonable,' Soapy said. He signed his name and almost floored the missionary by hauling out a thousand-dollar bill. 'That'll start you off,' Soapy said.

" 'My prayers are answered,' said the soul-snatcher. 'Maybe I can win you over to God's side.'

" 'Don't squander your money on that kind of bet,' Soapy said, walking away."

Mizner mopped his jowls with a napkin and almost choked with laughter. "Well," he went on, "what do you think? By nightfall the astounded prophet had collected *thirty thousand dollars* in cash! He was all hopped up with dreams of building a cathedral, instead of just a meeting-house. He sat up in bed when he felt a gun buried to the trigger in his umbilicus. It was Soapy's gun. Soapy not only got back his thousand-dollar investment, but twenty-nine grand *profit*. Yes, Soapy was a ripe character."

Although he was the best bartender in the West, if popularity is a criterion, Tammen's dreams were far, far beyond polished mahogany and ivory beer-skimmers. He was made of Napoleonic stuff—with a generous dash of Phineas T. Barnum, and a peculiar, though grand sense of humor. He could scheme and laugh simultaneously. When "Gassy" Thompson and "One-eyed" Riley fought a duel in front of the Windsor, it was Tammen who encouraged the fray, but saw to it that blank cartridges were put in the pistols. Both men—dead shots—threw their pistols down in disgust on failing to score.

Gassy Thompson was a neat schemer himself. One time he contracted to sink a shaft a hundred feet beyond its already existing two hundred feet. He found the shaft filled with water to a depth of nearly one hundred and eighty feet. The unwatering, of course, was a big item. Gassy hit upon a marvelous ruse.

At that time a man was mysteriously missing from camp. Rumors of foul play were abroad. Gassy came excitedly to town and reported to the officers of the law. He took them to the collar of the shaft. There in the snow was evidence of a titanic struggle, with bloodstains everywhere. The county commissioners installed a hoisting plant and unwatered the shaft. No body was found. Gassy rented the outfit and proceeded with his contract.

When Lou Blonger sold a gold brick to a clergyman, Tammen chuckled for a week. In the years that followed, Tammen tried to shield the personable Lou from prosecution.

Soapy Smith, a mixture of Cagliostro and Dick Turpin, was not an inimate of Tammen, but the latter always enjoyed his didoes and studied his methods. Tammen had no desire whatsoever to go outside the law in any of his numerous schemes, but he saw in the antics of confidence men a certain hilarious motif; he read plainly that the public not only likes to be fooled—but actually insists upon it.

One of Soapy Smith's triumphs aroused a good-natured envy in Tammen's breast. Before Soapy departed for the Yukon, a report reached town that two prospectors, digging in the hills near Florence, Colorado, had uncovered the feet and legs of a petrified, prehistoric man. Hundreds of citizens flocked thither to glimpse the careful unearthing of the rocky Pleistocenian. The press grew garrulously scientific with descriptions of the discovery. It was hazarded that anthropologists might have to rewrite Darwin's books because of this find.

Soapy was early on the ground. He negotiated a purchase of the "remains," and superintended the exhumation. He shipped the stone-man to Denver (it weighed seven hundred and seventy-two pounds) and placed it on exhibition. It cost a spectator twenty-five cents to look on this wonder. After Soapy had gone North, it was learned that he himself had commissioned a destitute, bibbing sculptor to construct this stone-man from cement, a few teeth from a dental clinic, a supply of hair from a barbershop floor, and other physiological bric-a-brac, such as cow-horn finger- and toe-nails.

This hoax gave Tammen an urge to open a curio store, there to display sundry relics of the romantic West, to sell to tourists things symbolic of a rugged region.

"We must be pretty smart," Tammen said, "for Barnum himself was trimmed plenty here—and he was the baby who said a sucker is born every minute."

He referred to the costly visit of the great circus man in the '70's. Spry real-estate enthusiasts had managed to sell P. T. a huge "suburb" to the southwest of town. Later it was found that the "suburb" comprised only a few shanties with no water supply It still is called Barnum, in honor of the great museum and elephant entrepreneur.

Barnum had one solace, however. His only daughter, Helen, a guest of honor at President Grant's first Inaugural Ball, became happily married to a rising Denver physician, Dr. William Hermon Buchtel. He had doled out pills to Sherman's army on its march to the sea. Mrs. Buchtel lived a long and charitable life in Denver, working constantly to better the condition of orphans.

Dr. Buchtel was a brother of the Rev. Dr. Henry A. Buchtel, Chancellor of Denver University, later Governor of Colorado. I remember Chancellor Buchtel best for his Belasco collars and the fact that he was responsible for Grandpa's loud severance from the Methodist Episcopal Church.

Grandpa, at most, took his church with a sort of castor-oil bravado. On this Sunday it happened that the good Chancellor monopolized the sermon-hour with an almost lascivious plea for University funds. He became so enamored of his theme that tears as large as worn dimes slid down his cheeks, pelting the pulpit books with a tropical rain.

Grandpa rose from the Amen Corner—where Grandma virtually had forced him to sit—and stomped loudly down the aisle.

He turned at the varnished pine portal and shouted:

"Hey! Somebody put him out of his misery!"

There were adverse comments on this impious scene—although Colonel Bob Ingersoll sent Grandpa a note of congratulation.

"The years have dealt kindly with your judgment," the note read.

Grandma confided to her pastor, the barrel-chested Dr. Merritt. "Someday God will strike Bob Ingersoll dead."

If Tammen had a stimulus to establish a curio shop—such as Taylor's Free Museum in Larimer Street—that urge became final when "Spec" Lyons brought a bag of "Peacock Ore" to town. The meeting of Tammen and Spec occurred under robust conditions, involving an incident at Madam Jennie Rogers' house of forbidden pleasures.

Gentlemen, a toast.

4

A Big Rock And A Brainstorm

EVERYONE OF a sound judgment—in matters concerning women or mining—agreed that Madam Jennie Rogers had been physically superb. There was a difference of opinion, however, as to her intellectual excellence. Indeed, this moot point carried dozens of her former patrons into a stormy second childhood.

The town's superannuated petrels were arguing about Jennie's cerebral worth as recently as the outbreak of the World War. It required the magnitude of that event to quash the crutch-shaking debates of these veterans of the aphroditic G.A.R. Discussion of the Marne and Verdun eventually dwarfed the bedroom Antietams and Bull Runs of reminiscent roués. Presumably a gentleman cannot be conversationally amorous and patriotic at one and the same sitting. What a pity!

"A woman of dignified bearing," Colonel Jameson declared on his death-bed. "Cultured, refined, and I know for a fact that Madam Rogers graduated, *cum laude*, from an Eastern college. It was the year I helped lay out the first baseball diamond in Philadelphia. She was an omnivorous reader. The Classics."

The old Colonel rallied amazingly. The doctors had been trying to administer oxygen. "Just wheel that gadget away from me," the Colonel said. "I don't want any gadgets."

No one (including Buffalo Bill) could outshout Grandpa in regard to Madam Rogers' schooling—or lack of it. Allowing for a congenital prejudice against book-learning (General Wallace being cited as an example of what excessive brain-effort does to one), still we cannot brush Grandpa aside.

Grandpa claimed that "truth was truth." He could not see Jennie as a Madam de Staël. "A fellow," he would say, "should lie *to* a woman, if absolutely necessary. But he should never lie *about* one. Now that you ask me, Jennie Rogers never was 'in fifty miles of a college—and more power to her. She was the best-built lady in town, though—maybe the best-built anywheres. And that's what men mean when they speak of a woman as 'refined' and 'cultured.' Yes, she was the greatest of the Madams; tall, dark, beautiful, and ignorant of everything except her trade. As for old Colonel Jamieson, he actually was only a Corporal, and got feeble-minded in Libby Prison."

The Jezebels of Madam Rogers' rendezvous behaved circumspectly in public, and never spoke to a man if he were out walking or driving with his wife. They acted with social restraint while traveling to the race-course in a coach and six, with trumpeter and other fashionable necessities. At home, they dressed in gowns of the Directoire period, sipped wine with a Versailles elegance, and pretended they enjoyed the parlor verbosities of lieutenant governors, the newly gilt mannerisms of mining magnates. No rough talk was countenanced this side of the boudoir.

Not a few alumnae of the Rogers' Finishing School became wives. Their divorce rate was almost *nil*—which may, or may not, point a moral.

The Rogers' place was in Holiday (now Market) Street, a co-educational campus which flourished until George Creel became Police Commissioner of Denver. He locked the zone tight with a chastity belt of civic ordinance. Thenceforward, pleasure was boot-legged.

The tenderloin comprised four blocks of whoop-de-doo, with tremendous rental fees to politicians for cribs—shuttered sties, where a Port Said atmosphere obtained. There were occasional parlor-houses, such as the temples of Verona Baldwin, Madam Fay, Vesta King, Leona de Camp and Mattie Silks. But Jennie Rogers, the pioneer, out-queened all the Western hetaerae.

In the lesser and ruder houses, you might come across remarkable stories, were you a reporter or a person addicted to collecting data of a robust sort. There was Madam Carrie, for instance. Two hundred pounds of out-moded flesh. She maintained a place near French Lizzie's and on a side street. She sold beer for a dollar a bottle, keeping the brew in a closet, the key anchored to a piece of two-by-four cypress, which she used as a truncheon in time of stress.

When Carrie had reached the sentimental fifties, she became infatuated with a dopey McGimp, who played the pianoforte by ear. He would sit at the seldom-tuned upright and oblige with slow, sallow fingers and whiskey-tenor voice. No one knew him by any name other than the "Professor." He served as house musician from eight o'clock until eleven-forty-five o'clock each night. At that time, Madam Carrie would rise in the parlor to make a never-varying speech.

She would walk among the guests, the Professor's black derby in one hand and her lumber-laden closet-key in the other. As she moved, she spoke:

"Gents, a little something in the hat for the Professor. He's had a long, uphill struggle, Gents. Drove from home by his cruel stepfather, he has rose from the slums. If he had of become discouraged and crooked, he wouldn't never got to where he is at. Gents, the Professor is *a self-made man!*"

Then the Professor would rise from the instrument, where he had been playing and humming "Silver Threads Among the Gold," bow, receive his derby and its contents, and repair immediately to Chinatown's Hop Alley, there to take his nightly long-draw on an opium pipe, a self-made man.

One night the police surgeon found the Professor, his skull crushed. The cypress key-block was on the floor beside him. And so also was the Madam, dead, her throat seared by carbolic acid.

Spec Lyons arrived in Denver on a five o'clock evening train. He carried with him a carpet-bag. In it was a large chunk of "Peacock Ore." This is a gaudy, but valueless blend, containing copper, galena, and an abundance of iron pyrites.

Spec was a dour fellow, with set ideas. One of his ideas was that "Peacock Ore" was lucky. Just why he chose to cart around so huge an amulet is not clear—but he did.

It was difficult to find a cab at the Union Depot, the rush of tourists being great. Spec finally shared a vehicle with a venerable fellow, a stranger, who wore a plug hat and seemed opulent.

"My friend," the stranger said, "shall we go to the hotel, or drive directly to Madam Rogers' place?" The stranger fared well there, but not so Spec. A few hours later, he was standing at the Windsor Bar, pouring forth his troubles to Tammen.

"This here Madam," Spec said, "put me out, for no good reason. I ordered some wine, and when I wouldn't pay ten dollars for a quart anybody can get for three, she set me outside."

"Forget it, and have one on the house," Tammen said.

"I won't forget it," Spec said. "I won't forget it to my dying day."

"Then don't forget it," Tammen said. "What's that you're playing with?"

"Oh, this?" Spec held up a fragment of rock. "It's 'Peacock Ore.' Awful lucky stuff, too."

Tammen examined the rock. "Not worth a dime a ton," he said. "Got some more of it?"

Spec nodded sadly. "I *did* have, but this here Madam kep' my baggage. There's a big hunk of it in my bag."

"We'll go get it," Tammen said. He was interested in the "Peacock Ore."

"She'll call the cops," Spec said. "She claims she is entertainin' the State Legislature tonight. No foolin'."

"What of it?" Tammen said. "I want to buy that ore. How much?"

"I carry it for luck," Spec said.

"I know that. But how much for your bag?"

"Would five dollars be too high?"

'Yes," Tammen said. "But here's the five."

"Sold," said Spec.

"I'll pay on delivery," Tammen said.

They drove behind Tammen's black horse to Madam Rogers' place. The Madam finally came to the door. She was not pleased to see Mr. Lyons.

"Don't you dare come in," she said.

"Give this gentleman his suitcase," Tammen said.

"It's a carpet-bag," Spec corrected.

"Both you crooks get the hell out of here," said Jennie.

"The bag now belongs to me," Tammen said. "And I want it or else . . ."

Madam Rogers slammed the door. Spec looked appealingly to Tammen. "She's inhuman," Spec said. "Imagine, ten dollars for wine!"

"Never mind," Tammen said. "Come with me."

"What you goin' to do?"

"Just get in the buggy, and do as I say."

They drove to a livery stable. There Tammen had a talk with the proprietor. Before Spec could learn the purport of Tammen's actions, he found himself beside the Little Dutchman, and on the seat of a hay-wagon, behind a span of plodding grays. Tammen held the reins. He steered the horses patiently to the Rogers' house. Then he ordered Spec to help him unload the hay on the Rogers' door-step. There were sounds of discreet merriment inside, with a bit of song. The State Legislature was in session over its brimming glasses.

"I don't understand it," Spec was saying. He worked hard, however, when Tammen urged him to keep moving. "What we doin', eh?"

Tammen was holding a lighted match. "Now listen," he said, "when the excitement begins, you run inside and grab that bag. Don't hesitate."

"By God!" Spec said. "The Legislature's in there!"

"They'll move to adjourn," said Tammen, setting fire to the hay. It was like a Fourth of July exhibition. Tammen and Spec crouched behind a neighboring piazza. Suddenly there were screams, and in the distance the clangor of fire bells. And now, plunging from the building, came the ladies, and a sprinting, struggling array of solons, some of whom were wrestling with their suspenders.

"Now!" Tammen said, giving Spec a shove. "Don't come away without that bag, or I'll push you back into the flames."

Inasmuch as Jennie's place was mostly of cold stone, there was little damage done. The firemen were somewhat disappointed, as revealed in a subsequent conversation between one of the laddies and his superior officer.

"Chief," he said, "you know how they talk about Jennie's library? Well, I went in there with Mac and Nate. We rummaged through the place, thinking we'd find some interesting books, you know, and pictures."

"And?" said the Chief.

"Hell! All she had was some prayer books and a picture of a shepherd with some Bible stuff printed under it. Ain't that a funny one?"

From the big chunk of Peacock Ore grew Tammen's first fortune. Sickness, however, delayed his start. He took ill with brain fever. On recovering, he resigned as bartender at the Windsor. Then he got out his dearly won rock, cracked it into bits, labeled each piece with a grandiose description, and began selling the specimens for souvenirs.

Later he bought out Taylor's Free Museum in Larimer Street. He began creating one of the largest curio businesses in America.

"I started out," he would say, "with a big rock and a brainstorm."

Tammen also built up a brisk mail-order trade. To publicize his wares, he devised a pamphlet, which he called "The Great Divide." It was his first sally into the field of journalism. He devoted much time and many adjectives to composing this catalogue. In later years, and after he was a rich and powerful owner of the *Denver Post* and the *Kansas City Post*, Tammen continued to publish a weekly newspaper, purely because of sentiment, calling it *The Great Divide*.

I served as managing editor of *The Great Divide* in my youth, and I can say authoritatively that Tammen never gave a damn whether it made any money or not—nor did I.

Tammen's curio business was a success from the beginning. He sold arrows "direct from the reservation." These weapons were made by schoolboys, vacationing in the basement of Tammen's store. His moccasins never came from the wigwams of aboriginal manufacturers, but were sewn and beaded by gentle old ladies of the city, who wished to earn pin-money.

A romantic tourist could obtain at Tammen's font almost anything suggestive of the Red Man or the Wild West. There were "authenticated" head-dresses and war-bonnets of mighty chieftains. You might purchase War Cloud's baby-bonnet for five dollars, or the pugnacious millinery of Sitting Bull for fifty dollars. He sold annually as many as eighty "authenticated scalps" of foes "slain by Geronimo," and goodness only knows how many times Geronimo's own, personal scalp (concocted on the premises) was sold for a high figure. Tammen was one of the first dealers to supply Navajo blankets (made in Eastern textile centers) to tourists. And certain of his factory-chipped arrowheads, it is vouchsafed, are in museums of natural history.

"Sometimes," he once said, "I am led to believe that our workmanship surpasses that of the Indians themselves."

Tammen pretended to be hard-boiled, to be paying no attention to the sufferings of his fellow men, to be laughing at the gullibility of "suckers"—yet already he was anonymously, almost slyly, contributing to the support of indigent tuberculosis patients, to homes for aged people and to numerous other charities.

To throw snoopers off the scent of his philanthropies, Tammen would jest about the very institutions he was contributing to in secret. For example, he composed and circulated a sizzling joke concerning a certain sanatorium.

"Poor old George Carver is dead," Tammen told a friend. "Died at the F—— Sanatorium last night. Suffered like hell. Before he died, he called for B——, the owner, and for Dr. S——, the head physician. He asked them to stand, one on each side of his bed, and to hold his hands, which they did. Then he addressed them. 'Boys,' he said, 'I want to die like Christ—between two thieves!' "

5

Moon-Eye, The Indian Maiden

TAMMEN BUILT a house in 1890. He moved in before the workmen were quite finished—an impatience to exult in new toys being one of his characteristics. He sat at the bay-window, a pencil in hand, a pad of paper on his knee. He was very happy, indeed; and rightly so. The *objets d'art* of his flourishing musee had been safely transferred to a new address, near the post-office, where tourists and other people trooped. Furthermore, he had acquired a sensational "attraction."

This "attraction" was an anciently embalmed lady, salvaged from the effects of a bankrupt undertaker. The mortician—he had drunk up his River Styx profits, and himself was on the brink of one of his own graves—had been unwilling to part with this evidence of pioneer mummification.

"The Egyptians never done a better piece of work in all their lives," the indigent undertaker said. "Just take a look at *that* torso."

The doughty valet of death shed a few manly tears on selling his handiwork to Tammen. "Just keep her in a cool place," he said. "It was my masterpiece."

Tammen carted the nameless, lifeless lady to his new home, there to conjure an atmosphere for her world première at the Curiosity Store. The Soapy Smith influence was at work as Tammen scribbled an advertisement for Tom Patterson's newly purchased *Rocky Mountain News.*

The Little Dutchman whistled a cowboy song, "Bury Me Not on the Lone Praire-e-e-ee," and wrote his blurb:

VISIT ME SURE—IT'S ALL FREE

Indian Relics, Game Heads, Navajo Indian Rugs, Fine Minerals, etc., Large Assortment of Opals and Agates, Tigereye, Topaz, etc. and Beautiful Souvenirs.

MOON EYE, the PETRIFIED INDIAN!
DON'T FAIL TO SEE HER!!!

Yes, Life was God's greatest invention.

Tammen had all the tailor-made clothes he desired—memories of ill-fitting homespuns and funny Dutch hats doubtlessly influencing his taste for expensive garb. Despite the man's flamboyant air, his devastating speech, he was somewhat conservative in dress. He preferred quality to sartorial show. He wore white, custom-made shirts, with soft, roll collars and button-cuffs (this in a day that demanded chokingly high throat-boards, with starched wings for its dudes). He liked the feel of linen, but also enjoyed the soft gloss of silk, so he imported linen from Ireland and silk from Japan, ordering his shirt-makers to fashion his chest-garments with *linen backs* and *silk fronts.*

The Little Dutchman wore blue ties with polka-dot designs, sometimes four-in-hand, more often bows. His shoes always were shined and in excellent repair. He claimed he could read character from footwear. Under no circumstance would he employ a man whose heels were run-over, no matter how talented or how highly recommended the applicant.

As for hats, however, Tammen never achieved a technique in keeping them presentable. He would buy a costly fedora—usually the shade of old ivory—and the next day it would look as though a football team had used it for scrimmage. He carried his hat, crumpled up, in his hand. He was an immaculate fellow, and always looked scrubbed.

He smoked good cigars and drove speedy, black horses. His contemporary, Silver Dollar Tabor, had gaudy stables and gingerbread coaches lined with crimson satin. The town talked much of Silver Dollar's steeds. But Tammen, quietly, selected horses as pedigreed as those of any Western stud-farm. He drove much, inasmuch as he was a lifetime foe to any sort of physical exercise. Walking seemed a nuisance.

Tammen sat at his window, repeating the words of the Moon-Eye advertisement, as though it were a prayer, a namaz. He chanced to look through the window-glass and toward a place where his lawn soon would be planted. Then he became deflated, almost sad.

A tom-boyish lass of sixteen, fair-haired and rosy, was taking some laths from his lumber-pile. Now Tammen was not stingy. He would have given almost anyone a part—even all—of his lumber, were a decent request made at his door. But he was constitutionally unable to play the rôle of "sucker." It was about the only thing that made him sulk. It took weeks, sometimes months, to get over that sort of a calamity: such as the time he bet on the Chicago White Sox, and on a friend's advice, and lost his wager through that baseball scandal. And again, when he promoted a prize fight in the East, a "sure money-maker," and it turned out a fizzle.

Although he detested exercise, Tammen now leaped from his chair, dashed past the recumbent Moon-Eye, and jerked open his front door. He began shouting to the girl at the lumber pile:

"Drop those boards!"

The young woman had her arms full of laths. "Why should I?"

The simple, defiant question unhorsed Tammen. He advanced ominously. "So you won't do it, eh?"

"No," she said, moving swiftly to the gate. "And you can't make me drop them."

"By God!" he said, chasing after her. "You're plenty fresh!"

He pursued her for a block. Finally he decided he was getting too much exercise. He was panting as he turned back, mumbling to himself. So far as is known, this was one of the three times when Tammen exerted himself to the point of panting. Another time, he was seeking safety from a tiger, which had broken loose in circus winter quarters. On a third occasion, he was playing catch at the insistence of Elbert Hubbard. Tammen threw down the indoor baseball—a sphere fashioned by the eminent Roycrofter's own hands—amazing Fra Elbertus with the pronouncement:

"To hell with it; it don't prove anything!"

Tammen brooded for a time, but Moon-Eye soon re-occupied his attention. He gazed with a paternal sort of pride on Moon-Eye's weathered lineaments, then retired to bed with forty schemes in his ever-busy skull.

Next morning, he stood on his front porch, the copy for Moon-Eye's ad in his hand. Then he saw the same fresh young woman of yesterday, riding a bicycle. There were some school books strapped to the handle-bars. The girl saw Tammen as he squinted menacingly from his piazza. She waved rather impudently.

"Hey, Sis," he called. "Come here a minute."

He thought her exceptionally fair and beautiful as she dismounted and wheeled her bicycle to his gate.

"Well," she said. "I'm not afraid."

"Who said you was? Want to see a mummy?"

It was a slightly fantastic invitation. "I'll be late for school."

"Never mind that," he said. "I never went to school a day in my life, and look where I've got. Come in and see the best mummy God ever made. It's tremendous."

They went to the parlor, Tammen bursting with pride. He gestured importantly to the open packing case in which Moon-Eye lay in stoical calm.

"She'll have a beaded dress tomorrow," Tammen said. Then he launched into a spiel that had a slightly nasal quality, after the manner of medicine-show orators selling pigeon-milk and tiger-fat with guarantees of renewed vitality to the purchaser. Then he stopped short in his harangue. His audience had paled.

"What's wrong with you?" he asked. "You ain't sick?"

She was swallowing hard. "Kind of."

He seemed astounded. "It *couldn't* be the mummy. Something you ate for breakfast . . . ?"

She was backing away. "It's . . . it's not a real mummy."

Nothing could have given him greater affront than to have the authenticity of his wares questioned. "The hell it . . . I mean, it sure *is* a real mummy! Don't *you* try to tell *me* what it is! I got a certificate . . . Say! You *are* a little under the weather. We'll get some fresh air."

On the porch again, the girl breathed with obvious relief. Tammen was perplexed. "I don't understand it," he said. "It's a genuine Indian mummy. A girl, 'Moon-Eye'. Petrified, and a princess. She died three hundred and four years ago, and her tribe . . ."

The girl interrupted. "I'm going home."

Tammen held the bicycle for her. "Now if you're really sick, I'll send for a doctor. What's your name?"

"Agnes Reid."

"Live near here?"

She pointed. "Up the block. We came from Virginia last week."

"Not much of a State," Tammen said. "Can't compare with this. I'm a Southerner myself, a Dutch Southerner. You look kind of English."

"Scotch," she said. "Well, thank you, sir."

"You can forget the 'sir' stuff around me. People who need to be 'sirred' haven't got much else to hold 'em up."

She was on her bicycle now. "I'm sorry about taking the laths yesterday. Some boys dared me to do it."

"That's all right," Tammen said. "Want some more?"

"They were building a hut, and they dared me."

He laughed. "I know. The fellow who's got no nerve always dares somebody that has. Well, Sis, after I perfume Moon-Eye, she won't affect you. Come to the Curio Store. It's free."

During two years following the Moon-Eye introduction, Tammen watched Agnes Reid grow into womanhood. He fell in love with her. And she with him. They took drives behind his fast horses. Often they went to the higher prairie sweeps, there to watch the sunset and to hold hands. One of these rises in the plain adjoined a cemetery, which later became Cheesman Park.

Tammen never tired of seeing the sun sink behind the blue hills to the west of Denver. "The sunset colors are grander than the sky of sunrise," he would say. "It makes you dream."

They were married. Their life-long love became a legend of beauty. This woman understood the Little Dutchman from the earth up, from the outside in. She stayed in the background, always, permitting him to shout, to perform, and to play without restraint. They were complete in each other.

To the world, Tammen might pose as Robin Hood—and often as Yorick—but in his home he was kind, loyal, thoughtful, and forever young. No matter if he clowned abroad, nor if he boasted repeatedly that he "really was a confidence man, an ex-bartender, and out to trim the world," Agnes Tammen never questioned his behavior.

Sometimes, and in the homes of so-called people of importance, Tammen would make shocking pretense that he did not know what finger-bowls were for, or inquired "which fork do we use for this fish." The truth was that in his own home, Tammen enjoyed the finest viands, served in expert and proper fashion. He enjoyed his "act," and was his own audience.

Tammen built still another home for his bride. After it was completed, the front walk began to sag. The trench dug for installation of the water system caused this depression. Tammen said he would help the workmen fill in the trench with earth.

"It'll take about three hundred bucketfuls," the foreman said.

Mrs. Tammen, amazed by her husband's sudden decision to do *physical labor*, came out to witness the phenomenal event. She saw Tammen take off his hat, coat and vest, roll up his sleeves, fill a bucket and carry it to the trench. Then, on emptying the pail, Tammen set down the utensil, calmly put on his vest, his coat and hat, and walked into the house.

"It was the first, and *very last* stroke of real work Harry ever did," Mrs. Tammen said. "He became a guest in his own house. Always a guest."

Most of Tammen's fortune was tied up in his business. He was unable immediately to furnish his second home in a manner suitable to his taste. It was said his store and stock now were worth in excess of $150,000. He began to outfit his home slowly, a piece at a time. There were Navajo rugs from the Curiosity Store, and Western gewgaws. He purchased a music-box, and, although he knew little of music, would sit for long, listening to melodies of the day. The only other contrapuntal sortie by Tammen was made much later in life, when he sang to the diva, Mary Garden, in return for her having sung to him.

"It was a good trade, too," he said. "Mary complimented me highly."

In 1893, and a few days after Tammen had paid a down-deposit on a rococo table, the panic came. He lost half his business, the other half going to creditors in covering an eighteen-thousand-dollar debt. He easily could have gone into bankruptcy, but whether it was pride, or an inner honesty (which he loudly denied ever having) Tammen refused to be known as a bankrupt.

"Never mind," he said to his wife. "The table we've been eating off of is good enough. I'll make another stake, but I want you to promise one thing."

"What is that?"

"Never buy anything, whether a thimble or a rope of pearls, if you can't pay spot-cash."

She made the promise—and kept it to the letter.

6

Bloody Bridles And Bull Hill

TWO COLORADO gentlemen, George G. Merrick and Harley B. Morse, invaded the Philadelphia Mint on January 3, 1891. They brought with them a brick of silver, weighing 514.8 ounces, troy. They made a somewhat extraordinary demand in the office of the Director.

"Coin this silver brick into dollars for us."

The Director was annoyed. "Just take your silver brick back home. We won't do any such thing, and you know it."

"Uh, huh," Morse said, "we *thought* you'd take a stubborn attitude."

"Yes," Merrick put in, "and now we're going to the Supreme Court."

The Director rose. "That's as good a place as any."

The United States Supreme Court denied it had jurisdiction in this matter. The two Colorado gentlemen brought home their argent brick to a state seething with silver-talk. The Nation, too, was resonant with debates on the white metal. Senator Henry M. Teller was Colorado's chief expounder at Washington of that doctrine which called for "free and unlimited coinage of silver." There were two hundred and twenty silver clubs in the Centennial State.

These were pre-Bryan days. Actual advocacy of a free silver policy dated back to the convention of 1876, which had nominated Peter Cooper for President. The modern espousal, alarming Eastern money, was announced at Cincinnati, May 19, 1891. This unequivocal pledge harried Colorado's two veteran parties; the silver platform appealing to Labor, as well as to independent political groups.

When Colorado Republicans renounced silver, and the Democratic plank was evasive, the Populists elected David Hansen Waite as Governor.

There was a man for you! One who played 'cello music on a mouth-organ. He was well-intentioned, moral, but given to alarums and excursions. Waite stepped into one of the blowiest gales of State history, and began to dance a highland fling on a rug of nails.

There was the silver question, the panic of '93, labor troubles, strikes and political rape. This lean fellow grew sore beneath the goad, mounted a rostrum and bellowed:

"I should rather see our men riding through the streets in blood up to the bridle-bits than surrender our liberties to the corporations."

From that time on, he was called "Bloody Bridles" Waite. Born at Jamestown, New York, in 1825, Waite successively had been a lawyer, merchant, state legislator in Wisconsin, principal of a Missouri high school, a newspaper publisher in Pennsylvania, and then a combination lawyer-editor-politician-school board chief at Aspen, Colorado.

On coming into gubernatorial harness, Bloody Bridles set out nobly to pull the wagon of forward-looking reform. He is credited with laying the foundation for woman suffrage in Colorado—three sopranos reaching the State Legislature in 1894. He also sought to check graft in Denver's municipal government by transferring the appointment of Fire and Police Board members from the Mayor's province to that of the Governor. He thereupon named two of his Populistic subalterns, Orr and Rogers, and one Democrat, Martin, to the public safety board. Hardly had these men begun to function than Bloody Bridles charged that Orr and Martin were "unfit." He moved to oust them but the nearness of the panic diverted him temporarily.

On June 26, 1893, the mints of India closed to the coinage of silver. That far-away happening was a blow to Colorado, a leading silver state. The metal fell within the week from eighty-three cents an ounce to sixty-two cents.

Operators shut down their silver mines. It was then that Bloody Bridles made his famous utterance concerning horses wading to the throat-latch in human gore. The oration was misconstrued by the hysterical East as a threat of revolution. Money-changers of the Atlantic seaboard refused further issuance of capital for the West. On July 17, three Denver savings banks broke. Nine others suspended within three days. Business ceased.

In a moment of high spleen, Bloody Bridles resumed his fray of State versus Muncipality. He ordered a trial of Orr and Martin, declaring they had been protecting gambling and encouraging other nasty didoes. He appointed two new men to the Fire and Police Board.

"Just try and put us out," said Messrs. Orr and Martin, whistling for help. "He's a King Lear in a rocking-chair."

Three hundred pals, many of them gendarmes and fire laddies, answered the commissioners' tocsin. They took station in City Hall doorways and windows. That architectural abortion became a fortress overnight.

The City Hall was a porcupine of rifles. The Governor had one of his schoolmasterly brainstorms.

"Call out the National Guard," he yammered. "We'll reduce that iniquitous sink-hole to ashes! Tarsney! Where in hell are you, Tarsney?"

Tarsney was Adjutant General. Tarsney yelled for Brooks. "Where in hell are you, Brooks?"

Brooks was Colonel of the militia. Brooks began yelling for his captains; and so on, until everyone was yelling: "Where in hell are you?"

"There's apt to be a shortage of arms," Bloody Bridles told Tarsney.

"I don't think so," Tarsney said.

"Commandeer all the rifles of high school cadets," he said.

"They're not much account," said Tarsney.

"Get 'em anyway," quoth Bloody Bridles. "They look all right to me. And the Grand Army weapons, too, in their arsenal."

Thousands of citizens gathered in lower Fourteenth Street to attend the antics. Denver always was the rootingest town for blood-thirsty assemblages. The citizens never—to my knowledge—failed to risk life or limb in behalf of witnessing a lynching, a gun-duel, a foot-race between an adulterer and his beloved's husband, or any other spicy drama involving mayhem and assault on the person. After a fray, there usually were three lists of casualties given out: one for each of the two contending forces; and a third list—quite often the largest one—that of innocent bystanders maimed or slain.

The glamorous Chaffee Light Artillery, a symphony in blue, with triangular forage caps on their noggins, hepped along Larimer and into Fourteenth Street. They trained their cannon on City Hall. Colonel Brooks was a cool-headed fellow in the field; otherwise . . . well, the casualty list of innocent bystanders undoubtedly would have been as large as the Biblical roster of begats. Finally, his skull simmering down to a slow boil, Bloody Bridles decided not to risk the lives of non-combatants. Anyway, the courts were stepping into action on the matter of ouster. The National Government had sent troops, now bivouacking at the Union Depot, in case the whole town began to shoot.

"To your armories, boys," said Bloody Bridles.

In a few weeks the courts sustained Waite's ouster, but hardly had he drawn a long and bloodless breath than trouble came from the Cripple Creek gold-mining camp.

The Western Federation of Miners had their union men working an eight-hour, three-dollar-a-day minimum schedule in forty mines of the Cripple Creek District. Nine other mines were permitted to operate on a nine-hour basis for a minimum wage of $3.25 a day. Among the latter was the "Independence" Mine, discovered by Winfield Scott Stratton, an ex-carpenter who could hardly read or write—the same dashing fellow who once tried to register with a blonde lady at the fashionable Brown Palace Hotel; and when refused, went out, bought the hotel for $800,000 cash, returned to the lobby, fired the clerk, and moved into a bridal suite.

On January 17, 1894, the owners posted a notice at the shaft-house of the "Pharmacist" Mine. It proclaimed that all miners would have to labor for ten hours a day—taking lunch on their own time—and for $3; or else, work eight hours for $2.50. Similar ukases appeared at the "Isabella", at the "Victor" and "Anaconda" properties.

"Why the longer hours and reduced wages?" asked a spokesman for the union miners.

"The properties are not producing enough to pay the present wage," was the reply. "Overland haulage from mine to mill eats up the profits."

The union men cited a recent quarterly statement, in which large dividends were quoted as having been paid to stockholders.

There was a strike. Strikebreakers worked the mines. The union men gathered their forces on Bull Hill. Sheriff Bowers of El Paso County appealed to Governor Waite for help.

"Call out the guard," Bloody Bridles told Adjutant General Tarsney. Tarsney again ordered Brooks, now a Brigadier General, to take the field. The militia arrived at Bull Hill the morning of March 18. There were parleys. Then Bloody Bridles decided no emergency existed. His men had marched up the hill, and now they marched down again.

Once the boys with rifles and forage caps had departed, the miners entrenched themselves on the hill. Good old Sheriff Bowers was fighting mad at this sign of war. He swore-in 1,200 deputies and advanced on the miners' Gibraltar.

"It was a fine sight," said Grandpa, who, of course, was one of the boys on Bull Hill. "For two whole days we had 'em buffaloed. One of our comrades, Larkspur, a squarehead Swede, was a great hand at making wooden things. He got a big log, lathed it out to imitate a cannon, then mounted it and painted it. When Bowers saw that big gun, he said: 'How in hell did they get *that* piece up there? Move careful, boys!' "

While Bowers was brooding, a Duke-of-Wellington frown on his forehead, the miners suddenly descended from Bull Hill. They took position near the mines of Battle Mountain, above Victor. Then they captured the "Story" Mine in bloodless battle, confiscating the arms and ammunition of deputies guarding that claim.

"Everything was going peaceable," Grandpa reported, "till some firebrand amongst us" (and he winked knowingly) "set off a charge of giant-powder at the shaft-house. Well, a lot of machinery was made second-handed right off, and the engineer and foreman were imprisoned under the wreckage for a day and a half. Later we rescued 'em and threw 'em in the caboose."

Bowers' deputies deployed to Walker Station. On May 25, some three hundred strikers set out to attack Bowers. Unexpectedly they ran into ambush, and Bowers' shooting irons began to spit. Several were killed on each side. Bowers captured six miners. He released them in exchange for the engineer and foreman of the "Story" Mine.

Bloody Bridles boiled over again on May 26. He issued a proclamation, calling for the laying down of arms. Then he took a special train from Denver, arriving at the Bull Hill sector on May 28, and presiding as sole arbiter at a miner-owner conference. Just as he was about to reach an agreement, there was interference by outside parties. The disarmament conference of '94 ended, as most of them end, in tons of words and ounces of constructive results.

Barely had Bloody Bridles risen, deflated, from his peace parley, then he received word that other camps were about to send strike sympathizers to Bull Hill. In fact, a party of one hundred miners that very hour had seized a Rio Grande and Southern railway train, and had steamed a hundred miles thither.

"Stop that train," yelled Bloody Bridles. "Telegraph them. Tell them to return that train to its righful owners."

There was another conference in Denver, June 4. This time, Waite rose, beaming, from his task. He had put through an eight-hour day, plus twenty minutes for lunch, and at $3 *per diem*. There was to be no discrimination between union and non-union men.

That night in Cripple Creek the citizens were happy. The strike had hurt business worse than had the panic. Also, it had been unpleasant to be rousted about, whether by deputies or by strikers. The citizens heard speeches, and demonstrated in saloons and thoroughfares of the picturesque mining town. Buildings bloomed with Fourth-of-

July bunting and torches. Bands played. The strikers were relaxed. They mingled good-naturedly with the townspeople. Suddenly a sinister note sounded.

The merrymakers heard—disbelievingly at first—that Sheriff Bowers, with an army of more than a thousand deputies, was engineering a coup; that he was advancing from Divide, traveling toward Bull Hill in forced marches.

"Hell!" said the usually non-profane Bloody Bridles, when word of the apparent abrogation of treaty reached him in Denver. "Oh, hell! Call out the Guard!"

Adjutant General Tarsney got out of bed at the Hotel Alamo and instructed General Brooks to dust off the epaulets and whet the sabre.

There were heavy rains and washouts, delaying the boys in blue on their way to the field. Bowers' men had arrived at Bull Hill on June 6, slinging a bit of lead at pickets. Then they had withdrawn to Grassy, to await the troops. On June 7, there were skirmishes, and I regret to admit that Grandpa was ignobly wounded in the rear of his left hip.

"It's a damnable lie," he said. "I wasn't retreating. Just looking around. I'd dropped my chewing tobacco."

By a coincidence, this was the second wound of that character sustained by Grandpa. In 1859, and while lost from his wagon-train near Leavenworth, Kansas, grandsire caught an Indian arrow in his hip (the right, rear, *that* time). I forget his alibi, but it wasn't tobacco.

When General Brooks' bugles blew reveille the morning of June 8, he was encamped between the lines of deputies and miners. He looked from his tent and saw the deputies moving in three columns toward Bull Hill, as though to assault. He assembled his staff and galloped after the columns.

"What's the meaning of this?" he asked Bowers.

The Sheriff was sad. "I got no control over my men."

"You'll regain control," Brooks said, "or we'll open fire."

The deputies retreated, to camp near Beaver Park. Next day, Bloody Bridles telegraphed General Brooks to "go up the hill and receive the peaceable surrender of the miners." This was done. But Bowers' men broke camp, went to Cripple Creek and began to demonstrate. They thought it conducive to their morale to capture citizens, kick, slug and make them run gantlets.

"Just boyish pranks," said Bowers, when questioned.

On June 10, there was another conference, this time at Altman. The owners and miners reached an eight-hour settlement, with suitable wage-scale. The owners agreed to withdrawal of all deputies. The latter moved to Colorado Springs, there to receive their pay and honorable discharges.

There were trials of several miners, charged with committing acts of violence during the strike. Adjutant General Tarsney, a lawyer by profession, represented several of the defendants.

At midnight, June 23, and during the course of one of the trials, a group of fifteen masked stalwarts called at the Alamo Hotel. They compelled the clerk to summon Tarsney from bed.

"What's up?" Tarsney asked sleepily. "Is Bloody Bridles on the war-path again?"

The masked men began to belabor Tarsney with their pistol-butts. Then they trussed him, put him in a carriage, drove to a remote section of prairie, where a fire was going beneath a foul-smelling kettle. The men stripped the Adjutant General naked, then smeared him with tar and feathers. They turned him loose in the dark of night. He wandered, half-fainting, to a farm-house, where a rural couple relieved the yammering Adjutant General of his painful cocoon.

Such was the time and the scene in which Harry Tammen lost the fortune he had accumulated by dispensing liquor at the old Windsor and by retailing the authenticated scalps of Geronimo.

In these storms he read opportunity for anyone who might appear as a "Peoples' Champion," provided that one had courage, foresight—and showmanship. But how to become a champion?

A struggling little newspaper, the *Evening Post,* was for sale. Tammen coveted that newspaper. He could envision a day when, armed with a powerful organ, he might rise to the heights. Timber line!

But right now he was "broke." He would need capital to take over the *Evening Post*; and capital was the thing that Colorado then had the least of; so he decided to go East, and not to come back empty-handed.

7

Napoleon Of The Cornfields

TAMMEN INVADED Chicago's Midway with a portfolio of Jackson's Architectural Views—tinted photographs of the World's Fair. He managed to rustle coffee and cakes; that was about all.

His roommate at the Great Northern Hotel was a brilliant young Celt, Thomas H. Malone, later a scholarly priest.

"Tomorrow," said Tammen, "I'm going to throw my photographs into the Chicago River. Then I'll *have* to get an idea. A man without an idea might as well be dead."

The next morning, Tammen tossed his portfolio to the river. Then he began walking. He had come to the old post-office and was about to pass a news-stand window when he caught sight of an engrossed Declaration of Independence. He went inside.

"One dollar," the clerk said.

"Fifty cents," Tammen said.

"Sold for seventy-five cents."

Tammen showed his purchase to Malone. The future priest regaled him with a history of the Declaration of Independence and embellished it with tales concerning the signers.

"Why," Tammen said, "it's romance, this Declaration. It ought to be passed along to the children."

He found a print shop and interviewed the boss. "I want this etched for cheap reproduction."

While scouting the press room, Tammen chanced on a book of peculiar tickets. "What are these?"

The boss-printer glanced up from the Declaration.

"Oh, those? Some lottery tickets."

"Who for?" Tammen asked.

"A fellow named Bonfils. Know him? Fred G. Bonfils."

"No. What's his game?"

The printer chuckled. "He's a lulu! Kansas City. We do all his printing. This Bonfils has a million dollars salted away."

"Banker?"

The printer shook his head. "Not so's you could notice. Gambler!"

Tammen pricked up his ears. "A million, eh? How old is he?"

"I'd say about thirty-three or four. Went to West Point. Says he's a relative of Napoleon. Sounds kind of crazy."

"He's not crazy," Tammen said, "or he'd claim he *was* Napoleon. How'd he make his million? Lotteries?"

"He got his stake in the Indian Territory land rush of '89."

"A prairie-dog trapper," Tammen said. "Then what?"

"Well, he helped lay out the town of Guthrie. Then there was a little commotion. When the boys began peddling high-priced lots in Oklahoma City, what does this here Bonfils do? Why, he puts big ads in all the newspapers, offering Oklahoma City parcels for one-third the market value; but it seems the ground Bonfils sold was *not* in Oklahoma City, Oklahoma, at all, but Oklahoma City, Hemphill County, *Texas!*"

"A genius!" Tammen said. "Is that all you know?"

"Maybe I oughtn't to gab so much," the printer said. "I don't know who you are."

"I'm the fellow who's going to trim this Bonfils," Tammen said.

The printer laughed loudly. "Any old time! Why, just let anybody try and get a dime out of him. It's like childbirth. . . . How big you want this plate? Same size?"

"A little bigger," Tammen said. "Let's not short change the people of Chicago. . . . Say, what kind of looker is this Napoleon of the Corn Fields?"

"A dude," the printer said, "handsome and military, if he didn't wear such loud check suits."

"Big fellow?" Tammen asked.

"Say, you seem interested!" the printer said. "You ain't a copper by any chance?"

"No, a con man."

"Oh," the printer said. "Well, you see the authorities are after him. They claim Bonfils' lottery is crooked."

"They're narrow-minded," Tammen said.

"That's why them lottery tickets are still here. Bonfils was supposed to get 'em two weeks ago. But I guess he's laying low. Some new laws was passed."

"Laying low where, partner?"

The printer hesitated. "Well, it's none of my business, but he works his lottery from Kansas City, Kansas, and lives in Kansas City, Missouri."

"Skips out of one State into the other on short notice?" Tammen asked.

"Search me. Now, you was askin' about his size. Taller'n you, I'd say. About five feet ten and a half. Well set up. Like a boxer. Moves sure and fast. They say he's got a punch like a mule. Anyway, you wouldn't want to cross him. He's got the damnedest set of blazing black eyes that ever looked at five aces. A sharp-pointed moustache and curly hair, black as coal . . . How many of these you want printed?"

"I'll be back," Tammen said. "If Bonfils shows up, tell him Harry Tammen is out to trim him."

"You'll be the first one, then," the printer said.

At his hotel, Tammen said to Malone: "There's a big sucker named Bonfils in K.C. As soon as I get a stake, I'm on my way to take him."

Tammen interested the editor of the *Chicago Tribune* in his Declarations of Independence. That newspaper had a daily circulation of 84,000.

"Give one to each subscriber," Tammen said. "It'll promote your paper."

The reprints cost Tammen $2 a thousand. The *Tribune* paid him $7.50 a thousand and ordered 100,000 documents. With his profits, Tammen set out for Kansas City.

"You stay here," Tammen said to Malone. "I'll bring the sucker into camp in a week or less."

Tammen did not find his man in Kansas City. Many other persons were looking for Frederick Gilmer Bonfils, government officers being anxious to question him concerning the *Little Louisiana Lottery*, which they claimed he had been operating under the alias of L. E. Winn. That interesting enterprise, said the sleuths, was fraudulent, Mr. Winn offering a large capital prize and several lesser ones, himself winning the big plum through members of his own ring.

The more Tammen learned of Bonfils, the more determined he became to contact the Napoleon of the Corn Fields. Information was plentiful. The *Kansas City Star*, the farmers' Bible, had been attacking Bonfils. The owner of that great publication, William Rockhill Nelson, was determined to rid the municipality and State of the swarthy Corsican. The Nelson publication declared Bonfils had operated under several aliases, such as L. E. Winn, E. Little, Silas Carr, and M. Dauphin. The last-named, said Kansas City journalists, was a fabulous person, who "had died conveniently" during some scheme or other.

As for Bonfils' biography, Tammen learned that he was a native of Troy, Missouri, the son of a Probate Judge, whose name originally had been Eugene Napoleon Buonfiglio. The family—a respected and well-to-do clan—had changed the name for purposes of simplification. Fred Bonfils' grandfather, it was said, not only had been closely related to Napoleon, but as a boy in Corsica had played with the future First Consul.

The Judge, on retiring from the bench, became an executive of the Triple Alliance, an insurance company of Troy. He had military ambitions for his son. He was delighted when Fred was nominated for West Point. The young man entered that war-foundry in the late '70's, a classmate of John J. Pershing and Enoch H. Crowder.

Stories varied concerning Bonfils' leaving West Point. One had it that he was demoted from the third to the fourth class for deficiency in mathematics; that he tried for three years to meet the academic requirements, but finally was dropped in 1881. Another version was that he resigned voluntarily to marry Miss Belle Barton, a young woman of Peekskill, N.Y. At any rate he did leave West Point before graduation.

It was said that he labored briefly in New York City, merchandizing gas mantles from door to door. Gossipers averred he was ejected from a Knox Hat Store when giving a sales talk, and was so lastingly enraged that, during his journalistic days, he ordered a virulent editorial attack on Senator Philander Knox, who was not related in any way to the Knox Hat people. Bonfils hated the very name, Knox.

Still later, he worked in the Chemical National Bank, learning something of finance. Then, after his wedding in 1882, he moved to Canon City, Colorado. There he became drill master and instructor in mathematics at a military school. After a short sojourn in Canon City, Bonfils returned to Troy, where he engaged in the insurance business of his father, Judge Eugene Napoleon Bonfils. He heard so much of Napoleon that eventually he came to believe himself a *lineal descendant* of the Little Corporal.

Tammen gleaned all the reports he could concerning Bonfils and went back to Chicago. He had an inside tip that Bonfils not only was "resting" in Chicago, but, wonder of wonders, in retreat at Tammen's own hotel, the Great Northern.

As to what occurred when the Little Dutchman met Bonfils for the first time, one must rely upon Tammen's own account.

"I walked smack in on him," said Tammen. "He eyed me up and down, and said:

" 'Who are you? and what do you want?'

" 'Kid,' I said, 'I hear you've got a million dollars in safety-deposit boxes, and I'm going to shake you down for half of it.'

"This sort of floored him. He said: 'Sit down, and we'll talk it over.' Then he imitated me. 'Kid,' he said, 'if you get a nickel out of me, you'll be doing more than anyone else ever did, or ever will do.'

"Well, I guess I'm a pretty good salesman. I told Bonfils about the new crop of millionaires out in our country. Of

Stratton, Tabor and the rest; Dave Moffat and the boys of the gold fields. Then I told him about the *Evening Post*, a little sheet that could hardly pay its office rent. And we went into partnership then and there. And there never was a scratch of a pen between us. There were a lot of things he didn't know. What he didn't know, I did. What I didn't know, he knew. And between us, we knew everything."

Tammen and Bonfils took lunch together. The Little Dutchman had a first taste of Bonfils' shrewd and everlasting guardianship of that which belonged to him, when time came to pay the check.

Bonfils swiftly added the figures. Then he said: "We'll each pay for his own."

Tammen studied the lithe, handsome gambler, who began a lecture on giving large tips to waiters. "It only spoils a man," Bonfils was saying, "to overpay him—to give him something that he doesn't earn. Never do it. It means you are really afraid of something—of looking ridiculous perhaps—when you give away a big tip."

They walked out, Tammen holding Bonfils' arm. Malone saw them in the lobby. Tammen winked at Malone and made funny faces, pantomiming that he had "landed the sucker."

It was the beginning of as strange and lasting a friendship as the West ever witnessed. The obscure meeting of these men of opposite temperament, of opposite behavior, was apparently unimportant to anyone but themselves. Yet, it was to affect the lives and thoughts of hundreds of thousands of persons, shape the destiny of a city, a state—and, within certain limits, a nation. For, from this friendship was born a blatantly new journalism, called by some a menace, a font of indecency, a nuptial flight of vulgarity and sensationalism; by others regarded as a guarantee against corporate banditry, a championing of virtue and a voice of the exploited working man. The important thing was that everyone would have *some* opinion of the product of this union.

Tammen was a dreamer. Bonfils was haughty, dynamic, publicly taciturn (except when his Peer Gynt temper broke through and words or fists commenced to fly).

Tammen was blond, plump, blue-eyed, bubbling with

ideas, slogans, acrobatic phrases, resourceful under fire, swaggeringly indiscreet, a showman, an upsetter of custom and tradition, a foe of stupidity, a born heckler, a Puck with both hands full of firecrackers, a lover of indoor comforts and physical luxuries. Bonfils was dark, leanly athletic, black-eyed, austere, a watch-dog over his bank-roll, believing that money meant Power, and that Power was God Almighty, a brooding Sphinx, a man with a metallic quality in his rare laughter, a fierce and untiring worker, a Spartan in his habits, imposing on himself plain living and near-privation, a male Hetty Green, a Manfred, yet a preacher of outdoor sport, a man whose striking face sometimes betrayed, despite its practised set of jaw, a vague unrest, a thirst, a bitterness, a frustration.

Neither of these men ever seemed to grow old in mind or body. They didn't have time. And as diametrically opposite as they were in most respects, they had this in common: both were strong of purpose, both were aggressive fighters.

In later years, each would speak bluntly, often "disrespectfully," of the other. Yet, neither permitted an outsider to criticize his partner. And, when reunited after occasional business separations, these two fighting, crusading, rambunctious characters would meet in public, throw their arms about each other, and *kiss!*

The initial agreement between the partners was that Bonfils provide the capital—$12,500 for the purchase of the *Evening Post*—and Tammen "the publishing brains."

The partners arrived at Denver in October, 1895. They bought the *Post* property on October 28 of that year.

"It's a piddling little paper now," Tammen said to Bonfils as they walked arm-in-arm up Curtis Street. "But we'll wean it on tiger-milk."

On the day of purchase the partners entered the red sandstone Chicago Block, next door to the old Curtis Theatre. They mounted the stairs to the second floor, turned to the left, passed the "Gents" room and turned into the editorial offices of the *Post*. The newspaper lads were smoking and playing cards, their feet on derelict desks. They didn't pay much attention at first to the newcomers.

"Don't let us disturb you," Tammen said. "But we've just taken over this paper."

The card-playing ceased. Feet came down from desk tops. If husbands are the last ones to hear of their wives' amative strayings, newspaper men are the last to learn of the sale of a paper over their heads. And the eventual revelation seems to carry the same shocking quality in either case.

Bonfils called out in his best military baritone: "Go ahead and just keep on working."

In addition to the *Post*, Tammen and Bonfils acquired the equipment of the defunct *Denver Democrat*, founded in 1892 by George Herbert, later a political writer on the *Kansas City Star*. They changed the name of their paper from the *Evening Post* to the *Denver Post*.

When the first week had passed, Tammen applied to Bonfils for funds to meet the editorial pay-roll.

Bonfils looked at his partner wonderingly. "Why, no, Harry," he said, "I won't advance another cent."

Tammen was deflated. "Then how in hell are we to keep on publishing?"

"That's for *you* to determine, Harry. You *say* you have the brains. I *know* I have the money."

"You won't put up a few hundred more?"

Bonfils embraced Tammen. "Harry, when we bought this paper, I expected it to make money from the grass up. And it will have to do just that. Not one cent more do I put in from *my* capital. The *Post* has got to earn its own way."

"Great!" Tammen said. "It will."

He went from the editorial room to various railroad offices. It was before a day when the Interstate Commerce Commission shut down solidly on passes. Tammen obtained from each office a number of passes—whether on promise of advertising space or with veiled threats is not known for certain. He took these gratuities to lower Seventeenth Street, near the Union Depot, where ticket scalpers had their stands. He disposed of the passes to cut-rate brokers at bargain prices.

With the proceeds of these sales, Tammen met that week's payroll.

"Harry," Bonfils said, "you are a man of foresight and dependability."

The first of Bonfils' many newspaper crusades was launched—it was a drive against the *operation of lotteries!*

8

The Bucket Of Blood

DENVER'S JOURNALISTIC history began in 1858, when W. N. Byers of Omaha saw some gold dust being transported East. Byers, Thomas Gibson, John L. Dailey and Dr. George C. Monell set out for the West. As an afterthought, they put an old and small printing press in an ox-cart and brought it along.

These men arrived at Denver on April 17. They were amazed to find the town excited about the "newspaper." They rented the attic of a log cabin owned by a man named Dick Wooten. They began setting type, and on the afternoon of their arrival published a "dodger" for some now-forgotten citizen who had lost his dog. That same evening another paper called the *Cherry Creek Pioneer* appeared. It was issued by a man named Merrick. He promptly sold out to Byers for some flour and bacon—the city's first newspaper merger. The Byers' organ became the *Rocky Mountain News.*

The paper was popular but passed through strenuous times. The visit of Horace Greeley—the Arthur Brisbane of his day—was a stimulus to Western journalism. Greeley arrived to investigate high-sounding tales of Colorado's mineral wealth. And despite his hectic experiences in the wilds, Uncle Horace wrote laudatory articles in his *New York Tribune* concerning the Eldorado. His "Go West, young man, and grow up with the country," was looked upon by pioneers as words from Holy Writ.

Horace went on a prospecting hike. Mounted on a mule, and with an old white hat of copious brim pressed to his Socratic forehead, Horace came to the Platte. The spring freshets had made of that river a roaring tide. Uncle Horace plunged in and began to buffet the torrent. He was washed from the back of his palfrey. The awestruck settlers saw the mule emerge and the white hat go sailing down stream. The celebrated maestro of the quill was nowhere in sight. Finally, the Westerners espied editorial bubbles rising to the surface, and then the almost albino noggin of the learned writer.

Now a boat-hook is one of the rarest of utensils in Colorado family life, yet someone appeared with a boat-hook and made thrusts at the floundering leader of the Fourth Estate. It so happened that the rescuer's boat-hook got a purchase in the seat of Horace's trousers. But no matter, he was drawn ashore, half-drowned, but gaspingly grateful.

Another untoward happening occurred when Horace was visiting Palmer Lake, and decided on a shave. He was stopping at a small boarding-house conducted by Catrina, Countess Marat, and her husband, the Count, Greeley didn't know the rank or former station of these poor people, nor that the Countess one day would be called "The Mother of Colorado," because on July 4, 1859, she cut long strips from her prized red marino skirt and a field from the count's blue night shirt, and with some white silk (a memento of a court ball) fashioned the first American flag manufactured in the Rocky Mountain region. The visiting journalist was more concerned with barber-shop ministrations than with the future history of Colorado's Betsy Ross.

The porter-barber-count placed Uncle Horace on a three-legged stool and lathered him. The journalist moaned beneath the razor. When the operation was concluded, the count solemnly charged Uncle Horace five dollars! It is said that this costly tonsorial indulgence led Greeley to shun barbers thereafter and to grow that immortal crop of chin-whiskers which one may see in bronze, if one cares to look from the doorway of Barney Gimbel's New York store.

When Greeley, Henry Villard (of the *Cincinnati Commercial*) and A. D. Richardson, a noted newspaper correspondent, visited Denver and signed a verifying statement as to Colorado's gold discoveries, the *Rocky Mountain News* issued an extra edition. It was published Saturday, June 11, 1859. There was a lack of newsprint. The edition appeared on brown wrapping paper.

The *News* became a daily evening paper on August 27, 1860. In November, telegraph reports entered its columns. These articles did not come the whole way by wire, for the nearest telegraph station was 500 miles distant. Pony express and stage coach picked up the dispatches and relayed them to Denver. Thus the first news of Lincoln's election left St. Joseph, Missouri, by pony express the afternoon of November 8. The *News* published it on November 13.

With the *News* becoming a daily paper, Thomas Gibson established a rival evening journal, the *Rocky Mountain Herald*. The Civil War now was on, and both publications organized pony express services to obtain latest advices of the conflict. It cost a subscriber $24 a year to get either of these papers.

Cherry Creek, which flows through Denver to the Platte, always was an erratic stream, a puny, trickling thing in late summer and autumn, and in the spring often a furious flood. The swollen creek washed away the *News* building on May 19, 1864, and the paper didn't go to press again until June 27, of the same year. In 1866 a new building was erected at No. 369 Larimer Street, and when the railroad came in 1870, the *News* became a morning paper.

The *News* had been a Republican mouthpiece. When it was sold to W. A. H. Loveland in 1878, it changed to a Democratic daily and enlarged to eight pages. John Arkins bought a third interest in June of 1880, and with James M. Burnell and Maurice Arkins, took over the entire holding on March 15, 1886.

Thomas M. Patterson, one of the outstanding attorneys of Denver and a man of political ambition, purchased a third interest on August 9, 1890. In 1892, Patterson got control by buying another third, that belonging to Burnell. John Arkins remained as managing editor until 1894, when he died. Patterson's son-in-law, Richard Campbell, became associated with the enterprise. In 1902 they purchased an afternoon newspaper, the *Denver Times*, from David H. Moffat and his associates. That organ had begun as a theatre program in 1870. Moffat was a leading banker, railroad builder and capitalist.

There was another morning paper, the *Denver Republican*, an outgrowth of a merger with the old *Denver Tribune*, on which Gene Field had served both as managing editor and a pioneer Walter Winchell. The *Republican* had gone a conservative and laggard way under the ownership of Nathaniel P. Hill, the mining magnate. Still, it had been a school for many famous writers, beginning with Field, who, while editing that newspaper, wrote "Modjeksy as Cameel" and other classics. Frederick J. V. Skiff, later head of the Field Columbian Museum in Chicago, and O. H. Rothacker, an authority on the life of Disraeli, were contemporaries of Gene Field on the *Tribune*.

The day Disraeli died, Rothacker was absent from his desk. Gene Field set out to find him. He located Rothacker at Mattie Silks' bawdy house; the thin, bright-eyed authority on Disraeli was seated at Mattie's board, sipping wine.

"Your friend, Disraeli!" said Field.

Rothacker set down his glass. "Dead?"

Field nodded. Rothacker thereupon wrote three fluent columns which were copied all over the world.

It was Rothacker, spurred on by Field, who played a joke on Oscar Wilde. That literary gentleman was due in Denver to deliver a lecture. He was billed to drive through the streets with an escort of local brain-barons before going to the Chautauqua hall. Field dressed Rothacker in a rather alluring, lace-cuff outfit, a sunflower in his lapel. He rode in a phaeton with Rothacker, the latter impersonating Oscar. Rothacker responded with a tired, artistic smile and restrained doffings of a plumed hat to titterings of unmannerly crowds, two full hours before the genuine Wilde came to town. When Oscar found no one to welcome him, he was offended. His sponsors with difficulty persuaded him to fulfill his lecture engagement. It was then that Field originated the pun:

"That's what made Oscar Wilde."

Not the least famous of writers schooled on the *Republican* was Arthur Chapman, author of *Out Where The West Begins.* A sterling journalist in all branches of the craft, Chapman gained wide recognition for his daily verses.

The *Denver Post* began publication (as the *Evening Post*) on August 8, 1892. It was incorporated as the Post Publishing Company by Hugh Butler, George D. Herbert, Caldwell Yeaman, R. G. McNeal, M. C. Jackson, I. C. Cross and M. J. McNamara. The directors were W. P. Carruthers, A. B. McKinley, Charles J. Hughes, Jr., and Platt Rogers. The offices of publication were at No. 1744 Curtis Street. The paper did not prosper. Its owners were Cleveland Democrats. The state was for silver and Bryan. On August 29, 1893, the paper suspended publication. It was resuscitated in 1894, by the Post Printing Company, which incorporated for $100,000 with H. J. Anderson, William Kavanaugh and Frank Medina, Jr., as incorporators. The directors were E. E. Dorsey, D. W. Shepherd and J. J. Cronan. It was this company and this paper which Tammen and Bonfils acquired in the autumn of 1895.

Bonfils and Tammen immediately began sensational attacks. Almost anyone in authority was "unsuited to the people." Nearly all Governors "were bad for the state and unfit for office."

Tom Patterson, owner of the *Rocky Mountain News*, had exercised a dominant influence on Colorado affairs. In 1901 he was elected to the United States Senate.

"That's our man," Bonfils said as he sat in the background, while Tammen played the showman for the benefit of a growing public. "We must keep hammering at Patterson."

Brilliant as he was in the law and politics, and shrewd editorially, Patterson lacked a sense of humor, and directly felt the barbs of the upstart *Post* and its nervy owners. He tried to ignore the attacks at first, but eventually attempted to answer in kind. He was overmatched, surely.

Bonfils, under a standing-head, "So The People May Know," wrote editorials flaying the bespectacled and serious Senator-to-be. They were masterpieces of invective.

The partners were fortunate in obtaining the mechanical services of William Milburn, a gentleman of many qualifications in the newspaper business. He had been a printer on Bennett the Elder's *New York Herald*, on Bennett the Younger's *Evening Telegram*, and on William Cullen Bryant's *New York Evening Post*.

Milburn was one of the older Bennett's few favorites, largely on account of a prank he had played on Horace Greeley. Greeley, it appears, was gloriously absent-minded, but vehemently denied that failing. References in rival newspapers as to his mental lapses made his whiskers curl.

Uncle Horace, when snipping paragraphs from other publications for editorial reference, had a habit of pinning them to his lapel. Thus he wouldn't be so apt to mislay his cuttings during a spell of wool gathering. Often he took such work to Hitchcock's restaurant, near the old *Tribune* building in Park Row, later emerging from a repast, his lapel decked with so many paper ribbons that he resembled a heavy winner in a dog show.

Milburn and other *Herald* men were eating at Hitchcock's one noon when Horace and his work got together at table. Now, in that era it was customary at Hitchcock's for a patron to be seated and served without ordering. The meals were "as is" for the day: soup, beans, bread, potatoes, coffee and apple pie. A patron paid on his way out. This day Horace was over-busy with the shears. The water brought his viands, but the eminent clipster didn't attend them at once; he just kept on clipping, pinning the cuttings to his lapel for safe-keeping.

Milburn and his cronies stealthily appropriated Horace's food and drink for themselves. At length the hard-working Greeley was ready to refresh his editorial duodenum. He seemed mildly astonished at the sight of his empty plates and drained cup. He studied the ceiling, as though reading the stars, pursed his lips, then sighed, dabbed his mouth with a napkin, rose and went out, paying his tab.

Milburn wrote an account of this absent-mindedness for the *Herald*. Greeley was outraged. He penned in self-defense that it was "a shame and a vicious libel, this canard concerning my alleged absorption in work to the exclusion of mundane needs," and added: "I solemnly aver that I not only ate my full meal at Hitchcock's, as any other normal citizen might, but actually *ordered and enjoyed a second portion of beans.*" Next day, the *Herald* contained a facsimile of affidavits sworn to by Milburn and his cronies.

Milburn had an affection for Greeley, however, because of the latter's quality of repartee. One episode in particular delighted Milburn. Greeley was riding on the steam-drawn elevated in New York when he saw a man reading the *Sun*. Always eager to ascertain why anyone consulted a paper other than the *Tribune*, Greeley spoke politely to his fellow-passenger, first about the weather, then adroitly inquiring:

"My dear fellow, why don't you read the *Tribune?* It's a better paper, more informative than the *Sun*."

New Yorkers of old must have been about as polite as they are now in replying to strangers. "Oh," said the fellow-passenger, "I take the *Tribune*, too. I use it to—— my —— with."

"Keep it up," said Greeley, "and eventually you'll have much more brains in your —— than you have in your head."

Fifty-seven years ago, Bill Milburn went to Denver to die. He had tuberculosis. He arrived on a stretcher. When it became apparent that he had only a short time to live, he threw away his tonics, got off his cot (to the horror of a physician), began to keep late hours and to take a few highballs. Bill Milburn, now in his eighties, is one of the sturdiest of Denver's gentlemen, walks five miles every day and weighs about 210 pounds.

Gene Field, Bill Nye and Milburn were a busy trio. Nye, then publisher of the *Larimie Boomerang*, virtually had been discovered and "promoted" into the public eye by Field. Milburn worked in the composing room of the *Tribune*.

Although he was the gentlest and most amiable of souls, Field sometimes was as absent-minded as Horace Greeley, when it came to money matters. His wife, therefore, had to watch his expenditures. She one day entered an understanding with Charles Raymond, business manager of the *Tribune*, whereby he promised to keep Field's pay-envelope intact until she could call for it. Field learned of this. He took a messenger boy to the *Tribune* office, gave him a dime to carry a note to Raymond. The missive read that "Mrs. Field is waiting outside, and would you please send Mr. Field's envelope to her?"

Raymond obliged. Field spent that night in Taylor's Museum, where he paid $40 for an Australian emu. It escaped and lost a decision and its neck to a fire engine. Field wore a brassard of mourning for a week.

In those days, Milburn relates, Field was tall, slim, angular and beginning to get bald. He was full of tricks and was an enemy of boredom. He arranged a special chair for dullards. There was no seat to it. He piled papers on the frame, concealing the caneless bottom. When a bore called on him, he would say: "I'm busy. Won't you sit down?"

He became over-solicitous as to the condition of a victim's sprained rear.

Milburn became head of the *Post's* composing room. He was more than a mechanical supervisor, however. Tammen's experience with newspapers had been confined to his ballyhoo pamphlet, the *Great Divide*. Bonfils knew little about newspapers; they were something you found on the breakfast table. Milburn advised them frequently.

"Let's make 'em sit up," Tammen said. "Half the town is good, and half is bad. The good ones will read the *Post* to congratulate themselves on being so holy; the bad ones to see what we've found out about 'em."

The paper proceeded to out-Hearst Hearst.

"Yes," Tammen said. "We're yellow, but we're read, and we're true blue."

He formulated slogans that caught popular fancy. He called the *Post* "Your Big Brother" and "The Paper With a Heart and a Soul." Enemies added the words "and a Price." There were whispers of "blackmail" and "strong-arm methods" in procuring advertisements. But in the entire history of the *Post*, no case involving its hinted blackmail was substantiated in court.

It is likely this sort of rumor gained strength through Tammen's own proclamations. He would declare that he and Bonfils were "dangerous babies," that they had ways of knowing things, that it would be good business for merchants and others to put their ads in the *Post*. Bonfils, it is said, did not relish this brand of cacophony, *per se*, but inasmuch as it brought money into his drawer, he permitted it.

It must not be inferred that the *Post* made great headway in advertising during its fledgling years. Bonfils insisted that every cent of profit should revert to the enterprise until it could pay dividends without strain, and then only should the partners draw a percentage of the gains for their separate usage. It was thirteen years, to be exact, before Tammen received a dividend—but he prospered with his reorganized curio company, utilizing the *Post's* columns to sing the dubious scalps and imitation arrows.

Bill Milburn showed the partners how to "type" a paper. Tammen wanted huge headlines, and when black ink didn't seem effective enough, he ordered the boys to sprinkle *red* headlines among the black. The result was (and is) startling to an unwarned eye.

Tammen knew nothing about writing headlines—from a technical standpoint—but the fact remains that he was one of the greatest "natural" composers of sensational and arresting streamers the newspaper business has known. One day he criticized a lifeless banner line and substituted one of his own:

"JEALOUS GUN-GAL PLUGS HER LOVER LOW"

"But,Mr. Tammen," the copy-desk chief objected, "your headline doesn't fit!"

"Doesn't fit what?"

"We have no type that can take care of your line."

"Then use any old type you can find. Tear up somebody's ad if necessary."

The copy-desk man fidgeted. "If you don't mind my saying so, your head isn't good grammar."

"Well," Tammen said, "that's the trouble with this paper—too God damned much grammar. Let's can the grammar and get out a live sheet."

The prospering partners decided to move from their dingy old sandstone building to quarters more symbolic of their talents.

Bonfils, the business genius, had three prospective sites for future endeavors. He was holding back until landlords bid one another into the gutter, hoping to pay as little rent as possible. He wanted to go to Arapahoe Street, where the rent seemed "reasonable." Milburn, however, pointed out that another address—although more expensive—was preferable. This was a building across from the opera house built by Silver Dollar Tabor. It also fronted the post-office.

"A hundred people pass there where one passes the Arapahoe place," Milburn pointed out. "The *Post* should be where people can see it."

"And *hear* it," said Tammen.

So the *Post* moved to Sixteenth and Curtis Streets, a corner location. Tammen had a huge electric flag done in twinkling bulbs, to hang outside the editorial rooms. The walls of the partners' new private office were painted a flame color. They called it "The Red Room," but it became known to the citizens as "The Bucket of Blood."

The *Post* assailed almost everyone of prominence—were that person not amenable to the newspaper's tactics or useful to the *Post* politically or in business. Bonfils happened to write one of his masterly "So The People May Know" blasts concerning a distinguished but crotchety old architect named Edbrooke. Edbrooke had served under Col. John McArthur in the Civil War. He was not afraid of anyone—not even of "Tam" and "Bon," as the citizens now called the two publicists.

When several polite warnings by Edbrooke failed to stave off Bonfils' editorial lambastings, the architect hit upon a more efficacious plan. He sat daily at his window, across the street from "The Bucket of Blood," and whenever Bonfils looked that way, he saw the Civil War veteran *cleaning a rifle*. The attacks stopped at once.

It was in this room, "The Bucket of Blood," where the

partners became immensely wealthy, where there were dramatic meetings with powerful political allies, announcements of assaults on their corporate foes—for the *Post* took up the "people's battle" against the "vested interests"—and where the intrepid publicists met threats, and survived at least one murderous attack.

There were three large windows looking to the street, and two desks with a table between. A swinging door led from the editorial room to the red sanctum of the partners. On Bonfils' desk was a globe of the world, at which he often gazed with a proprietary stare. Within his reach was a sawed-off shotgun.

9

The Benign Masters

TAMMEN TREATED the staff with an almost fatherly kindness; Bonfils with a reserved, but just indulgence. There was a family spirit about the editorial room. Whenever a staff-member was in trouble, he could cry on Tammen's shoulder and receive some consoling philosophy, such as:

"Don't worry. Everybody thinks his own fire is the biggest."

So long as a man had ability, discharging his duties competently, there was no loud complaint concerning moral laxities. Bonfils himself was a temperate man—on rare occasions he would take a glass of wine—and Tammen was actually more moral than most of the town's "goodies," although he yammered to the winds that he was "A bad one under any and all circumstances." The staff could do almost anything it pleased—except ask for a raise in salary.

"We mustn't commercialize those artists whose talents are entrusted to our care," Bonfils said.

"Don't ask F. G. for a raise," Tammen would counsel. "He's the tightest guy in the world. The Shylock of the Rockies. Whenever I get a nickel out of him, I have to mortgage my immortal soul. I often wonder how I bamboozled him into buying this paper."

Tammen once borrowed a thousand dollars from Bonfils. Bonfils (who was his own bookkeeper) entered the loan in a little black volume, kept in his desk drawer. In recalling this incident, Tammen said:

"After Bon had gone home, I sneaked into his office and got the little black book. Sure enough, he had entered my loan. I took an eraser and rubbed it out. A week later,

I took another look at the book. Bon had put the entry back; so I rubbed it out again. A week after, I looked in the book, and there it was, the thousand-dollar entry. I got tired and gave up. He never forgot anything like that—but neither of us ever said a word about it."

In later life, Bonfils had a habit of delivering morality sermons to members of his staff. I think he really was sincere, growing genuinely patriarchal, but sometimes his lectures had a bird-call quality. When a dashing, fierce-eyed warrior in a checked suit suddenly bursts into a Billy Sunday hosanna, the victim's soul may be saved, but his mind is baffled.

I do not mean to infer that the *Post* paid niggardly wages. Bonfils and Tammen rewarded their "stars" with salaries comparable to metropolitan standards. What I do wish to say is that no one might expect an *increase* without besieging Bonfils' hermetically sealed pockets with every artifice known to science, and every struggle within the compass of brute strength.

Witness the case of James Barton Adams, a prolific versifier. Adams adopted a ruse of writing telegrams to himself, purporting fabulous offers from Eastern publishers. He would sign the name "Hearst," "Pulitzer," or "Watterson" to the wires. He had obtained one or two salary increases in this manner, but Bonfils spotted the sham.

One day Adams came in with a glowing "offer from Hearst." Bonfils bestowed on Adams' talent and work more praise than Lord Tennyson ever received from the critics. He ended by saying:

"And much as we hate to lose you, Jim, Harry and I are too big to stand in your way. So good-bye, my boy, and God bless you."

It was with tears and contrite heart that Adams decided he "could not bear to leave the *Post* family." So he stayed on, but *at a much lower wage* than had been his when he entered the "Bucket of Blood" with his spurious telegram.

The first ready-made writer to grace "The Paper with a Heart and a Soul" was Julian Hawthorne, son of Nathaniel. He accepted the then fabulous offer of $250 a week to labor for the "People's Big Brother," and for a month had the time of his life.

There was one mystery concerning the Hawthorne bide-a-wee. Few of his Denver works emerged on newsprint. Instead, Hawthorne's by-line appeared at random above stories which he never had seen, let alone written. He condoned that fact, however, because the *Post's* proprietors fascinated him. He couldn't make up his mind whether they were pirates or evangelists—or both.

Bonfils liked Hawthorne, and when he favored anyone, he became that person's moral guardian, with special emphasis on physical welfare. He decided that Hawthorne should become—like himself—a child of the great outdoors, one who communed with Nature.

Bonfils (when not angered) had only two loud characteristics. One was his raiment, the other his pæans in praise of Mother Nature. Hunting and fishing were two of his most insistent themes. Bonfils actually was not the best hunter in the world, but conversationally he seemed a one-man *safari*. Indeed, when it became unlawful to kill elk, he obtained a permit to slay one "for the museum, and in the name of science." He went to Wyoming, and for a week stalked a big and almost tame elk, wore out a guide, felled the elk, then collapsed from his exertions.

As to fishing, he offered a yearly prize of $50 for the biggest trout caught with a fly in Colorado waters. In later years, it is declared, Bonfils invaded a preserve, tried with a hook to entice a fat and lazy trout from a well-stocked breeding pond, and, failing in this, captured the lethargic monster with a wire loop, *winning his own prize*. Nasty rumors were current that he sometimes used worm-bait when no one was looking, a rather serious flouting of the fly-fisherman's code, a sort of below-the-belt blow at Izaak Walton's ghost.

When a new century arrived, Bonfils turned to golf. The snooty Country Club declined to elect him to membership—he had flayed too many of its wealthy members in print. So Bon and some associates formed the Lakewood Club, where he golfed, talked golf and lectured his putting comrades on the virtue of controlling one's temper.

"Where would I be if I gave way to my temper?" he would say.

If ever you had heard him "give way" to his temper, you might wonder how a man could survive such skull tornadoes.

In the Hawthorne era, Bonfils was consistently lyrical about fishing. He bemoaned Tammen's lack of interest in this divine pastime.

"There's nothing so thrilling as feeling a trout strike in the clear mountain stream. It would make a new man of you, Harry, my boy."

"Let's go fishing with him, for God's sake," Tammen said to Hawthorne. "But let him go on ahead; we'll pop in and surprise him."

Bonfils wasn't expecting callers at his camp the day Tammen and Hawthorne descended on him. He was upstream, his hipboots in the musical waters, his hook fluttering.

He occasionally affected a short-stemmed pipe during outdoor pursuits, and now he held a briar between his strong white teeth. He thought he heard someone calling, but the purling waters sometimes offered an illusion of merry voices. Finally he looked up and was thunderstruck.

He saw a city cab, drawn by two exhausted white horses. A driver and a red-faced gentleman were on the box. Halloas of a college-yell quality came from inside the conveyance. Then Bonfils saw a head poking from each cab door. One belonged to Tammen, the other to Hawthorne. The vehicle did not stop at the bank, but kept coming right into the stream! Tammen was leaning far out his window, yelling:

"Hey, Fred! Here we are."

Bonfils was chagrined. Hawthorne was getting out a rod and tackle. Tammen indicated the man on the box. "Meet our chef," he said. "The red-faced fellow. If I catch a fish he's going to cook it. The other guy's the cabby. If we catch a fish, he'll eat it."

Tammen began to "fish" from one side of the cab, Hawthorne from the opposite door. Bonfils moved farther upstream to brood over this sacrilege. He didn't have too great a sense of humor. Life was no joke—nor was fishing.

When Bonfils returned to camp that night, somewhat mollified by his good luck in landing a dozen beauties, Tammen was nowhere to be seen. Hawthorne had been napping; he couldn't say where Harry had gone. Bonfils was worried, so they began a search. At a cabin half a mile down the valley, they found Tammen. He had *bought* a feather-bed from the cabin owner and had paid a day's "room rent." And this moment he was asleep. He had *cut open the feather-bed* and was inside the ticking, among the feathers, *à la* Oliver Goldsmith.

"This is what I call roughing it," he said on being awakened. "Have my carriage ready at nine in the morning."

Near the close of his short-term contract, Hawthorne asked Tammen to assign him to some worth-while story. "I'd really like to earn my money," he said.

"If you're that worried," Tammen said, "there's some railroad officials going to make an inspection of the narrow gauge. Suppose you go along with them."

Hawthorne returned with a story called "The Dancing Mountains." Tammen proclaimed it a "marvelous work," but it didn't appear in the *Post*. A few days later, Hawthorne bid Tammen good-bye.

"Now that I'm going, Harry," he said, "will you enlighten me on one thing? How on earth could your paper afford to pay me a thousand dollars for delivering practically nothing?"

"Let your conscience be at rest, Kid," Tammen said. "Remember that yarn about 'The Dancing Mountains'?"

"You didn't print it," Hawthorne said.

"Of course not. It was so good I sold it to the railroad for an advertising pamphlet. We paid you a thousand dollars for a month's time. The railroad handed me twenty-five hundred for 'The Dancing Mountains'. Catch on?"

"Good-bye," Hawthorne said, "nothing can defeat you."

Frederick W. White, a facile essayist of New York, became drama critic of the *Post*. He also wrote Sunday articles, graceful and exceedingly erudite monographs, which lent prestige to the paper. White was well connected socially and an intimate of leading theatrical figures of the day, Belasco, Bernhardt, Mrs. Fiske, Mansfield and the rest. He was a member of the Lotos Club of New York and of one or two London clubs. He was the one person of whom Bonfils stood in awe. Tammen stood in awe of no one, living or dead, but had an affection for the cultured, talented White.

White signed his articles with the initials, "F.W.W." No matter where the *Post* was banned or its owners excommunicated, F.W.W. was welcome. There was an old-world manner about this stocky gentleman, a Washington Irving charm to his writings. He was the father of four children, one of them Lilian White Spencer, a leading authority on Indian folklore, and a poet of outstanding merit.

When the partners visited Mr. White at his Evergreen summer lodge, they did so in a special train. They did not pay for this train, but buffaloed the railroad into putting it at their disposal on a Sunday, a day when no regular trains ran to that mountain resort. Seemingly they could get anything they went after, were it a million dollars or a private locomotive.

F.W.W. was guest one evening at the Country Club—the organization which had deemed it inexpedient to admit Bonfils to its privileges. One of the diners, a Judge of the Federal Court, was charmed by White's conversational powers.

"Tell me, Mr. White," the Judge said, "how it can be possible for a man of your proven talents and culture to be on such a newspaper as the *Post?*"

F.W.W. lifted his brows. "Judge, there must be at least one gentleman on the paper."

The *Post's* drama critic wrote in long-hand, typewriters reminding him too strongly of a machine age. There was only one linotype operator who could decipher that handwriting with ease, and he had gone home late Saturday night. F.W.W. arrived at the composing room after having attended a performance at the Tabor Grand Theatre. (He was one of few critics anywhere on earth who remained until the final curtain, holding it unfair to players and public alike for a commentator to dash away at the end of Act II.)

He handed his article to Bill Milburn, the foreman. Bill was sad as he contemplated the critic's handwriting. He gave the copy to a substitute linotyper, a young man who was a constant playgoer.

"See if you can make this out," Bill said.

"I saw that play before I came to work tonight," the youngster said.

"Great," said Bill, "but be careful, F.W.W.'s as touchy as a bride's father."

White sought out Milburn the following Monday. He asked who had set up his article Saturday night.

"Why, wasn't it all right?" Milburn said.

"It was splendid," White replied, "but only the first line was mine. It was an excellent piece of work. What's the man's name?"

"Robert Burns Mantle," said Bill.

Mantle today is an authority on the drama, a leading New York critic. His late brother-in-law, Gene Taylor, also a newspaperman of prominence, and a city editor of the *Post*, had been a linotypist. The two young men went one time to a Chicago typesetting tournament. Taylor won the world's championship for speed, with Mantle the runner-up.

One of White's sons, Frank E. White, succeeded his father as drama critic on the *Post*. A most amiable and modest fellow, Frank's autobiography follows:

"When I was in my fourth year at East Denver High School, I was ordered to Principal Smiley's office for a reprimand. It was the good Professor Smiley's custom to keep a culprit waiting for hours—a period for much squirming and pangs of conscience. While I was thus shivering in this den of doom, a virile and bouncy fellow came into the prisoner's dock. He, too, had been sent up for moral repairs.

" 'What did *you* do?' he asked.

" 'Caught whispering,' I said.

"He studied me for a time, then said: 'Gee, it's a great day outside. Let's go out doors.'

" 'Hardly that,' I said. 'Think of the penalty.'

"He laughed almost too loudly. 'Think of nothing,' he said. 'Want to go with me?'

" 'I've got no hat,' I pleaded.

" 'Well, neither have I. Good-bye, my lad!'

"With that he dived, hatless, through an open window and was gone—gone to wealth and fame, for the young man was Douglas Fairbanks, Sr., and now look at him. And then look at me! I wonder if I made a mistake in not leaving my hat and going with Doug that day?"

Despite Mr. White's sally when questioned by the Federal Judge, there were gentlemen a-plenty on that fast-moving, colorful staff at the *Post*. I know of none finer than the Hon. Lyluph Ogilvy, younger son of the sixth, and uncle of the ninth Earl of Airlie.

The Earl and his second son visited Colorado in the wild and wooly '80's. Young Ogilvy fell in love with the country. A man of superb education and aristocratic background, he possessed nevertheless an earthy quality that endeared him to all. His knowledge of agriculture and livestock was second to that of no one in America. To accompany him to a stock show, to hear him talk of cattle and horses, was like listening to a reading of Homer. He was a Walt Whitman in build and dress, but with something less of beard—sporting a sandy, spade-shaped whiskerage which later had a path of gray down its middle. He was (and still is) tall, straight, handsome, with muscles like wire cables, and blessed with the highest type of human fire and daring. His legend is a glamorous and honorable one.

"To hell with calling yourself 'Lyluph'," Tammen said. "I wouldn't know what it meant in a thousand years. You're the son of an Earl, ain't you? Well, you're going to work for me, and you're going to be 'Lord Ogilvy' to me, whether you like it or not."

Lord Ogilvy maintained an estate near Greeley, Colorado. There he hunted to hounds and entertained with baronial generosity. He occasionally came to town, putting up at the Windsor.

When Ogilvy's sister, Lady Maud, visited him, he led a team of unbroken horses to meet her at the station. He waited until the train was about due before hitching these unbroken two-year-olds to the surrey. His grooms were at the bridles until Lady Maud and Ogilvy got in. Then there was action. The horses—as Ogilvy had foreseen—took a short, whirlwind route homeward. Lady Maud didn't say a word as the vehicle careened madly, with Ogilvy at the reins.

When they had finished their wild mile run, the spent horses gave a final lunge. The whiffle-tree broke loose, the surrey almost turned over, a cottonwood tree preventing. Lady Maud stepped serenely to earth and said:

"What a charming place you have, Lyluph!"

During her stay in Colorado, Lady Maud went on the box with a stage-coach driver—a place reserved for an honored guest. Ogilvy had put this Jehu up to driving as fast as his six could take the coach. Finally, the brakes smoking on a perilous hill, the admiring driver turned to Lady Maud and said:

"Ain't you just a little bit afraid?"

"I'm used to driving with my brother," she said.

Ogilvy is credited with having been the first person to take a taxicab ride in Denver. He was visiting with Buffalo Bill Cody at the Windsor in the '80's and had missed his train home—he lived some fifty-two miles in the country. As he left the hotel, he saw a steam roller. He appropriated it, fired the boiler, and started for the prairies. Whenever necessary, he stopped to re-fuel, to re-water his juggernaut, completing his journey with great satisfaction. Most of the wooden bridges between Denver and Greeley were condemned for three weeks following this ponderous visitation.

Lord Ogilvy was extremely witty, capable of heroic jests, but in no sense a clown. His pranks provided lasting merriment, but sprang from sheer spirit, a gay and driving spirit, never from buffoonery. He was slow to anger, but God help an antagonist if that anger did arise!

He and a friend were in the old Arcade, a gaming house of renown, when the killer, Jim Moon, swanked into that ornate rendezvous. Moon had just left a restaurant, where he had terrorized the occupants by slitting a tablecloth to ribbons, then shattering all available chinaware with pistol shots.

Moon had a grudge against the Arcade and its clientele, fancying they thought themselves superior to him socially. He strode between Ogilvy and the latter's friend, interrupting the conversation.

"Get out of here, Moon," Ogilvy said.

"Who the hell are you?" Moon asked.

"Get out of here," Ogilvy repeated, "and leave in as gentlemanly a manner as is possible for a mongrel."

Moon made a reach for his gun, but not quickly enough. Ogilvy's good fist alighted on the killer's chin, and Moon kissed the tile floor. Outside, and when he came to on the sidewalk, Moon swore he would shoot Ogilvy at sight. Ogilvy heard of this threat next day: that Moon was waiting at the Arcade.

He went straightway to the gaming house and found Moon at the bar. He said nothing, just stared at the killer. Ogilvy had a pair of blue-gray eyes that could look through any brand of bravado or hypocrisy. Moon couldn't withstand that diagnostic stare. He slunk away. Once out of focus, he became so incensed that he ran upstairs, where patrons were busy at a roulette table, and shot down two men! He escaped to the hills.

Ogilvy was a consistent friend and adviser of the farmer—and it is a tough task to farm that mountainous state, save on the western slope, where the soil is rich and fruitful. Ogilvy was wont to jest with farmers on occasion. One day he was dining at Tortoni's, Denver's challenge to Delmonico's, when a farmer resented a charge of twenty-five cents for a baked potato.

"Well," Ogilvy said, "didn't you enjoy it?"

"No. To tell the truth I didn't. Imagine charging twenty-five cents for a baked potato, when that's all I get for a bushel of 'em!"

"My friend," Ogilvy said, "all I can say is that the farmer is on the wrong end of the potato."

Ogilvy still works at his desk on the *Denver Post,* and he has not changed much in appearance or spirit. He was in rare story-telling form only recently, and remarked that the younger generation is "getting a bit soft." His son, Jack, had come home for the week-end.

"Now my Jack is all man," Ogilvy said, "but I was astounded last night when he asked for an *extra pillow*. It shows how civilization can affect even an Ogilvy. It made me recollect tales concerning the early members of our clan. We're Scotch, you know, and a braw people. Our clansmen fought among themselves a lot, mostly for exercise and from sheer joy of living. But whenever the English came up, the clan would quit fighting one another and get together to lick the whey out of the British and send 'em back to England. Then the Ogilvys would begin fighting among themselves all over again.

"Now it was the custom of the family warriors to travel lightly in the field. Just their weapons and a bit of oatmeal. They slept wherever the battle of the day had left them, never worrying one whit whether the ground were hard or soft, damp or dry.

"The first Lord Ogilvy had a seventh son about eighteen. They were engaged with the English, and at the close of the opening day's battle, His Lordship was making the rounds of the camp to see how his youngest cub was faring after his first major conflict.

"He came upon the boy, slumbering happily on the ground, his head pillowed on a rock, his claymore at his side. The old gentleman was infuriated. He kicked the rock from under his boy's head and shouted: 'Let it never be said that a son of mine was reared in the lap of luxury!' "

General Frank Hall, painstaking historian of Colorado (a four-volume stint), was mining editor of the *Post*. Although he had an outstanding list of journalistic feats to his credit, the dear old General lived in a state of constant alarm lest he be "scooped" on some vital mining story.

I don't think any of us was lacking in appreciation of the General's abilities, or wantonly disrespectful, but the fact is we used to watch him on his way hither and yon, and made bets as to how long he would retain his trousers. The General's pants continually hung at half-mast, in a baggy, zouave fashion. But he never lost 'em, nor was he scooped—two points in support of Tammen's theory that the only thing to fear in life is fear.

The General was a devotee of peace and quiet, but bore up beneath the noisy, riotous behavior of staff-members. His office-mate in late years was Fay King, the gay and brilliant pen-and-ink artist, now on the New York *Daily Mirror*. Fay followed Nell Brinkley on the *Post*.

This little girl, a whirlwind of laughter, curiously didn't disturb the peace-loving General. She had an almost child-like faith in the ore-critic's ability to retain his trousers to the last.

Fay was temperamental, capable of fainting when one of her numerous canary birds met with an accident, or of fighting a bob-cat if necessary. Whenever the General dozed off while penciling a prediction of "another bonanza in the Crede district," Fay would become alarmed, fancying the Grim Reaper was catching up with the snoozing General. At such times, Fay would run into the corridors, unable to repress a belief that the General was dead, announcing his demise with heraldic overtones.

"I wish you wouldn't report my death," he said to Fay one day, after nearly everyone except the coroner had responded to her alarm. "When my competitors learn it is not true, it shocks them that much nearer the grave themselves. I'll outlive them all—but I want to win fairly and squarely."

Field, Cy Warman and Chapman were the most highly rated of Rocky Mountain versifiers, but the leading minnesinger of the *Post* staff was Walter Juan Davis. Unshaven, unkempt, reefed in with an old imitation chinchilla overcoat buttoned only with the top button and swelled out at the bottom like the flaps of a wedge-tent, Davis for long years warbled in *Post* type.

His favorite drink was the "Bing Bing," half Bourbon and half water in a tall glass. He was author of:

"When your heels hit hard,
And your head feels queer,
And your thoughts arise
Like the froth on beer,

When you sit around
The whole night long,
And laugh like hell
At some damn fool song,
You're drunk, by God!
You're drunk!"

Davis died while working at the *New York Morning Telegraph* copy desk, the last port of call for so many brilliant news writers. He had started on the old *Sun,* gone West, then returned to New York—a voyage so often taken by the merry travelers of a merry business.

In preaching his gospel of a fairly good mind in an awfully sound body, Bonfils sometimes visited prize fights. He frequently attended the Coliseum, a temple of fistiana presided over by Reddy Gallagher, a quondam athlete of Cincinnati, and of whose right first the saying originated: "Let 'er go, Gallagher."

It was in Reddy's arena that Abe Attell and Young Corbett readied themselves for pugilistic titles. Young Corbett (born William Rothwell) was named after his manager, Johnny Corbett, and not, as was popularly supposed, after Gentleman Jim Corbett. Johnny Corbett was an immaculate fellow of much personal charm. He continued to dress in the styles of the '90's, no matter what sartorial improvements came to the market. He was as loyally proud of Young Corbett as was Sam K. Harris of Terry McGovern, and when Corbett defeated McGovern at Hartford, Johnny was as happy as Harris was blue. Young Corbett was the only ring-man I ever saw who could hit with terrific power and accuracy while stepping back, going away from an opponent, instead of toward him—a fistic paradox.

Among Corbett's trainers was Frank Newhouse,. Frank also was "keeper" of the erratic baseball virtuoso, Rube Waddell, left-handed Philadelphia pitcher. Newhouse later wrote amateur baseball notes for the *Post* and promoted state baseball tourneys in the name of that newspaper.

These tournaments were so popular that Bonfils decided to share in the prize-moneys which formerly had gone entirely to the competing teams. One year Bonfils held up the play, revising the schedule There was much confusion and adverse criticism. He had extended the tournament an extra day to glean added gate receipts, a matter of perhaps $2000 or less, and at a time when he was reputed to be worth in excess of $10,000,000.

It was Newhouse who unraveled the strange weight problem of Young Corbett. The fighter was inclined to take on fat. To enable him to make the then lightweight limit of 133 pounds, his manager sent him on the road for two-hour runs. Corbett would return to camp, perspiring profusely, and worn out, but the scales would indicate he had *gained* a pound or so.

Newhouse spied on the pugilist. He learned that Corbett would lope out of camp for half a mile, then rummage about in a pool, where he had cached numberless bottles of beer. Then he would sit down and drink for an hour, sleep another hour, then get up, have a parting couple of bottles and then sprint madly to camp, breaking into a sweat. In all, he took only one mile on the road—the rest on his back and at the business ends of brew bottles.

Gallagher eventually sold the Coliseum to the government for a post-office site, and at a price said to have been $300,000. Bonfils admired Gallagher.

"He knows how to keep his money," Bonfils said. "He doesn't mind if his critics call him close-fisted. It's better to be criticized for saving money than to be damned for becoming a hobo. It is immoral to throw money away, betraying a trust. It is an unforgivable sin. Gallagher doesn't have to smoke, drink or spend money to be successful and happy. He is a model citizen."

Otto C. Floto was sports editor of the *Post*. He had been a bill poster, subsequently a saloon man. He knew almost every one of importance in the sports and theatrical fields. He met Bonfils at the James J. Corbett-Robert Fitzsimmons championship fight in Carson City. Later he became acquainted with Tammen and began a journalistic career of great color.

Floto ranked with Bill Mizner and Wilton Lackaye as a raconteur. If he had seen fit to write as simply and as entertainingly as he talked, he would have reached genuine literary ground. As it was, he became an internationally known authority on boxing and wrestling. He seldom sat at his typewriter without a Thesaurus, and harbored a stubborn belief that polysyllabic words were trumps.

For example, when his pal, Bill Naughton, the foremost boxing commentator of California, died, Floto wrote:

"As we stand upon the threshold of grief this melancholy morn, there is an increased secretion of our lachrymal glands."

A prodigious effort by his intimates sidetracked Otto to a more simple vein. Then he wrote one of the finest tributes a sportsman ever composed in memory of another.

"Floto admitted when he took the job that he couldn't write," Tammen once said, "and the truth of the matter is I hired him because he had the most beautiful name in the world. It fascinated me; 'Otto Floto'. He begged me never to tell what his middle name was, and I don't blame him. It was 'Clement'. Parents will have to answer in the next world for putting handles like that on a kid."

"How do you account for Floto's success? Hard work?"

"Nuts with hard work!" Tammen said. "Any man is a genius, if you can only place him."

Floto was married twice. His second wife, Kitty Kreuger, was foremost bareback equestrienne of her day. Floto was sitting in a circus box when he first saw Kitty, "The Girl in Red." He rose and shouted his admiration while she was dancing tiptoe on the wide beam of a pink-eyed horse. This experience annoyed Kitty, but on meeting the personable Floto, she decided the best way out was to marry him.

Floto and Bat Masterson (the late boxing authority of the *New York Morning Telegraph*) were life-long enemies. Both were past masters in appraising pugilists, being America's foremost critics of pancratia. Let a gladiator make one lacklustre feint, the slightest error in leading, the least violation of rhythm in footwork, timing of punches or coördination of brain and fist—and these Dr. Johnsons of sport would reprove the offender with galleys of bitter type.

Yet, I remember, as a lad, an encounter on the street between these two interpreters of *le boxe*. Did they indulge in fancy steps, neat left hooks, graceful fiddling? Nay. They advanced like any two charcoal burners of the Black Forest, and began *kicking each other in the groin!* That event was richly symbolic of the critical poohbahs of any art, men who know every move, whether of pen, brush, violin-bow or naked fist, and yet themselves can find no bridge from the academic to the practical.

As a character, Floto was lovable, generous and loyal. It was a memorable experience to have worked for him in the *Post* sports room. He was proud of the boys who had served an apprenticeship there. Among those stars were Damon Runyon and Charles Van Loan. Runyon, born in Kansas, was a precocious newspaper cub in Pueblo. He went to the Spanish-American War as the youngest soldier in the service. Nicknamed "The Colorado Kid," he wrote camp verses of merit, then returned to work on the *News* and *Post*, and became just about the best all-around journalistic writer of our generation. The pranks Runyon played, in company with that elfin little cartoonist, Doc Bird Finch, were such as utterly belie the stoical photographs one glimpses of El Runyon in the Hearst newspaper columns. When he is in reminiscent mood, "The Demon," as Tad called him, can spin fantastic yarns.

I believe every newspaperman looks back on his sports-reporting days as the best of all. A reporter finds in that sphere a work that is play. The characters are of raw material, the games filled with direct action; it is a world which reflects one's own gay youth, without the intrusion of old men's defeatist mottoes and old women's dismal taboos. There are friendships which are life-lasting, and familiarity that does not breed contempt. And, again, a sports reporter has leeway in his writing, a privilege which sometimes makes for glib, sloppy composition, but more often makes possible the best stories in a paper—largely because one is free of the pussyfoot censorship of managing editors, a burned-out breed that too often stifles the energies, ambitions and talents of a young reporter. A pox on the old tsetse flies!

There were several notable women writers on the *Post*. Polly Pry was the first, joining the staff in 1898. She was extraordinarily beautiful as well as artistic.

Daughter of a Kentucky horseman, Polly married the president of the Mexican Central Railroad when she was but fifteen years old. She became bored with private cars and ambassadorial *soirées,* and at the age of seventeen went alone to New York City to seek a position from her family's friend, Editor Cockerill of the *World.*

"Give you a job nothing!" said the editor. "I ought to spank you and send you home to your husband."

He finally gave Poly a job. Later she worked for Street & Smith, writing a fiction story every week. She was traveling in the West with her mother when she met Bonfils. They were on a train through Kansas. She saw a copy of the *Post* on a seat, and the effect of its huge, red bannerlines was so startling that Polly gasped. Bonfils asked her what she thought of the paper.

"Does one *think* about such a monstrosity?" she asked.

"I'm the owner of that newspaper," Bonfils said. "Will you have lunch with me in the dining car?"

Polly found Bonfils a man of enormous ego, but not offensively so. He talked much of himself, his paper, and a little of nature as expressed in the bake-oven fields of Kansas. Polly was impressed by Bonfils' large, handsome head, his soldierly carriage and his checked, brown suit, with its wide black braid. She went to work for him, eventually becoming inured to the red headlines. Her daily articles were an immediate success, and there was no story she couldn't handle.

Another outstanding woman journalist was Winifred Black, who possessed a vitality in writing that fitted with the *Post's* style. Winifred had been (as she is again today) a Hearst journalist. Originally she had been a member of the Black Crook theatrical company. At the time she came to Denver, Bonfils' younger brother, Charlie, seemed to be on the verge of getting married.

In his rôle of moral and spiritual arbiter, Fred Bonfils decided that Brother Charlie must not surrender his status as one of the city's handsomest bachelors. It is said that Bonfils envisioned the fine-looking Winifred Black as one who might make Charlie forget brides in general. He never dreamed that Winifred would consider jeopardizing a most promising career by marrying less than a Duke.

Charlie fell in love with Winifred, and she with him. When Fred learned that his master-minding was to eventuate in Charlie's *marriage,* he cut up no end.

There was one child, a boy, born to the couple, and he became the darling of his Uncle Fred. The latter now had two daughters, Helen and May, but the prospect of a male to bear the Bonfils' name was a joy indeed. And with the coming of that child, Bonfils began to soften; that is, to grow more lenient, somewhat tolerant and less icy.

But a tragic star seemed to look upon anything humanly tender this man sought to keep for his heart's sake. The lad died, and something that had come for so short a time to leaven this man's darkling spirit died with the boy. His bitterness was all the more evident because he sought so savagely to conceal it—he was prone to deny that anything, love even, was capable of mastering him.

Once, when asked why he had quit eating ice cream, he replied:

"Becaue I like it too much"

Asked to amplify his remark, he continued: "Whenever I find I am craving anything, liking it beyond a merely normal appetite, I quit it entirely. A man never should allow anything—not even ice cream—to become too needful to his desires."

He drew away from so many things in life—and yet, he seemed forever to be seeking, thirsting, and not finding. On the other hand, his partner, Tammen appeared never to be seeking anything, but always finding something entirely acceptable, enjoyable to his soul.

So many other compelling characters on the *Post!* The alert and gymnastic George Creel, Wilbur Steele, the cartoonist, Paul Gregg, the painter of Western scenes, Paul Thieman, Hugh O'Neill, and later the gifted writer of fiction, Courtney Ryley Cooper, Leola Allard, Al G. Birch, and the famous daughter of the famous Judge James Burns Belford ("The Red Rooster of the Rockies") Frances Wayne Belford. Bide Dudley, Lute Johnson, Martin and Joe Dunn, James R. Noland, Robert Seymour, W. C. Shepherd, Frank Webster—the list is long, and there are anecdotes concerning each, such as that of Robert Emmet Harvey and his horse.

Harvey, the son of wealthy parents, was a reporter and a "fire bug." He kept a stud horse in the basement of his mansion, and in a fire stall. The bay stallion was trained to be hitched in three seconds. The city obliged this enthusiast of the flames by connecting its fire alarm system with a box in Harvey's bedroom. Beside the bed were Bob's trousers, tucked into his boots in regulation fire-fighter fashion. There was a greased pole from Harvey's boudoir to the stable of the oft-disturbed horse. When an alarm sounded, Harvey would leap into his clothes, slide down the pole, drop the breast-strap on his horse, snap the bits and holdbacks and then streak off to the blaze. There was a gong on the dashboard of his "buggy" which sounded like a Polish wedding.

Jimmy Noland, of police-reporting fame, once was Harvey's house guest. An alarm sounded during the night. Noland went with Harvey. As they turned a corner, the buggy struck the curb, capsized and catapulted Noland to the gutter, breaking his collar-bone.

Harvey, also a bit dazed, got up, cut the traces, and, leaping to the back of his horse, continued to the fire. He shouted back to the fractured Noland:

"No time for sentiment!"

10

A Simon Legree Of The Paste Pots

Of all the gay and salty characters who sailed under the *Post's* Jolly Roger, none surpassed the irascible Josiah Mason Ward in typographical skill or journalistic passion. He joined the paper in 1897 and for fifteen indefatigable years cracked the whip as city editor. He performed almost entirely behind the scenes, a pudgy, nearly bald, gray-moustached hetman, with a squealing voice that vibrated like the whine of a firework. He carried his thick pencil and over-long scissors as though wearing sidearms. He was habitually careless with these stork-billed shears, but never wounded a disciple—severely.

Born in 1858—God knows where—Ward eventually appeared on the journalistic horizon in California. He was going great guns in the Far West until fouled during a comedy of errors. That experience transmuted him from a plumply generous human into a savant, a wheezing cynic, and finally a slave driver. He wore his spectacles on the end of his short nose, and when he snorted with cynicism the glasses jiggled. Whenever his harvest-moon brow grew an apoplectic scarlet, his blue-button eyes mounting in candle-power, I think the condition was induced by a re-opening of the old psychic wound, a smouldering heartache re-kindled, with ancient devils dancing a can-can in his reminiscent skull. His tongue became a knout.

A lady, of course, was the goddess of the machine in Joe's California decline and fall. It is well authenticated that he did *not* love her, nor she him. There's where the tragi-comedy germinated.

The lady—let's call her Pepita and have done with it—was beloved and pampered by an editor, Joe's superior officer on a Far Western newspaper. The doting wooer lavished on Pepita's limbs and digits a plethora of gifts, such as bracelets, necklaces, rings and other gim-cracks. A love-nest and bolts of rose-point lace from Parisian marts were editorial tokens of the quill-driver's yen.

Pepita (the scented hussy!) was what the boys call a chiseler, a two-timer and a baggage. The while she was accepting amatory emoluments from her daffy swain, she was giving of her heart to another, gratis. The recipient of Pepita's Sunday kisses was a homely, gawky, Ichabod Crane of a copy-desk lieutenant, a glutton for rum, a sulky sponger, a limping malingerer who claimed his head ached with Byronic genius, a snaggle-tooted Casanova, four-ply alimony dodger—in short, the kind of barnyard stooge that only a blind mother—or Pepita—could love.

Although the editor adorned his Delilah with constellations of precious rocks, stuffed her with hothouse victuals and plied her with social-register wines, he never allowed her a privy purse. A subconscious censor prevents the crossing of palms with silver, in love idylls of this ilk. Hence, Pepita was unable to give more than her body to the dog-faced boy of the copy desk. And, if she pawned any of her jewels—which, alas, I am compelled to admit, was proposed by the shameless cuckolder—the editor would be sure to learn that he was wearing horns as big as any bull moose.

A barnacle, attacking himself to the nearest hull, Burt, the copy reader, became friendly with Ward. Not only did he borrow from Joe's salary, but used Ward as a male chaperon. That is to say, he and Pepita would take Joe to semi-public drinking parlors to lend an appearance of innocence to the dastardly liaison. This is an almost medieval dodge, which I have christened "The Three-legged Stool" or "Farmer in the Dell."

On the night of debacle, Joe accompanied the lovers under protest. His own girl, a lass of incendiary temperament, was cooking a Bavarian dinner for Joe. He had promised to be home not later than seven o'clock. But Burt was a fluent scoundrel, no matter how drastically nature had short-changed him as regards appearance. Joe agreed to convoy the couple—shivering, however, when he thought of his own sweetheart's temper and of her thwarted Tyrolean dinner.

They visited one of the better cafés and preëmpted a booth, curtained by green baize hangings festooned for drinking and other privacies. Joe, of course, bought the liquor until his purse went limp. Then Burt began running up a bill, the proprietor finally refusing to risk further service.

"Jus' nev' mine," said Burt. "T'hell with it. Jus' you pe'pul stay here w'ile I go borra s'money."

Burt left Joe and Pepita at table and went into the night. He did not return. His idea, perhaps, was to permit Joe to wiggle out of the bar debt as best he might. Suddenly Pepita had an idea. (They always do.)

"Why, Joey," Pepita said, "why, this is the very place Mr. Blank and I come to a lot of times. Let's drink up and *charge it to his account.*"

"A pious thought," Joe said.

Pepita called the manager. "Just put the whole thing on Mr. Blank's bill," she said. "And bring us some more drinks. What're you having, Joey?"

Joey wanted Bourbon. Pepita was disgusted. "Joey!" she said reprovingly. Then to the manager: "Wine. Keep bringing us wine. Mumm's *Cordon Rouge.*"

As the night progressed, Pepita grew playful. Joe forgot the Alpine dinner and his waiting sweetheart. Pepita draped Joe with gems, taking her chemise ribbons with which to tie rings to his pink ears. As a grand finale, she insisted that he sit in her lap, which the squatty, gay fellow did. They engaged in loud baby-talk and sang *Münchener* drinking songs, with yodels. There was all manner of innocent fun until the green baize curtains parted.

And, standing there, pale and almost inarticulate, was *Mr. Blank, the editor!*

Pepita tried to laugh it off (you know the kind) and shoved Joe from her lap. She began to dance a fandango toward her benefactor, but ye editor stiffened and said:

"I'm sorry I disturbed you." Then he gazed with a Banquo-like stare at the plump Bacchus, who had pearls instead of grapes on his brow, and departed.

This visitation sobered Joe and he fled Pepita, now weeping like a widow in Wall Street. He suddenly remembered the Bavarian dinner and his *Eidelweiss* sweetheart —one more river to cross. He had no money for cab hire and had to walk up and down sad municipal hills. When he arrived at the home of his lamb-love, he endeavored to weasel his way inside. But Fate had yet another uppercut to deliver before the count of ten, a second shot from a double-barreled fowling piece.

The lights flashed on, Joe saw his sweetie standing on a balcony (but wholly unlike Juliet). What was she holding aloft? A large silver tray, heaped with dishes. Before Joe could duck, down came the tray and all.

"There's your God damned dinner!" yelled Joe's heart-of-hearts.

He did the only possible thing—two things. He left his sweetheart forever, and resigned his job. With the money due him at the paper, he started East. He got as far as Denver, and there he stayed to become famous. He shunned vivacious women as though they were the bubonic plague, for not only does a burned child dread the fire, but a strangled puppet avoids the strings. He spent his time between city editing and writing a learned book on the ancient East, called "Come With Me Into Babylon."

Joe Ward's fringe of hair had grayed, but life was big within him. He was the greatest journalist I ever knew, and could smell a news story farther off than an ambitious mother can scent a movie director. He was a stickler for typographical accuracy, and if a reporter misspelled a name or misstated an address, Joe would simulate an attack of angina pectoris, and then brandish his shears with the threat:

"I'll cut your guts out and strew them all over Sixteenth Street!"

He was the only person I ever knew whom a reporter could love profoundly and hate entirely at one and the same instant. He could detect a lie and disembowel an excuse, and his sarcasms were of Voltairean proportion.

"You ought to be arrested for impersonating a human being," he would yell.

He wrote a hand which none but a student of cave-dweller runes might decipher. It was his custom to scribble an assignment on a small bit of paper, put the note between the points of his shears and thrust it toward a reporter, who felt as though he were ordained to pluck a hot coal from tongs. If the victim were trembling from a hangover, Ward would hold out the shears, then, as the culprit reached for the assignment, withdraw the scissors quickly.

"Look!" he would pipe shrilly. "Just look at your hands! Trembling! Shaking! Drunkard's palsy! God, Must I be at the mercy of quaking dipsomaniacs?"

One day he gave such an assignment slip to a new reporter, a fellow who said he had labored for the *New York Times*. The chap took the little paper to a corner of the room to read it. Joe roared:

"Get out on that story! I'll give you a book to read when you come back."

The new reporter went to the Press Club, an historic retreat, and submitted Joe's slip to the veterans there.

It was beyond the powers even of Ward-trained scholars to decode. It read:

"M. M. E. & C. 2nd F."

"Damned if I know," said Jimmy Noland, the police reporter. "It must be one of Joe's tests."

"He's crazy," the reporter said.

"No, hardly," said Jimmy. "Better keep trying to puzzle it out. Have a drink?"

The reporter had several. He brooded and stewed. He got drunk. He stayed drunk two days. On the third day, his courage fortified by three more shots of Bourbon, he returned to the *Post*. The other reporters were getting their assignments, via the shears. The new fellow edged into the queue at Joe's desk.

"You'll have to wait," Joe said.

"Wait, hell!" the reporter said. "I'm quitting. Nobody in God's world could make out the assignment you gave me."

Joe's spectacles jiggled. "Got it with you?"

The new man threw the riddle to Joe's desk. "Yes. And you stick it!"

Joe pulled his spectacles up from the tip of his nose, where they seemed to fit best. He began reading aloud: " 'M. M. E. & C. 2nd F.' What's hard about that? It means 'Man murdered, Ernest and Cranmer Building, second floor.' " Joe looked up from the cryptogram. "Now go to your desk and quit bellyaching. Lucky I sent another reporter on the story; you couldn't have gotten it anyway."

Bonfils once wished a cub on Ward—owners of newspapers so often bedevil their editors with friends of friends of theirs. This lad came to Ward's desk on the second day almost frothing with excitement, his eyes bulging.

"Mr. Ward, I've got a sensational story!"

Ward had been writing Babylonian paragraphs during a lull. He pushed his glasses up, snorted and licked his lips. "Well, what is it? Murder?"

"No, sir. I was walking last night past the East Denver High School, when I looked . . ."

"Is this a story of your own life?" Ward asked.

"No, sir, but I was walking . . ."

"Well," Ward said, "quit walking and get down to facts."

"Well, I saw a lot of young people, and they were doing . . . they were . . . I don't know just how to put it . . ."

Ward folded his arms and leaned back in his chair. "Like Adam and Eve?"

"Yes, sir. I counted fifteen couples. Right in the shadow of the high school."

Ward cleared his throat. "Son," he said, "this thing has been going on for thousands of years, maybe millions. Evidently your parents didn't enlighten you, so I'll try to describe the effect—if not the cause—of this phenomenon. You can preach against this practice, denouncing what you saw, in every pulpit of the nation. You may write books attacking it; you may create plays condemning it; you may call out all the armies of the Lord, as well as of the United States Government, to suppress it—but, son, mark my word, you never will succeed in making it unpopular with the masses."

Mr. Ward then returned to his Babylonian chapters.

Joe was a competent drinker, and he patronized Tortoni's bar. Ordinarily he carried his liquor with a defiant waddle, but sometimes (usually after Bonfils had come to a Vesuvian boil and decided to shake the ground) Joe would become overserved. Never, though, did he neglect his work. His legs might bow beneath the liquid burden, but the keen, knife-like mind seemed undulled, and he would be even more apt than usual to ferret out misspelled names.

"A man," he would say, "will stand for almost any insult sooner than have his name garbled in print."

Once, and after he had left the *Post* to become editor of the *Republican,* Joe took too much firewater aboard. The same could be said for that beloved cartoonist, the late Clyde Spencer. Spencer had been trying for three swaying hours to draw a cartoon. Finally he brought his bristol-board to Ward and began to shed large tears.

"I'm licked," Spencer said. "I'll never be able to draw again."

"What's wrong, Spence?" Joe asked.

"Everything," Spencer said. "Just look at this cartoon. I've tried and tried, and now just look at it. A botch."

"Well, don't cry," Joe said. He examined the picture. "Who is it?"

"Uncle Sam. Something's gone wrong, but I can't figure out what."

Joe pondered the drawing. Finally he said: "Yes, he's dressed like Uncle Sam all right, same clothes and build, same hat, but it doesn't look like him." He called in a sub-editor, MacLennan. "Mac," he said, "Spence here has drawn a cartoon. Does it look all right to you?"

"Who's it of?" Mac said.

Spencer and Ward spoke in unison: "Uncle Sam."

"Well," Mac said, "it may be Uncle Sam, but *where in hell's his whiskers?"*

The news policy of the *Post* was unconventional; some

critics have described it as unethical. The owners encouraged editorial comments and opinions in the news columns, so much so that they discontinued orthodox editorials, with the exception of those captioned "So The People May Know," grape-shot attacks that depended on Bonfils' vacillating moods.

Ward, although a competent, shrewd, agile news man, allowed himself to participate in editorial-laden news reports.

The *Post* once saw fit to assail David Moffat, the best-known of Denver capitalists, railroad builder and banker. Moffat was of J. P. Morgan reticence when it came to newspaper contact. Ward sent a reporter to "interview" Moffat, instructing him to pose extremely personal questions concerning Moffat and his concerns.

Joe was laboring over his Babylonian opus when the reporter returned to the city room.

He looked up, pushed his spectacles back from his nose-tip and squeaked: "Well, what did the old buzzard say?"

"He didn't say anything except 'No'," the reporter said.

Joe went back to his literary grind. Without looking up, he said: "Write it, but be brief; we can't use more than two columns."

Tammen (a constant visitor in the city room) was at Ward's desk one morning when a young reporter reeled in.

"What's wrong, son?" Tammen asked. "Drunk?"

"No," the man said. "Sick. I just saw a murder. It was awful. Right before my eyes. Awful! And right after I had breakfast."

Tammen waited for the young man to talk himself out. Then he said:

"Go ahead and write it. Never mind the story about the murder. There are lots of stories like that. Just write about how you felt when you saw it, and how you almost lost your breakfast. Write about yourself, not the crime. The readers will get *that* out of your story. Write it just as you told me. If I like it, it's a good story."

Joe Ward, the slave driver, fooled not one of the old

hands on the *Post* with his snarls, his bellyaching and his hetman antics. Underneath that pudgy chest beat a sentimental heart. The great old man would sneak into his office at the *Republican* (where he worked a year away from the *Post)* and, thinking himself unobserved, would remove a bit of bread or cake from his pocket, crumble it and place it on a window sill for the birds. He also kept a pan of water there for the callers from the sky.

In 1902, Joe's book, "Come With Me Into Babylon," was published. It received praise from archeologists the world over. This grand old reporter never had been within five thousand miles of Babylon, but he knew how to smell out news, even if it were buried beneath the winged bulls of ancient palace gates.

11

Fife And Drum

A KITE, IT seems to me, is the most symbolic of toys—a child's bid to the gods. So frail an implement, and so slender its string—as man is frail and mortality a slender cord—yet it rises bravely, for all to see, its rag tail an insolent whip at the cloud's flank, or again, remotely soaring, an exclamation point in the sky.

What disillusionment when the once intrepid visitor to heaven runs afoul of chimney pot or telegraph wire, and is revealed, shorn of mystery and challenge—an unheroic crumple of paper, a few pitiful sticks.

A newspaper is a kite, doomed to keep moving or perish; tricky to navigate; quick to fall at the slightest interference, and once down, soon forgotten. Like any celebrity.

The *Post* flew high and kept moving. Its advertising revenues grew with increased circulation. When the partners acquired the paper in 1895, its circulation was in the neighborhood of four thousand copies. On July 2, 1898, the masthead claimed for the *Post* a circulation of 27,136 in a city of 133,000 inhabitants. It sold for ten cents a week, including the Sunday morning edition, and Tammen devised a slogan:

"Delivered anywhere on earth. Kick like a bay steer if you don't receive your *Denver Post*. Call Mr. Bonfils or Mr. Tammen personally."

Tammen was a man of spontaneous action. He arrived quickly at decisions, depending on his showman-like instincts for results. Bonfils, contrawise, took each day's edition home, sat up until nine or ten o'clock, checking and re-checking the columns, meditating on what had been left undone. Next morning he was ready to tear his editors and their work apart, filling them with awe and inspiring a belief that he could see the smallest detail of journalism at a glance.

Tammen paid little attention to the mechanical departments. Bonfils was here, there and everywhere, trying to catch the typographical union napping. Although he installed up-to-date machinery to supplant old equipment, once it was in place, Bonfils hesitated to spend money on upkeep.

Foreman Milburn had a "trouble-shooter," a mechanic who stood by in case of breakdown of machinery. When Bonfils saw this man *sitting* in the composing room, apparently doing nothing, he complained.

"We ought to get rid of him," Bonfils said. "All he does is sit. We've no time or money for sitters."

"Whenever you see that fellow in his chair," Milburn explained, "it means that we are in shipshape. When he is not in his chair, there's trouble."

Milburn urged that an emergency electrical system be installed, in case the regular one got out of kilter. Bonfils refused. One morning he went to the composing room, finding the "trouble shooter's" chair unoccupied.

"Where's that sitting fellow?" he asked.

Milburn was rather dejected. "We're having trouble. If you had put in that emergency system, we wouldn't be held up now. As it is, we're liable to miss the mail edition."

"We can't miss it," Bonfils said. "No matter what."

"The trains won't wait," Milburn said. "The thunder storm has got us down; if we fix the wires by one o'clock, we have a chance. But it means the linotypers will have to break a record."

"Let them break a record," Bonfils said. "We mustn't miss the mails under any circumstance."

The machines were in operation by one o'clock. "Boys," Milburn told his operators, "I know it's asking a lot, but if you'll pitch in and work doubly hard, it'll be appreciated. We *must* make the mails."

The men put on full speed, "railroading" type to the make-up stones without stopping to correct proofs. A few minutes past three o'clock, Bonfils entered the composing room.

"Will we make the mails?" he asked.

Milburn was proud of his men. "Yes, by breaking our neck."

"I want to see you in my office," Bonfils said.

Milburn followed him downstairs, very happy to think he would receive a compliment for the unprecedented two hours of typesetting—perhaps a cash bonus for the record-breakers.

Bonfils drew the printer's time-sheet from his desk. "Milburn," he said, "I want you to take this pay-roll and cut it in half."

Milburn was astonished. "I don't understand."

"Well," Bonfils said, "you may not understand, but I do. The men today have proved they can accomplish in two hours the same work it has been taking them eight hours to do. So cut them in half."

"I'll be damned if I will," Milburn said. "They may break their backsides for me, but they won't for you."

Merchants of Denver tried to boycott the *Post*, objecting to the high advertising rates asked by that paper. In reprisal, Bonfils and Tammen launched a child labor campaign against leading department stores of the City. Bonfils released daily thunder in his "So The People May Know" column, flaying the dry-goods barons.

"Little girls, poorly dressed, and with pale faces are employed in these department-store sweat shops," he wrote.

Bonfils took up the cudgels against each of the boycotting merchants in turn. For example:

"As you entered Monarch's Fair, a little, pale-faced girl opened the door, standing in the dangerous drafts, for people kept coming and going. This method of employing child labor at starvation wages, which are all the way from $1 to $1.50 a week, for which they work from sixty-five to seventy hours a week, may be fashionable in Baxter Street, New York, but we desire to serve notice upon these establishments that it will not go in the Great

West. Let every lover of childhood, let every person who has children of his own see to it that such conditions cannot find favor. In this battle, the *Post* is right, and 'thrice armed is he who hath his quarrel just'."

No matter what commercial reason prompted the attacks, the *Post* was thereby enabled to push through decent and humane child labor laws. In this phase of activity Bonfils excelled—benefitting himself, exercising a revenge on corporate interests, and at the same time bettering conditions of the poorer people of the region. In the child-labor campaign, the *Post* had the support of Judge Benjamin Barr Lindsey.

It is unnecessary to write at length of Judge Lindsey's life and work, for he became Colorado's best-known citizen, widely acclaimed abroad and utterly unappreciated in his home town. Persecuted, ridiculed, lied about and hounded (ending with his abominable disbarment in recent years) this little fellow had more courage than any fifty of his detractors. It is impossible to know how many boys he advised, guided, kept from criminal careers or prison, how many girls he safeguarded from public shame.

His Juvenile Court is his monument. His Companionate Marriage premise has been garbled, misrepresented and falsely analyzed, but it is not unlikely that the Judge has more than a glimpse of the future—and may prove to be a century ahead of his time.

The war with Spain afforded the *Post* an opportunity to exercise its brightest red inks and loudest eagle screams.

A perusal of bannerlines and subheads during the tiff with Spain is a somewhat interesting history of that war. Interspersed with the war-lines were other "human interest" local headings, for Bonfils had a gospel in favor of home news.

"A dog fight in a Denver street is more important than a war in Europe," he would say.

Juicy crimes in states other than Colorado received due attention, and pictures (especially of women) were used whenever possible. When attacked in pulpits or women's clubs on the ground of sensationalism, of catering to the mass moronic mind by playing up crime and criminals, the *Post* owners said they did this to show that "Crime doesn't pay." As an earnest of good faith, they sprinkled little black-face lines of type throughout the paper, usually closing a tale of morbidity with the grace note:

"Crime never pays."

When William Henry Theodore Durant was hanged at San Quentin Prison, January 7, 1898, for killing girls in Emanuel Baptist Church, the *Post* headline was:

**DEMON OF THE BELFRY
SENT THROUGH THE TRAP**

The *Post* headlines of 1898, samples of which are here given, form a motion-picture series of lap-dissolves, showing the progress of the war.

CLINK OF HANNA GOLD HEARD AT COLUMBUS

UNCLE SAM A CAT'S PAW

AMERICANS NOT SAFE AT HAVANA

WARSHIPS READY FOR THE SIGNAL

MARTIAL LAW RULES CUBA'S CAPITAL

WAR'S ALARM MAY SOON BE SOUNDED

UNCLE SAM TO TAKE CUBA UNDER HIS WING

**MAINE'S HULL WAS
PIERCED BY A SHOT**

PRESIDENT LEAVES FOR WASHINGTON
WHERE WAR'S ALARM WON'T DISTURB

GRIM-VISAGED WAR
IS LURKING NEAR

WALL STREET GUIDES
THE SHIP OF STATE

McKINLEY WILL NOT
UNLEASH DOGS OF WAR

WHAT McKINLEY HAS NOT
DONE IN TWELVE MONTHS

UNCLE SAM READY
FOR CASTILIAN FLEETS

BEARS SHOUT WAR AS WALL STREET
TREMBLES

IS IT THE LULL BEFORE THE STORM?

DONS MUST DANCE TO YANKEE DOODLE

PEACE AT ANY OLD PRICE

TRYING TO HOLD
McKINLEY STEADY

McKINLEY ALMOST PERSUADED

CUBA MUST BE FREED
OR SPAIN MUST FIGHT

CALL TO ARMS SOUNDED
AND BOYS IN BLUE ARE
MARCHING TO THE SEA

**THE WAR EAGLE SHRIEKS
ACROSS CUBAN WATERS**

**OLD GLORY NOW FLOATS
OVER THE PHILIPPINES**

AN EYE ON THE SPANIARDS OF DENVER

**SCHLEY'S GUNS BRUSH AWAY
DIPLOMATIC COBWEBS AT LAST**

**UNCLE SAM BEWILDERS DONS
WITH PYROTECHNIC DISPLAY**

**SHELLS FROM SAMPSON'S SHIPS
PUNCH HOLES IN MORRO'S WALLS**

**ADMIRAL CERVERA'S FLEET IS NOW
BOTTLED UP IN DAVY JONES' LOCKER**

**VICTORY CROWNS AMERICAN
ARMS AND EAGLE SCREAMS**

**GHASTLY PICTURE OF DEATH
IN THE SPANISH TRENCHES**

**NO POWER IN EUROPE
DARES TO INTERFERE**

IT WAS EASY TO REMEMBER THE MAINE

**THIS IS BEGINNING TO LOOK
LIKE CRUELTY TO ANIMALS**

SPAIN NOW CUTTING THE OLIVE BRANCH

**HAUGHTY DONS NOW DO
SEEK SOLACE IN PEACE**

DEATH LEERS FROM THE TOP OF EVERY HILL

DONS PLANNED TO SKEDADDLE IN THE NIGHT

YELLOW JACK AT SANTIAGO

BANNER OF FREEDOM IS FLYING IN TRIUMPH

With the war ended, the *Post* returned with redoubled vigor to its "So The People May Know" crusades.

12

So The People May Know

WHEN LOCAL coal companies refused to charge less than $5 a ton for their commodity (and wouldn't advertise too extensively) the *Post* founded its own coal concern. The fuel dealers howled with rage, for the *Post* began selling its coal for $3.75 a ton, and with Tammen's slogan:

"A full ton and an extra lump."

When the "Coal trust" shaved its tariff to meet the newspaper's figure, the *Post* leased several coal mines and listed their product at $3.50 a ton. The opposition newspapers, the *Rocky Mountain News, Denver Times* and *Republican,* nicknamed the *Post* partners: "Nut and Lump."

The *Post* attacked high water rates and recommended municipal operation of the privately owned plant. The tramway, headed by William G. Evans, came in for a sustained roasting in "So The People May Know" blasts, which dubbed the trolley-car chieftain "Napoleon Bill."

The corporations had a friendly organ in the *Republican,* and it was charged that corporation heads virtually controlled the *Denver Times.* While Tammen thought up ways to make people talk, such as sending the spiritualist, J. McIvor Tyndall, blindfolded, and driving two horses through the city in *successful* search for a hidden needle, Bonfils sat back and dipped his quill in high-powered acids.

Concerning the *Times,* he wrote as follows:

"The *Denver Evening Times* long cut a figure in our local history. Once it was a prosperous and even dominant factor in the Denver newspaper world, but it has long since fallen from its high estate, and the beginning of the end is now at hand. It has reached that stage when it is losing its own self-respect. In a few brief weeks it will find none so poor as to do it reverence. The fatal blight of monopoly has long been upon it; it is now in the business not to publish the news, but to suppress it. From an organ of public opinion, it has degenerated into a paid hireling of the political corporations, and the discriminating reader looks askance upon its opinions, even when they happen to be right. Inane, hysterical, a poor, suffering creature, that is blown thither and yon by every passing breeze, it is reaching that pitiable stage when neither excuse nor apology will justify its further existence. The poison had gone to the marrow, and it is now not a question of months, but of weeks, when the *Times* will turn up its shriveled toes to the daisies, and the haunts that once knew it will know it no more."

This diatribe was released by the *Post* on a day when the proprietor of the *Times*, Earl B. Coe, was absent from the city. When Mr. Coe returned, he was upset. The *Times* belatedly came out with headlines:

"F.G. BONFILS FORCED TO LEAVE KANSAS CITY BECAUSE HE SWINDLED POOR PEOPLE."

The article contained Bonfils' alleged aliases as a lottery operator, and continued in part:

"Patience has ceased to be a virtue with the *Times*, and this paper, after having suffered repeated lying, malignant and blackmailing attacks, proposes to inform this community as to the character of the person responsible for these criminal and blackmailing articles regarding the *Times*.

"Who is the man with so many aliases that thus seeks to traduce and malign institutions and citizens of Denver? Who is the man that, posing as a champion of the people through his paper, built his fortune swindling the laboring classes of another state out of an enormous sum of money, that he might operate at some other point for the purpose of adding to his gain? The *Times* has looked up and partially gives his history and character.

"The people of Denver are invited to read the career of crime and swindling practised by this person and as published in the papers of the place, from which he came to Denver."

The article went on to allege that Bonfils' departure from Kansas City was made under escort of the police department of that municipality, and that he had carried with him "a sense of deep relief from impending disaster, and behind him he left a score or more of ugly indictments not well calculated to fittingly adorn the escutcheon of a gentleman and an honest, self-respecting citizen."

The article continued:

"Once removed from these disquieting surroundings, and safely aboard a westbound train, he contemplated the future. Fortune had not been unkind to him, from a pecuniary view. He was generally credited with a bank account approaching the million mark, accumulated a dollar at a time by sundry and divers methods of financial diplomacy. He had been thrifty, as thrift goes in the purlieus of a great city. But, unfortunately for his peace of mind, his ambition did not cease at thrift. He yearned to shine as a financial magnate, a man of affairs and a social star. These privileges had been denied him in the community in which he had risen from poverty to affluence, and where the public persisted in recognizing him as a cheap political grafter, a fake lottery swindler, a tinhorn policy flim-flammer, an all-around bunco steerer and an enemy to society in general.

"The realization of his ambitions had also been hampered by the persistent complaints of hard-working laboring men, who had contributed their earnings to his numberless swindling schemes, of ignorant servant girls who had been made the dupes of his duplicity, and of widows and orphans who had been victimized by his dishonest craft. Finally the penitentiary yawned for him and he departed for pastures new, where his odorous fame should be unknown and his past be blotted out.

"By what lamentable mischance Denver became the final objective point of his wanderings is purely a matter of conjecture. But history records that to Denver he came, and upon the register of a prominent hostelry inscribed the name of Fred G. Bonfils. As a matter of prudence, doubtless, the numerous aliases by which he had heretofore been known were not added to the list, and so to local fame his full identity was not revealed.

"Had Mr. Bonfils' peculiar methods of gaining a livelihood ended here, this history would never have been written—at least not by the *Times.* Had he adhered to the worthy resolves which are credited with having accompanied his somewhat hasty and enforced departure from Kansas City, he might have lived an honored and respected citizen of this community, protected by the mantle of honorable charity. But this he did not do. No sooner had he safely escaped the clutches of the law, than his true nature again began to reassert itself, and he cast about for an opening in which to engage his talents as a swindler, a blackmailer and a public plunderer. He had not far to look."

The article then described Bonfils' meeting with Tammen in Chicago, saying: "Bonfils had money; Tammen imagined he had brains." It went on:

"The latter (Tammen) knew that there was no legitimate field for an afternoon paper in Denver, but he had heard of newspaper blackmail and was ambitious to try it. He knew there were rich men and corporations in Denver, and other large and important business interests which could be seriously damaged by unscrupulous attacks which should pander to a false and distorted public sentiment, already created by demagogical and designing politicians, and here, he conceived to exist, a rich field for the journalistic blackmailer and holdup.

"To a man of Bonfil's character this was, indeed, a congenial proposition. Birds of a feather flock together, and so the infamous compact was speedily made. The newspaper plant was purchased and the 'twin journalists' started on their blackmailing career. As is usual in the case of rogues who prey upon the public, a mask of virtue was assumed. Their sheet posed as the friend of the people, the enemy of all public corruption and the unselfish champion of reform.

"All the arts of demagogy were employed to deceive and mislead the public while at the same time the proprietors of the sheet were hawking the price of their silence wherever they imagined they could find a market. Public officials, corporations, railroads, bankers, business men and private citizens were indiscriminately attacked with misrepresentation and abuse of the most vile and unscrupulous character, in an effort to establish a reputation for virtue and fearless integrity.

"It was editorially proclaimed that the action of the sheet could not be controlled by any influence, while every man of affairs in the city knows that its silence or its support upon any occasion can be directly purchased for a price. Ex-Bartender Tammen has publicly proclaimed that fact time and again, and no attempt is made to conceal it except from the general public, which is used as a lever for the nefarious business."

The same article declared that "without Bonfils' support Tammen would be ruined and forced back to the slums from whence he came." The whizz-bang writer then went on to state:

"He (Tammen) is, therefore, becoming more desperate every day, and in his last extremity has thrown off every mask, and openly threatens the business men of Denver with personal vilification, slander and abuse and attack upon their business interests unless they submit to his blackmailing demands and furnish him an advertising patronage from which they (the advertisers) could get no returns. Such threats have been openly and publicly made within the past few days; and privately, nearly every business man and corporation in the city has been given to understand the same thing."

Incorporated in this article was the following charge concerning Bonfils' lotteries, of which it was said he had operated eight separate ones:

"Harry Lay, who now conducts a barber shop in the Grand Opera House block in Kansas City, was Bonfils' confidential man, and according to his statement, he drew all the capital prizes of $15,000 each. Lay and Bonfils were reared together in Troy, Mo., and when the latter became in need of someone to draw the capital prizes, he secured the services of his old friend.

"Occasionally some outsider was allowed to draw a

prize of a few hundred dollars as an advertisement. The lottery got out circulars purporting to be from 'M. Dauphin,' the deceased president of the Louisiana Lottery, and posted them so that the people would be deceived."

The polemic described gambling conditions in Kansas and Missouri in the '90's. There had been no law in Kansas prohibiting gambling, and fake lotteries and gaming houses flourished. They were compelled to cease operating by the Cubbison Law, which went into effect Feb. 5, 1895. The article continued:

"The sentiment of the law-loving inhabitants was pitted against the thousand of the gambling fraternity. Once the latter essayed to buy a Kansas Legislature. Bonfils was a ringleader in most of these moves and was the most notorious of the gamblers, by reason of his extensive operations and the vast fortune accumulated from the working people. On one occasion he was arrested twice in one day. One charge was that of vagrancy and the other that of maintaining a gambling device."

The article quoted a story from the *Kansas City Times* of Thursday, Dec. 20, 1894, which described Bonfils' alleged arrest and his "magnificent lair in the Husted Building."

A *Kansas City Times* item of January 18, 1895, also was quoted as follows:

"The new county attorney, Samuel Miller, has personally notified L. E. Winn, proprietor of one of the most successful of Kansas City, Kan., fake lotteries, to pull up stakes and move from the city. Winn says he will close up his business tomorrow."

The *Denver Times* article proceeded:

"When Bonfils was driven out of Kansas City, Kan., he went across the river and started a 'brokerage' business. . . . His game in the Missouri city was evidently a crooked one, also, as he was told to vacate by Chief of Police Speers."

The article then quoted from a *Kansas City Star* story of Sunday, June 24, 1894, telling of Bonfils' many alleged lotteries. After charging that he had paid out few prizes, it continued with an account of Bonfils' purported clash with Maurice Barrymore, as follows:

"Bonfils has the reputation of paying no prizes. Instance on instance where he has failed to do so could be cited. Notable among them is the case of Maurice Barrymore, the actor. Mr. Barrymore had invested in one of Eli Little & Co.'s tickets, and it drew a prize of $20. He could get no answer from the lottery by writing, so he gave it to a friend to cash for him. He (Barrymore) then went to Little & Co.'s office in the Husted Building and encountered Bonfils, who, for the occasion, was 'Eli Little' and was entrenched as usual behind his wooden partition. He calmly refused to pay the ticket, and when the actor protested, he said, with a wave of his hand: 'You have your way of doing business; I have mine.'

"The actor used strong language and offered to fight, but Bonfils merely laughed. He said if he were to take advantage of all his chances to fight, he would have to be a wooden man. The actor had to leave unappeased."

Was this ancient *contretemps* responsible for a virulent *Post* attack on Ethel Barrymore, the daughter of the sheared Maurice, when she appeared in Denver not long ago? Was it a visitation of Bonfils' perpetuated wrath on the daughter of an unpaid foe?

The gauntlet had been thrown. And now the town was to see an interminable newspaper war, with attacks, journalistic screams, fisticuffs and gun play. Bonfils grew more bitter, more glum, more violent with his "So The People May Know." Tammen laughed, shouted and claimed that he "loved every minute of it."

Meanwhile the people of the city read this whirlwind paper in increasing numbers. It became known everywhere—and feared.

13

The Man-Eater Emerges

THE ADVENT of a new century found the partners, Bonfils and Tammen, in constant motion. Although political attacks, critical barrages on corporate fronts and growls at their journalistic competitors were the rule, these men also "boosted" Colorado with all their might. Even the climate—which is indeed a bracing one—was extolled, rain or shine.

Bonfils was the great laureate of Colorado's weather. He hailed the sun each day with all the zeal of a pharaoh yelling his approval to Ra. If it snowed until trains were blockaded on the Great Divide and the traffic of towns bogged down, Bonfils was the first to point out that it "could be far worse," and without exception called such blizzards: "Million dollar snows," saying: "It's a godsend to the farmer." So also with torrential rains. They were "good for the farmer." And all these weather-warbles, from snowy lyrics to rainy madrigals, were headed:

" 'Tis a Privilege to Live in Colorado."

Yes, it was necessary to crusade *for* things as well as against them. So, as the old century died and a new one took its place, the *Post* campaigned for prison reform, betterment of conditions in City Jail and State Penitentiary.

The facile Polly Pry made the rounds of state penal institutions, poking her pretty head into this and that warden's preserves. She visited the penitentiary at Canon City and was being escorted through its somber cloisters when she saw a tall, solemn, silent fellow near the bake shop.

"Who is that?" she asked the warden.

"Packer," the warden said. "Alfred Packer. They call him the man-eater."

Polly "pooh-poohed" the idea. "He doesn't look like a man-eater to me. I don't think he would harm a fly."

"I don't know," the warden said. "It's been so long ago."

"I'd like to have a talk with him," Polly said.

The warden shrugged. "He hasn't talked much since he's been here. Only a 'Yes' or a 'No.' I don't think you'll get very far with him."

Polly Pry found Packer a morose fellow, but he finally began talking, and when he had finished, Polly told him she was going to interest her paper, the *Post,* in a movement to have him pardoned. She believed him innocent of having slain and eaten five prospectors in the blizzard of 1873.

She returned to her paper to lay the case before Bonfils and Tammen.

When the *Post* launched its first stories in behalf of Packer, there was broad protest in official circles. The Governor, Charles Spalding Thomas, said he wouldn't pardon Packer under any circumstance.

In a letter to William C. Blair of the *Lake City Silver World* in 1930, Thomas wrote:

"Regarding Packer himself, the *Denver Post* bedeviled the life out of me, during my whole term as Governor, to pardon him, and its proprietors were bitterly vindictive because I would not do so. It so happened that I was in possession, through former Sheriff Shores of Gunnison County, of a somewhat extended, one-sided correspondence, consisting of letters from Packer to his relations. They were the foulest compositions that I ever read and were filled with all sorts of threats against them, in the event he regained his liberty.

"I was under pledge not to make them public, but they were sufficient to justify a refusal of pardon even if he had any claim for consideration. I saw him several times during this period and talked to him at considerable length, but saw nothing in his attitude to change my opinion."

The refusal of Governor (later United States Senator) Thomas to grant the pardon stirred Bonfils' bile. At this time Governor Thomas was issuing several pardons, so Polly went to the Parole Board files to look up the records of convicts receiving executive clemency.

Polly copied only the cases of persons who had been guilty of such jaunty crimes as murder and rape. The *Post* printed these histories in full, asking: "Why not Packer?"

Two days later the man-eater was paroled. He became a sort of hanger-on, an unofficial bodyguard at the "Bucket of Blood."

In seeking amnesty for Packer, the *Post* had retained an attorney, W. W. ("Plug Hat") Anderson, to represent the man-eater. Polly Pry had promised Anderson a fee of $1,000 in the event he succeeded in freeing the notorious cannibal.

One afternoon Polly heard that Anderson not only was expecting the *Post's* fee, but that he had appropriated Packer's funds—about fifteen hundred dollars earned in prison from the manufacture of hair-ropes and bridles. Polly reported what she believed to be the fact to Bonfils.

"Fire Anderson!" Bonfils thundered. "And we won't pay him one cent, either. The crook!"

Tammen, sitting at the other end of the *Post's* Red Room, echoed: "We ought to sic Packer on him."

Polly went to Anderson's office, where she accused him of taking Packer's prison-savings without consulting Bonfils or Tammen.

"And *who* are Bonfils and Tammen?" asked Anderson.

"They have decided to discharge you," Polly said.

Anderson was a hot-headed fellow, but he kept his temper, rose and bowed Polly out. "I refuse to be fired," he said. "But tell your employers I shall call on them at once."

Anderson's offices were across Curtis Street from the *Post*. He put on his silk hat, buttoned a topcoat over his Prince Albert and left the building. Polly meanwhile had hastened to the Red Room. Both partners were at their desks. Polly warned Bonfils of Anderson's mood and sat down to await the call.

Anderson came into the room, his eyes blazing, his hands in the pockets of his topcoat, which was noticeably shorter than his Prince Albert. No sooner had he entered than Tammen began to revile him.

The short, strutting Anderson had taken his hands from his topcoat pockets and now was drumming with his fingers on the back of a chair.

"Sir," said Anderson, "I'm a Missourian and a man of culture."

Tammen shouted: "You're not a man at all. You're a low down son of a bitch! And a robber to boot!"

Anderson's face became white. He thrust his right hand to his topcoat pocket.

Polly Pry leaped to her feet. "Look out He's got a gun!"

As Polly screamed, Bonfils lunged at Anderson. He struck the attorney with great force, knocking him down. Blood oozed from the attorney's cheek-bone.

Bonfils now was on top of his man, ready to beat him to death. "Don't, Fred!" Polly said. "Don't! It will be a terrific scandal. Let him go."

Bonfils, breathing heavily, and very pale, listened to this counsel. He rose, stepped back, and then Anderson got to his feet, a very groggy barrister.

"Now get out, you bastardly thief!" Bonfils said. "Get out!"

Anderson started for the swinging door, leading to the City Room. Both partners followed him. Bonfils was hurling epithets; Tammen was exhausting on the worsted attorney one of the finest accusative vocabularies west of the Mississippi River.

"Please keep quiet, Harry," Polly was saying to Tammen. "He's been beaten. Let him go in peace."

"Let him go to hell!" Tammen said.

Bonfils pushed the door against Anderson's departing hips. Then, without warning, the door swung back; and into the room Anderson's gun was thrust. The three occupants could see the eyes gleaming behind the muzzle of the weapon.

"Look out!" Polly cried.

Anderson fired at Bonfils, the bullet passing through his

shoulder and ploughing upward to his throat. Then another shot, this time grazing Bonfils' heart. He sagged to the floor. Tammen dived behind the long table which separated the partners' desks. He knelt beside a suitcase, his hand on the top of it. Tammen was cursing and howling, believing his partner to have been slain. And now Anderson came into the room, snarling. He advanced on Tammen, firing as he leaned over him. He shot downward, a bullet almost shattering Tammen's wrist as it rested on the suitcase. Then he fired once again, the lead furrowing Tammen's shoulder.

Polly Pry, failing to summon help from the street or City Room, herself tackled the would-be assassin. First she stepped to Tammen, shielding him beneath her skirts. Then she grappled with Anderson, seizing the pistol. The barrel was hot, but she clung to it with both hands. Tammen's blood stained her dress, but she stood above him, holding to the gun, looking Anderson straight in the eye.

Joe Ward and several staff-members had heard the screams, the shots, the sounds of struggle, but they didn't come to the rescue. They stood outside the door, listening in a paralyzed fashion to the rumpus. The lady from Virginia fought a lone battle against the gentleman from Missouri. He threatened to kill her.

"Go ahead," she said. "And then hang."

Bonfils was unconscious, bleeding and barely breathing, lying on the floor. And Tammen was groaning beneath Polly's skirts.

Two men, hearing the screams as they were walking in the street below, now came upstairs. One was a merchant named Cook. The other was James Brown, son of the founder of the Brown Palace Hotel. But they didn't have to interfere now. Anderson had come to his senses and was walking calmly past the *Post* staff-members, putting his gun in his pocket. No one offered to molest him.

Anderson went immediately to Police Headquarters in City Hall. He handed his gun to a police sergeant. "Arrest me," he said. "I'm a murderer. I've just killed two snakes named Bonfils and Tammen."

When an attending surgeon began to remove Tammen's

shirt, the wounded man said: "Don't cut it, damn it—it's silk."

Tammen's wounds were more painful than severe. Although he had been winged in the wrist and shoulder, he maintained in later years that the shots had caught him in the hips.

"Most embarrassing," he would say. "I wouldn't have minded a wound anywhere else, but just think of it—the hips! Shot in the pants, by God! The work of a fiend."

If he had not possessed a frame of steel and an abnormal will to live, Bonfils might have died of his injuries. He was ill a long time. There is a doubt that his nervous system ever entirely recovered, and it is said that his subsequent grewsome nightmares, from which he suffered to an extent which caused him to dread sleep, to keep a man near his bed to awaken him from the weird visitation, were traceable to the Anderson wounds. One bullet he carried all his life; it was deemed inadvisable to extract it.

Anderson eventually was cleared of assault charges. It is said that while he was in jail, awaiting trial, he received a bouquet and a note, purported to have come from Governor Thomas. The missive read:

"I congratulate you upon your intention, but must condemn your poor aim."

When Bonfils recovered, he was unable to enjoy an occasional smoke—his throat having been peculiarly affected by one of the pistol balls. He enjoyed the smell of tobacco, and it was one of the bizarre sights of the *Post* to see this fierce-eyed descendant of Napoleon sitting still while some henchman or another puffed a cigarette and *blew smoke in Bonfils' face.*

Bonfils never entirely *forgave* Polly Pry for saving his life. She had done something for him which he had been unable to do for himself. He was under an *obligation,* a condition abhorrent to his strange, defiant nature.

He picked quarrels with her over small matters. Finally he accused her of lying to him. A woman who wrestles with a smoking pistol—a beautiful woman from Virginia—does not take the lie when passed. Nor did Polly. She left Bonfils and passed to other literary fields—a great and tender character, with courage unbounded.

14

Eagles And Hens

THE ROCKY Mountain region was a new empire. Yet the dawning century found it without a czar. Trail blazers had penetrated it, leaving their names embalmed in grade-school geographies, or memorialized on cigar bands and street signs. Treasure hunters had tapped its resources, sending themselves or their sons to the Senate, their prancing daughters to the Court of St. James's. Capitalists had exploited the region's material possibilities, played the cut-throat among themselves and then joined hands to throttle the laboring class from which most of them lately had sprung.

Yet the empire was awaiting—unknown to itself—a dictatorial conqueror, a tyrant even; and Bonfils had a Napoleonic awareness of the hour. If he were to prove a flank-ripping driver with a bull-whip, mark he was not dealing with a swarm of butterflies. Men kept their teeth a long time in the West; it was an age of dog eat dog.

Bonfils was not content to remain a mere journalistic satrap, dominating a pleasant town. Still, he saw the immediate necessity of becoming felt in the city of Denver, of making it his first battlefield and a whooping, imperial abode.

The Corsican carried his dark head high, the better to see far horizons, the better to hear the voices of a new century. In aspiring for power and influence outside the city limits, however, Bonfils never once lost sight of the smaller kingdom, the town. It was his key position, his royal residence. It was here he must stand to cast the ever-growing shadow.

He was relentlessly active with local onslaughts, unflagging in campaigns for civic betterment. His inky thunders caused utilities companies to twitch with St. Vitus' Dance. Although reputed to be a stockholder in the very companies he assailed most fiercely, Bonfils hurled stink bombs at water, power, telephone and traction interests.

He traduced, reviled, belittled his competitors. His editorial razmatazz surpassed in noise and rancor the pamphleteering excesses of the most strident invective-mongers. His ordnance never cooled while bombarding journalistic foes. United States Senator Thomas M. Patterson, owl-eyed lawyer and proprietor of the *Rocky Mountain News,* now served as principal target for "So The People May Know" projectiles. The aging Patterson, himself an "opponent of the vested interests" and hitherto "the people's stalwart friend," was at first astonished by the uncalled for shower of hot lard, then deflated, and finally as inflamed as a spaniel turpentined by uncouth boys. The *News* fought back gamely, but began to slip in circulation and influence, while the *Post* was definitely on the rise in both respects.

The *Post* mace bounced on the craniums of political boyards, irrespective of party lines. Befoozled aldermen, distrait Mayors and giddy Governors took to their bottles or their beds with nervous indigestion. Anyone that showed his head was this Caesar's meat.

When the churches and other fonts of spiritual guidance demurred to the *Post's* sensationalism, its show windows of scandal, its chronicles of left-handed love affairs, the partners did not reply with argument or excuse. They blandly invited a committee of church folk to edit the paper for a day, a week, or a month—should they care to undertake that task.

The servants of the Lord accepted the challenge. A minister of God and his (the minister's) committee put on eyeshades, armed themselves with pencils and wrestled with the public prints for one day. No crime news was permitted on Page One. The headlines were reduced in stature and greatly pasteurized. The result was that circulation did a nose-dive for the day. The pastor received such reproof from members of his own flock for "publishing a dull and silly newspaper" that he never again opened his theological larynx concerning "The Paper with a Heart and a Soul."

Similarly—and at a later date—some women's clubs were attacking the *Post*. When Bonfils heard that these good ladies "would never let that filthy sheet in their homes again," he offered them the privilege of editing the *Post*.

"Show us the sort of paper *you* would like," he said, "so we'll know what and how to print the *Post* for you."

A bevy of the town's scrupulous ladies undertook to produce an edition. Whether their labors would have suffered the fate that befell the pastoral essay into journalism, none can say. The ladies began bickering, quibbling, quarreling, caterwauling, shimmy-shammying, helter-skeltering, higgly-piggling, haggling over which stories should go into the paper, and which be left out; what was right, and what wrong; who had brains, and who had none, so that everything went snaffer. The edition was late, underset, the type lost or pied, galley proofs uncorrected, the copy readers became crazier than usual—in short, the editorial room was visited by a Black Plague of feminine whimsies, and the composing room resembled an obstetrical ward in Singapore. It was necessary for the *Post* pilots to leap to the helm and rescue the edition before the day's circulation and all advertising contracts went on the reef.

"It's a business for eagles, not for hens," said Tammen, coming up for air.

Tammen mocked Bonfils' grandiose drudgeries, describing his partner as a "slave to money"; but such epithetical screens masked Tammen's fundamental activities. He played the king's jester to the outside world, but under cover labored like a prime minister to sustain the Bonfils' legend. Still, he was always gay, blowing gigantic tubas and belaboring gargantuan kettle drums up and down the streets to impress the public, and thinking up such eight-column headlines as:

"DOES IT HURT TO BE BORN?"

The city of Denver was peculiarly situated in regard to commercial, industrial and transportational facilities. It was virtually lacking in all three elements usually identified with the growth of large centers. Its mile-high atmosphere was nearly free of factory smoke; its foundation stones rested in ground barren of mineral wealth. No trunk line of a major railway served this state capital, and yet, industrially unimportant and remotely placed as it was, Denver became the largest city between Missouri and California. For, next door to it, and on all sides, were regions of material wealth.

Bonfils appraised this field carefully, then moved boldly. He was conservative in thought and radical in action. He knew that the granite ramparts to the west were pregnant with gold, silver and the lesser ores. To the east were billowing prairies, ready for "dry farming," intensive cultivation, the tractor and silo. To the south were the steel mills, Pueblo, and "Little Pittsburgh," and the coal fields, Trinidad, a theater for labor wars, strikes, the future scene of the unholy Ludlow massacre. To the north, this Caesar envisioned a Gaul divided into three parts: Cattle, Coal and Oil. The last-named was to reach noxious blossom in the Teapot Dome-scandal, which—as only the ironical destiny of a Bonfils could permit—was to break sensationally and exclusively through Bonfils' own newspaper, and then to backfire, catching Bonfils in its aftermath of duplicity, as a *participator* in piratical benefits.

As he stood, at the turn of the century, surveying the Western scene, Bonfils decided to make of the whole region a "family," with the *Post* as patriarch. He would give his journalism an intensely intimate, personal quality, binding together the peoples of Colorado, Wyoming, Utah, Northern New Mexico, Arizona and as much of Kansas and Nebraska as he could win from the paper of his particular detestation, the *Kansas City Star*.

He began at once to address these citizens of the several states as "The Great *Post* Family." He told them, day after day, that the *Post* was their "Big Brother." When cattle were snow-bound in winter or gasping with July drought, the *Post* was the first to campaign for relief funds. Good roads were clamored for in the *Post* columns (and many were built by convict labor), all for the benefit of the *Post* family.

And despite the apparent eagerness to fill his own money-drawers and build his own imperial dream, I am naïve enough to believe that Bonfils had an actual, deep love for the region which made him a journalistic czar. The many sides of his character offered puzzling testimony as to his actual feelings on any subject. For example, there was his fondness for dogs.

He always kept one or two poodles himself, and they seemed to be the only recipients (other than Tammen) of this man's public petting. He loved all dogs, especially lost mutts. Whenever a stray hound was reported, Bonfils regarded it more important to restore the wanderer to its owner, or find the beastie another home, than to solve a great murder mystery. The *Post* usually was barking with dog news, with tear-jerkers concerning the travail of homeless curs, of canine bravery, intelligence and loyalty.

The *Post* was a daily evening newspaper, but it also published a Sunday morning paper. Thus, the staff worked all Saturday to get out the afternoon paper of that day; then, after an evening meal, which was shoved in hastily, landing, as Bill Nye used to say, with a "dull thud" in the editorial stomachs, the boys pitched in to produce the Sunday morning edition. Along toward 2 o'clock of any Sunday morning, the lads were a bit punch-drunk with labor; and on Sunday, the *day off,* a good bed was about all a fellow craved in this world or the next.

Consequently, the *Post* offices were deserted on Sundays—by all save Bonfils. That dynamo never ran down. He would come to his office for some reason or other, perhaps to punish himself for being rich. At any rate, he was visiting his office one quiet sabbath day when his telephone rang. He answered, and a lady reported that she had lost her dog. He promised to find it for her. The *Post* never failed in a case of this sort.

He ran through all the offices, but a watchman was the only person to be found. He knew virtually nothing about dogs except they had to be taught certain parlor restraints and might bite you if so inclined.

"Things have got to be readjusted around here," Bonfils roared. "I'll see to *that!*"

The next day Bonfils gave a standing order that a telephone operator and two reporters were to be on duty every Sunday in the *Post* editorial rooms.

"Yes," he added, "and keep a man at police headquarters, too, to look out for important happenings. Some poor lost dog, for example."

An alert sub-editor one day had an inspiration concerning Bonfils' dog fixation. He went to see F.G., who was very busy with a big deal. When he learned, however, that the sub-editor wanted to talk about dogs, he dropped everything and listened.

"My idea," said the sub-editor, "is this: you see we carry so much dog news all over the paper that it might be a good thing to group the stories, and to run a regular lost-and-found dog-column. Then anyone who loses or finds a dog can just report it and we'll print it."

Bonfils looked solemn. "No, my boy," he said, "we can't do that. Much as I love dogs, don't you see the fallacy of your suggestion? Why, if we did that, people with lost dogs would quit paying us *ten cents a line* for lost-and-found ads."

He loved dogs and he loved money. What was his love for the West?

In pursuance of his "family" policy, Bonfils bid for reader interest in neighboring states by emphasizing news of their respective localities.

The *Post* was partisan in almost all matters, were it a squabble between two Greeks concerning a kitchen stool, or the clash of kings with a throne at stake. Thus, when the sensational case of Tom Horn, Indian fighter, guide and stock detective, began with the murder of Willie Nikell of Iron Mountain, Wyoming, the *Post* "played it big" and took sides against the defendant, Horn.

The *Post* displayed this daily slogan as its masthead:

"Dedicated in perpetuity to the service of the people,

that no good cause shall lack a champion, and that evil shall not thrive unopposed."

It is possible the *Post* regarded Horn as a symbol of evil, but it would be difficult to fathom why the paper was so bitterly eager to have him hanged.

15

Talking Boy

TOM HORN was born in Memphis, Missouri, November 21, 1860. He was one of eight children, and of German parentage. He ran away from home at the age of fourteen, his father having fanned him beyond endurance with a harnass trace.

Horn traveled westward afoot, save when he could promote a lift from wagoners, and in 1875 reached Santa Fe, New Mexico. By the time he had arrived in the Southwest, he could speak Mexican and certain Indian dialects. He procured a job as Overland Mail and Stage driver between Santa Fe and Los Pinos. Soon he was known as an outstanding rifle shot and a horseman of parts. He possessed a clear eye, cool head, was powerfully knit, tall, and had an attractive personality.

Horn became a night rider and herder of oxen for wood haulers at Camp Verde when he was little more than sixteen. Next he was boss of a quartermaster's herd—the War Department's cavalry horses traveled overland from California in droves of as large as four hundred head.

Tom was not yet seventeen when he achieved the dream of most boys. He fell in with a chief of scouts, Al Sieber, a snowy-haired sage, who had been wounded twenty-nine times in Indian fights, but who kept on scouting, battling and umpiring frontier disputes despite his rheumatism and wounds.

Sieber took Horn to the San Carlos Agency as his Mexican interpreter. Tom cooked, tended packs, cleaned guns and studied the ways of men, both white and red. When there was no uprising or desertion from the Apache reservation—a tract one hundred and twenty miles long by sixty miles wide—Sieber spent his days looking for Apaches who made *Tis-win,* Indian whiskey.

Any redskin (and many white men, also) suffered delusions of grandeur after five or six swigs of *Tis-win.* When an Apache primed his æsophagus with this aboriginal vodka, his first urge was to wrap his tepee about his grandma's neck, then part his squaw's hair with a tomahawk. The scouts ferreted out the *Tis-win* makers as a preventive of massacre.

Horn's first observance of Western justice was on a day when old Al happened on an inveterate *Tis-win* distiller, one Chu-ga-de-slon-a ("Cockroach"). The latter was fermenting his primitive Old Parr in a gigantic earthenware pot. Sieber rode up and dismounted. The Roach produced a rifle, but blundered. The Happy Hunting Ground was his next stop, the Roach toppling beside his big clay pot, his head almost severed from his body by one swipe of Sieber's bowie knife.

Sieber then stuffed the dead *Tis-win* maker into the earthenware vat. Thus was Horn initiated into ways of bloodshed.

Sieber and his apprentice called at the camp of Chief Pedro, sachem of some four hundred fairly quiescent Apaches at White River. The old Indian listened to the story of the *Tis-win* maker. Horn did the interpreting between Scout and Chief. Pedro took a fancy to Horn, seeking to adopt him.

Sieber believed that Horn had a flair for frontier life and might prove useful to the Government. He urged Tom to stay with Pedro's people, there to observe the ways of the savage and to master the Apache idiom. Tom agreed to this.

Pedro (a rather virile old chap) chose from his forty sons a boy named Ramon, to be known as Tom's "brother." They took a lodge together. The girls of the tribe became interested in Pedro's new "son" and threw sticks at him. This, Tom explained in his autobiography, was a way the tribal débutantes had of encouraging emotional advances. Horn did not reveal what, if any, romance came of the kindling-wood gestures.

Tom and his "brother," Ramon, hunted deer and turkeys, and had a fine time of it. Pedro was fond of Tom Horn and gave him the name: "Talking Boy." He counseled him regarding Apaches and their natures.

"There are good Apaches and bad ones," said Pedro. "But the bad ones never get killed or grow old. You see, they are part devil and can't die."

Among Horn's friends was Micky Free, a young Scout, who had long red hair, a short red moustache, and a single blue eye. The other orb had been hooked out by a wounded deer. Free was half Mexican and half Irish. His parents had been slain by Indians and he had been reared a captive among the Apaches. He was married to a harem of squaws by the time he was twenty.

Before long, Tom was proficient in Indian lore and language. Sieber recalled him from Pedro's camp to serve as interpreter under Major (later General) Chaffee. The Major was the first Government agent appointed for the Apache country, supplanting a civilian agent, following a scandal involving the sale of rations.

Of 7,000 Apaches on the reservation, only 1500 braves and squaws would consent to appear each Friday for their food allotments. The majority preferred to live on the country, to plunder and rob. Thus, the civilian agents had a surplus of provender on hand, which they sold illegally to freighters and merchants at bargain prices.

After numerous adventures, Horn left the scouting life temporarily, an appropriation having failed to materialize at Washington. He and Sieber went to Arizona. One of Sieber's friends, a man named Scheflin, had found silver in that state but had been chased out by Indians. The savages had slain Scheflin's partner, Lenox. Scheflin barely had time to bury Lenox's body and escape. Now he was returning with a party of sixty from California, and Sieber and Horn were invited to join.

They reached the place where Scheflin and Lenox had been digging and saw a pit twenty-three feet deep. The old prospectors said it was a great claim. They held a meeting, distributed shares and voted to name the site Tombstone, because of Lenox's nearby grave.

Within a year there were 7,000 persons occupying the town of Tombstone.

In 1878, Washington got around to making an appropriation for scouts, so Sieber and Horn returned to active service under General Wilcox.

Word had come—via a squaw messenger—that the renegade chieftain, Geronimo, already the villain of two uprisings, was in Mexico and wanted to dicker for a return to the Apache reservation. General Wilcox assigned Sieber to hold a parley with the wily chief. The Scout took Horn along to interpret. Tom was excited over the prospect of meeting and talking with this notorious warrior.

They traveled across the border and to Geronimo's hideout on the top of a mountain. There were nearly a thousand Indians in camp. When Geronimo put his coal-chip eyes on the eighteen-year-old boy, he expressed misgivings as to Tom's ability.

"I sometimes speak very fast," Geronimo said to Horn. "Can you interpret as fast as I talk?"

Horn said: "You have only one mouth and one tongue that I can see. So let loose."

Geronimo's eye lighted. "Well spoken, boy!"

Horn long afterward wrote his autobiography in a Cheyenne jail. In it he recalled that Geronimo "certainly was a talker, and had more grievances than a railroad switchman."

Geronimo outlined conditions for a peaceable return to the reservation. Among his demands was the privilege of having two Mexicans make mescal for his own use, Geronimo was quite a rumhound and could drink any two bibulous chiefs under the blanket.

He also desired all the arms and ammunition he could use, calico for the women, children's shoes during the snowy season and some money. Geronimo, according to Horn, was "the biggest chief, the best talker and the biggest liar in the world."

He spoke for an hour, and then Sieber said: "Geronimo, you have asked for everything I know anything about except for these mountains to be moved into the American country for you to live in; and I will give you till sunrise to tell your people and see if you don't want these hills moved up there. If you are entitled by your former conduct to what you have asked for, then you should have these mountains, too."

Then Sieber walked away.

The next day Geronimo (having talked all night to his followers) declined to return to the reservation. Sixty-two of his people went with Sieber and Horn, however, and as they departed, Geronimo said to Horn:

"I'm glad they're going. They're growing old and are of no use to me. As for you, come to see me as often as you want. You are a young man and will always be at war with me and mine, but war is one thing and talking business is another. And I will be just as pleased to meet you in battle as in council."

Horn had many adventures between the time of that meeting with Geronimo and 1883, the year in which he again went to the Apache leader's Mexican habitat. This time Horn was serving under General Crook. The General had two Mexican interpreters and took Horn along mainly to spread peace propaganda among Geronimo's people. But Geronimo spied Horn and remembered him. He insisted he would do business through no other interpreter.

Other great chiefs, the aged Nana, Ju, Loco and a thousand braves gathered at this council. General Crook made a speech. Geronimo thought it over all night, called on Sieber (whom he trusted, although a foe) for advice, and then decided to go back to the reservation.

In returning, however, Geronimo secretly sent forty or more of his braves into the Mexican country to steal horses, knowing that he would have United States soldiers to escort his raiders and their plunder across the border. When the General heard of this ruse, he was furious.

"My God!" he thundered, "this makes me an accessory to a crime. I'll be court-martialed!"

By the time the troops and Indians arrived at the boundary line, the missing tribesmen had reappeared with more than a thousand head of stolen horses. Geronimo claimed the steeds actually belonged to his people, that his braves merely had gone to the hills to round them up for the return to America. Many excited Mexicans, however, were present at the border to claim their animals. The Indians refused to give them up. Crook had to pay off the Mexicans in cash before taking the horses across the line. The Mexican press denounced Crook, claiming he and his men were *encouraging* Apache raids in Mexico! And that the General was *fittingly named.*

Horn said that Geronimo would have "made a great confidence man."

Tom became chief of scouts in 1884 and after Sieber had been crippled by a rifle-bullet. He knew the ways of the Apache better than any other man now in active service. He was subject, however, to governmental caprice and lack of appropriations. So finally he changed to a career of mining. But not for long.

General Miles sent for him to help persuade Geronimo to return to the reservation—that liquor-loving rascal once again having made whoopee and fled into Mexico. Miles had objected to a civilian interpreter, which now was Horn's technical status, but Geronimo had refused to deal through a military interpreter. So Horn agreed to serve. There had been newspaper attacks on Horn, articles suggesting that he had an "in" with Geronimo, that he was in sympathy with the Apaches. There was, of course, no rightful basis for this surmise.

There had been a brush between soldiers and Indians, during which old Chief Nana was captured. The latter—now in his ninetieth year—was highly disgusted. Not because of his capture, but on account of Geronimo's order to his one hundred and ninety-five followers to "run for the river."

"Take me to the reservation or even the guardhouse," Nana said. "Gladly do I go, for I am ninety years old and I never thought I would see the day when my people ran foot races. Once there was a chief of this tribe, and his name was Nana. And he never said 'run for the river'. He was not a foot racer. He said: 'Fight', and his warriors fought. Yes, take me. All I ask is the privilege to sit around and growl. But do you expect me to talk against my people, for though they win foot races instead of battles, they are still my people."

Later, and while Horn was escorting Nana to the boundary line, the party was attacked by Mexicans. Old Nana appealed to Horn:

"Give me a rifle and some bullets. I want to help you fight. I shall not engage in a foot race."

Horn gave the old man a rifle and ammunition. According to Tom, the aged warrior had a delightful afternoon's shooting and bagged several Mexicans. Then he surrendered his gun and went along, a prisoner.

Horn officiated at the Geronimo surrender at Fort Bowie, the old chief saying he was tired of the war path and that his people were worn out by years of strife. He laid down his arms on the parade ground. General Miles packed the Apaches on a special train of the Southern Pacific, and the day of Geronimo was done.

With the passing of the bison from the great plains of the West and Southwest, a new "thundering herd" came to graze upon the wide ranges. Cattle barons took up vast tracts of land, imported pure blood strains and stock raising became a mighty Western industry.

So large were the herds that it was hard for a major rancher to keep account of his many thousand head of cattle. There were round-ups each year, at which time the season's calves were branded with the insignia of their owners. And here entered a fore-runner of modern racketeering.

Bands of cattle-thieves, "Rustlers," would cut into a cattleman's herd, steal unbranded calves and sear the young flanks with their own marks. They sometimes stole cattle and superimposed their own signs.

This practice led to long and bitter warfare between cowpunchers and rustlers. Many lives were lost. The cattle barons employed stock detectives to "get the goods" on the culprits. Sometimes, it was charged, the stock detectives actually were "killers," men hired to do away with rustlers whom the law was unable to apprehend or convict.

Such a war between punchers and rustlers was a progress during the spring of 1887 in the Southwest. In that year Horn was a deputy sheriff for Bucky O'Neil and acted as a mediator in the rustlers' war. In July of 1888, he was appointed as interpreter during the trial of the Apache Kid, an Indian who had killed a freighter. Horn, however, had gone to Phoenix, where he was competing in the rodeo of the Territorial Fair. While there he broke the world's record for roping, throwing and tying a steer with a riata, his time being forty-nine and one-half seconds. He heard, while there, that the Apache Kid had escaped with five other Indian prisoners and had killed a sheriff named Reynolds.

"Had I been there," Horn wrote, "this never would have happened, because I would have heard them talking in Apache and would have prevented the break."

In 1890, Horn sold his mine for $8,000 and accepted a position with the Pinkerton Detective Agency in Denver. He found the work "tame." He solved a train robbery case in August of 1890, the Denver & Rio Grande having been held up near Texas Creek, but was disgusted with methods used to pursue the robbers, writing:

"There were so many men scouring the country that I, myself, was being held up all the time; I had been arrested twice in two days and taken to Salida to be identified!"

Horn always preferred to be a lone wolf.

After the posses had become "tired," another operative, the famous C. W. ("Doc") Shores, and Horn were left on the trail of the suspects, "Peg Leg" Watson and Burt Curtis. Shores got Curtis. Horn tracked Peg Leg through many states, finally capturing him in Indian Territory.

An interesting side of the case was that the men received life sentences for robbing the United States mails, although the actual crime was an express company theft. In holding up the train they had compelled the fireman to break into the mail compartment of the express car. When they saw their error, they did not touch anything in that compartment but went on to the express company's cubicle. The Government, however, proved it technically to be a *mail robbery*.

Tom left the Pinkerton in 1894. In 1898 he served with the rank of Colonel in the Spanish-American War, first as chief packmaster for General Shafter's army, later as master of transportation. He was credited with salvaging supplies and ammunition for the Battle of San Juan Hill.

Horn took down with yellow fever and returned to America. His magnificent physique and strong constitution enabled him to survive this disease. He was in the hospital a long time, then went West to the ranch of John C. Coble in the Iron Mountain section of Wyoming.

Coble was a highly respected ranch owner, and among other things gave to the rodeo chutes the greatest outlaw horse of all time, "Old Steamboat." This ferocious bucker was the wisest, slyest, strongest, most active sunfisher ever seen. He had a roar like that of a lion, and for nearly ten years tossed champion broncho busters from his spine as though they were sacks of oats. Sometimes, when Old Steamboat was stiff-legging it, r'aring, snorting and pitching, it seemed as though it were raining cowboys all over the arena.

A younger generation has heard that Old Steamboat finally was ridden by one, Dick Stanley. Yes, but how? It was one of those cloud-bursting days and Steamboat was up to his hocks in mire. That critter was never ridden on *dry ground*. Later he broke a leg while the boys were loading him on a railway car, and had to be shot. It took *three bullets* from a forty-five to polish off this equine Tartar. A much tougher horse than the present broncho king, "Midnight." Ask T. Joe Cahill.

T. Joe became one of Horn's best friends. His father had been a buddie of Horn on the old Overland Stage route. Young Cahill now was county clerk at Cheyenne. He was a man to be reckoned with in peace or war. He became the guiding spirit of Cheyenne's world-renowned rodeo. (We used to call 'em "round-ups" in the days of Old Steamboat.)

T. Joe's start in wild-west carnivals was a rather inauspicious one. For a year or two he sold programmes in the grandstand. Later he acquired a concession. He took a canvas tarpaulin, cut a hole in it, stretched the canvas between two posts and hired a nigger to put his head through the hole. Then he permitted visitors at the

Cheyenne Round-up to chuck nickel baseballs at the nigger's head—a very ancient but popular game.

The second day of T. Joe's operations, a soused cowpuncher, named MacFarland, bought some balls and tried to bean the smiling Negro. Either the smile or his own poor marksmanship enraged Mac. At any rate, he leaped over T. Joe's counter, hurled a rock at close range and howled at the black boy:

"How'd you like *this?*"

T. Joe put the nigger in an ambulance, his skull cracked from stem to stern. Young Cahill's business for that year was utterly ruined, other Negroes refusing to serve as understudies for the baseball act.

At the time Horn went to work for Coble, the Iron Mountain rustlers were on the rampage. These men lived with their families in a district about fifty miles square. The ranches of that zone were relatively small ones, but the great holdings of the cattle barons were on all sides.

Not only did the rustlers steal mavericks, unbranded cattle, but they also kept sheep. Here was another problem for the cattlemen. Sheep ruin any grazing range. In the first place, the bunch grass has shallow roots; it grows in loose, dry soil, and the small, sharp hooves of sheep tread out and uproot the grass. Secondly, sheep crop close, and what grass they do not kill with their feet, they devitalize by their close cropping. Thus, sheep were anathema to the cattle owners. And between the sheep and cattle factions there was war, literally to the knife.

Horn became a stock detective on the pay-roll of several cattle barons. He rode the range in open season to discover who the thieves were. Part of his duties consisted in throwing the fear of God into the rustlers and sheep men. And there is no question that they feared him. He had a reputation as a killer. He did not hesitate to make threats, and, when drunk, was inclined to boast overmuch. Stealing would cease for months after Horn's "casual" visits.

Whenever there were killings, they were attributed to Horn. The authorities either did not want to accuse him or had no evidence against him. Horn did not deny these deeds; the mysterious slayings built up his reputation as a killer. Hence his dreaded influence with cattle thieves and sheep herders.

Among the embattled families in the Iron Mountain district were the Millers and the Nickells. Kels Nickell had brought sheep into the cattle country, an unforgivable crime in the eyes of his powerful neighbors. It became gossip that the cattle barons were "out to get Nickell."

Kels had two sons, Willie, a fourteen-year-old, and Freddie, eleven. On the morning of July 19, 1901, Willie had been gone from the ranch house for an hour or so, and Kels sent Freddie to look for him. Freddie came upon his brother about three-quarters of a mile from his home, dead, shot through the heart. A boulder was placed in a curious fashion beneath the boy's head. That was the only clue, the boulder. In several other unsolved slayings, boulders had been similarly discovered.

The authorities immediately suspected the Millers. It was said that Jim Miller had been lying in wait for "Ole Nick"—Jim's name for Kels—and had shot the boy by mistake. A coroner's inquest developed evidence tending to show that Victor Miller, not Jim, was the guilty one. It was testified that he had had many fights with Willie Nickell. But this evidence was not strong enough to justify an arrest.

Some one drove several hundred sheep from the Nickell homestead, across the public grazing range and onto Miller's deeded land on August 3, 1901. This was an inflammatory act. When the Millers found Nickell's sheep in their meadow, they made ready to fight. Finally the flock was withdrawn, but next day somebody took thirteen pot shots at Kels Nickell, wounding him. Nickell said his assailants were Jim Miller and one of the latter's sons. Mrs. Nickell told the neighbors:

"They will try to lay this on Tom Horn, but he never done it!"

There was a second session of the coroner's inquest on August 9. It was testified that the men who had shot at Kels had been seen riding away toward the Miller homestead, one on a bay, the other on a gray horse. Miller owned two such horses.

While Kels was in the hospital convalescing from his wounds, four masked men visited his homestead, chased away the herder and clubbed to death many sheep.

Nickell had had enough. He put a for-sale sign on his ranch and moved with his family to Cheyenne.

Suspicion began to point to Tom Horn as the slayer of the Nickell boy. It became known that he had stopped at the Miller ranch two days before the killing of Willie. But no one was able to bring the crime directly to his door. By the end of summer the sheep men had been run out of the district and Horn had completely intimidated the rustlers. Thievery stopped and Horn's employers dismissed him.

And now, although the Nickell murder threatened to pass into the West's long list of unsolved slayings, a deputy United States Marshal was secretly working on the case. His name was Joe La Fors. He cultivated the friendship of Tom Horn, and patiently waited the hour to trap him.

In December, La Fors told Horn's friend, Coble, that he knew of a stock-detective job in Montana. La Fors actually did land the job for Horn, who went as far as Omaha, got drunk and lost his outfit. He returned in January to Coble's ranch for a new outfit. While there, Horn received a letter from La Fors, saying that a representative of the Montana stockmen was in Cheyenne and wished to meet him.

"I also have a letter from Montana which I wish to show you," La Fors wrote Horn.

Horn set out for Cheyenne, but got drunk on the way. The morning he arrived at Cheyenne he was wobbly and unable to talk to La Fors. The deputy made an appointment for that afternoon, but Tom ignored it and went foraging for liquor. When La Fors located Horn, the latter was slumbering on a chair in the back room of a saloon.

"Wake up, Tom; it's important," said La Fors.

Tom was giddy with rum. He allowed La Fors to lead him to the office of the United States Marshal. And there Horn *talked*. La Fors had stationed a district court stenographer, Charles Ohnhaus, and a deputy sheriff, Leslie Snow, behind a door separating the Marshal's two offices.

The next day Deputy Sheriff Snow asked Horn's friend, T. Joe Cahill, to "go with him to get Horn."

"What for?" T. Joe asked. He was unaware of the "confession," or that Horn was wanted for a crime.

"Oh, just come along in case he's drunk and starts trouble. We want to talk to him."

T. Joe went with the officers to the Inter-Ocean Hotel, where Horn was waiting to have dinner. Cahill went up to him. "Hello, Tom."

"Hello, T. Joe," Tom said. "What's up?"

"I don't know," Cahill said, "but the Sheriff here wants to talk to you."

The officers had Tom covered. He was cool as they read a warrant charging him with the murder of Willie Nickell. When searched, Horn's pet weapon was found in his pistol pocket.

On his way to jail, he fixed T. Joe with his cold, staring eyes. "Were you in on this, T. Joe?"

"I certainly was not," T. Joe said.

"I believe you," said Tom.

There was great surprise at this arrest. Only the authorities knew what evidence was to be presented. The public did not learn of the confession until the preliminary hearing on May 10. Horn was arraigned on that day, pleaded not guilty, and his trial was set for October 10, 1902.

The *Post* made much of this case, naturally. But in its news stories there was a decided anti-Horn feeling, an inference that the *Post* not only wanted Horn to hang, but wished to involve "certain powerful cattle barons."

Horn claimed that his "confession" had been garbled, whole lines deleted from it, and the meaning of his talk with La Fors twisted. Horn was lodged in the Cheyenne jail, heavily guarded. The *Post* claimed that the cattle "interests" were planning to liberate him. In face, he did break jail one August.

It was a hot Sunday and the church organ was playing as T. Joe Cahill passed the collection plate. There were cries and shots from outside. Everybody made for the door, T. Joe dropping his offertory platter and in the van.

As T. Joe reached the street he saw his friend, Horn, loping along. The fugitive had a pistol in his hand. The Sheriff and some men were coming after him, full tilt. As Horn passed T. Joe, to duck into an alley, he was nervy enough to say: "Howdy!" and, "You stay out of it, T. Joe!"

But in the alley, Horn had trouble. Whoever had smuggled his weapon had provided a *modern automatic pistol.* This was an unhandy tool for a man schooled in the usage of an old-fashioned Colt's single-action shooting iron. Horn dropped the weapon with disgust. He was captured, and as he was being taken to jail, one of the officers delivered on his skull an uncalled for blow with a rifle-butt. Horn never forgave that "foul" and thereafter would not let certain deputy sheriffs enter his cell.

Horn had four loyal friends, all of them highly respected, honorable, and in no way connected with his professional life. They were Frank and Charlie Irwin, Coble and T. Joe. The Irwins were noted cow-men, and were Horn's favorite singers of old-time songs of the range. They called at his cell, and among their offerings were railroad songs, a kind of ballad beloved by cowpunchers.

Judge Richard H. Scott of the District Court of Laramie and Converse Counties presided at Horn's trial. He occupied a swivel chair in an old court-room of high ceiling and flower-papered walls. A tall stove, rusty and tobacco mottled, stood before the bar, its pipe rambling half way across the chamber. On the bench were two Edwardian bronze candelabra for gas lights. The jury box was behind a slender and glistening mahogany rail. Long, green-covered tables stood before the Judge's bench, extending from the jury box on one side to a cottage organ and a blackboard on the other. The ancestry or reason for that court-room organ never was divulged nor its presence questioned. Incongruities of that character were to be met frequently in the West.

The court-house itself was a square, red structure, the county jail within its walls.

It was a sharp, chill morning. Spring had been laggard in coming to the Wyoming uplands. The old stove was working like a donkey engine. The windows and doors were kept shut, the air stuffy, the chamber crowded with rugged, sun-bitten men of the range. Ranks of unsmiling fellows stood, their broad backs to the wall, their huge-brimmed hats held like Zulu shields at their chests. They were stoical, silent, watchful. There were but two women in that assembly: Mrs. James E. Miller of Iron Mountain and her daughter, Eva.

Kels Nickell, father of the murdered boy, sat in the front row, twitching and uneasy. By a freak of circumstance, he had been subpoenaed for jury service. The court officers were watching his hands, for he had threatened to shoot Horn on sight, in court or out.

The Prosecuting Attorney was Walter R. Stoll. The defense counsel were former Judge John W. Lacey, ex-United States Attorney Burke, and Attorneys Matson and Kennedy.

Judge Scott said: "Bring Horn into court, Mr. Sheriff."

There was a craning of bronzed necks as Deputy Sheriff Snow pushed through the crowd, bringing Horn before the bar. Horn was erect and handsome, though somewhat chalky from long confinement. He bowed to his attorneys, sat down and smiled slightly. His deep-set eyes were bright and sharp. As he leaned back, he occasionally twirled his small moustache. He wore a dark suit and corduroy vest.

The clerk called Kels Nickell among the first twelve prospective jurors. Horn looked steadily at Kels in a confident, half-amused way. The Judge, of course, dismissed Nickell from jury service.

The jury selected, Horn glared at each witness produced in turn by the prosecution. There was one exception. When Mrs. Nickell, mother of the slain lad, took the stand, Horn lowered his eyes in an almost gallant manner.

The trial lasted for days. More than all other evidence, the story of Joseph La Fors, containing the "confession," was the real body blow at the defense. When La Fors took the stand on October 15, there was almost half a minute of silence as he and Horn exchanged glares. Each had sworn to kill the other should Horn obtain his freedom.

La For's face reddened, but his voice steadied as he recounted in detail Horn's alleged disclosures, regarding the slaying.

" 'The dirtiest job I ever done,' Horn told me," La Fors testified. "When I said there was a job waiting him as stock detective, but that the Mountain people were not sure he was their man, he said:

" 'Joe, you yourself know what my reputation is. I can save these men witness' fees and lawyers' hire." Horn said he had to kill the Nickell boy, because he was waiting for the father, and the boy caught him waiting. When Horn came to the office, I gave him a letter of introduction to W. G. Pruitt of Montana, where he wanted to go to work. He said he didn't want to make reports of the shooting, and said: 'When it comes to shooting, you know me, but I have never gotten my employers into trouble by my shooting.' I then said: 'Yes, Tom, you're the best man I ever saw to cover up your trail. In the killing of Willie Nickell I could never find your trail.' Horn replied: 'Yes, by God, I left no trail!' I then asked Horn why he killed Willie Nickell, and he replied: 'You know how it is when a man is hiding and a boy rides up on him and then tries to run away.' "

"How far away did he say Willie Nickell was when shot?" asked the Prosecuting Attorney.

A—About three hundred yards, he said, and added: "It was the best shot I ever made and the dirtiest trick I ever done."

Q—What did he say about his trail?

A—Mr. Horn told me that when he shot the boy at the gate, he ran over the ground with his boots off. I asked him if he did not cut his feet, and he said yes. He said he thought at one time the boy would get away, but he said he had made a good shot.

Q—What was said about telling more of the killing?

A—I told Horn I wanted to know all about the shooting, and he replied: "It's too new now, but I will tell you all about it when I come back from Montana."

Q—How much did Horn say he was paid for the job?

A—He told me he got $500 for it.

Q—What did he say about his gun?

A—Horn told me he used a 30-30 Winchester for the killing.

Q—Did he say anything about a rock, a boulder?

It had been testified that a boulder was placed under Willie Nickell's head.

A—I asked him why he put a rock under the boy's head but he only said: "That's my sign."

Q—State whether or not he said anything about killing people as a business.

A—Yes. He said: "My specialty is killing people. I look upon it as my business, and I think I have a corner on the market."

On October 17, Horn was called suddenly to take the stand in his own defense. The court-room was electrified. Horn smiled as he was sworn in.

"My residence at present is the county jail," he said.

His attorneys asked questions tending to establish an alibi, also concerning his clothing. It had been testified that he had worn a navy-blue sweater the day of the crime.

"I wear a sweater only in winter," Horn said. "I wouldn't wear one in summer. And the only one I own is a white sweater."

It had been said that he rode a bay horse. He testified:

"The horse was a nearly dark brown named 'Cap', branded on the right shoulder and with a roached mane. I had a rifle, a 30-30 Winchester. On my saddle I had a field glass. I had no pistol."

Kels Nickell was leaning forward, his head on his hand, glaring at Horn. Horn was saying:

"I never did have anything to do with the killing of Willie Nickell. I never killed him, had no occasion to kill him and didn't kill him. La Fors commenced to josh me about killing Willie Nickell, and so I joshed, too. There was no reason for the concealment of those men in the next room. If they had been sitting with us, it would have been just the same."

Three witnesses had testified that Horn, while drunk in Denver, October 1, had boasted of the killing to a saloon group.

"Not one word of truth in it," he said. "It couldn't have been true. I was in the hospital in Denver the first week of October. I got my jaw broken Sunday night."

(His jaw was broken in a drunken fight with a Denver newspaper man. The journalist did not know at the time, nor did he care, *who* Horn was. When the latter became offensive, the reporter let go an uppercut and cracked Horn's jaw. But when the news man learned the identity of his victim, he *boarded a train for California,* where he *still* resides.)

The defense rested on October 20. The case went to the jury at 11:25 o'clock, October 24. At 4:20 o'clock that afternoon the jury agreed upon a verdict, having cast six ballots. They found Horn guilty of first degree murder, and Judge Scott sentenced him to be hanged.

Court appeals and repeated efforts to obtain a commutation of sentence from Governor Chatterton failed. Tom Horn spent his final year in the Cheyenne jail. His dark hair became thinner on his high, narrow temples, and his straight, long nose stood out more boldly than in other years. He lost weight, but kept his nerve. He studied the Bible, listened to songs by the Irwin boys, made hair ropes, bridles and quirts, handiwork with which he had become familiar when a boy on the Apache reservation.

The date for execution of sentence was set for November 20, 1903. Horn insisted that his friend, T. Joe Cahill escort him to the gallows.

"I want you to slip the noose over my head," Horn said.

"I'd rather not," Cahill said.

"But I want you to do it. It'll be a personal favor."

The week before the hanging, Horn wrote farewell letters. One of them was to the stenographer, Onhaus, asking him to change his testimony concerning shorthand notes taken at the La Fors "confessional." Another was to Coble, a résumé of evidence and a declaration of his innocence.

The day before the twentieth, a canvas tarpaulin was draped over Horn's cell. He heard thuds and a weird symphony of hammers and saws. Carpenters were putting up the scaffold. The hanging-platform faced Tom's second-tier cell. The drop extended down to the first tier.

Tom objected to the curtain. "What you fellows up to?" asked Tom through the canvas.

"We're fixing the gallows, Tom," said T. Joe Cahill. "Testing it out."

"Take this curtain away."

Cahill removed the tarpaulin. Tom stood at his barred door, cigarette in hand, watching coolly.

"How's she doing?" he asked.

Cahill was folding the curtain into a compact bundle. "Seems O. K."

Horn inhaled his cigarette smoke. "It had better be."

The men were testing the gallows now, tying bags of sand to a hemp rope. It was one of the old Jim Julian scaffolds. The prisoner steps on a platform, and the platform yields a fraction of an inch, releasing a trigger, in turn—attached to a cord—draws a cork from a water-box. That box, constructed on the principle of a toilet reservoir, contains a floating ball-and-arm mechanism. The water trickles forth with a ticking noise, spattering on the floor. When the water level sinks to a given point—a process requiring from forty to fifty seconds—the ball-and-arm arrangement springs the trap. The victim falls like a plumb-bob.

The theory is that a doomed man thus hangs himself, his own weight setting the Rube Goldberg apparatus in operation. Thus, the state saves its conscience for slaying one of its erring members, and all is well.

John Coble called on Horn the evening preceding the day of punishment. Horn said: "Keep your nerve, John, for I'll keep mine. You know Tom Horn."

After Coble had left, weeping, Tom put the final braids in his last hair bridle. His hands were steady. The jailor said "Good night" and turned down the corridor lights, but Horn kept on working in the semi-darkness, his strong but slender fingers moving whitely as he braided. From beneath him came the uneasy groans of Jim McCloud, an alleged train robber.

"What's the matter, Jim?" Tom asked. "A bellyache?"

"God, Tom, it's awful, ain't it? You hangin' in the mornin'!"

Tom was silent for a while. Then he said: "Just quit groaning, Jim. It'll be fairer all around."

The day of execution in the Old Cheyenne Jail found Tom Horn quite placid. He rolled cigarettes often. Then he switched to cigars.

He had warned Deputy Sheriff Snow not to come into his cell, or even into the adjoining high room where the scaffold stood before two tiers of cells. He wore a static smile and his eyes were bright. He seemed pleased when told that Snow was not among those present.

Most of the witnesses had climbed an iron ladder to the platform fronting the top tier of cells. Guards with carbines were there and elsewhere in the jail. Outside there was a great crowd of plainsmen. The Governor had sent militiamen to patrol the court-house grounds. A billboard abutting the building bore colored posters, advertising the drama: "Ten Nights in a Bar Room."

Charlie and Frank Irwin were at the foot of the scaffold. It was rumored they would ask Tom, as he stood on the trap, to admit or deny his guilt with a flat statement of fact, and to say if he had written any letters on this score. They were seen whispering to Jailor Proctor at 10:55 o'clock, A.M., at which time he had opened the door to witnesses.

"All right," he said to the Irwins. "I'll ask him." Then he said: "The newspaper men will come first and take their places by the side of the gallows. There will be no smoking, no chewing tobacco, no talking."

Six guards with carbines were pacing the corridor. The witnesses could see Horn, who was half reclining on his cot, reading from the Book of St. John. He was in his shirt-sleeves. He had on a red and black shirt and a tie to match. He held a half-smoked cigar between his lips. As the witnesses walked past the death-cell, Horn raised his eyes from the Bible, with a searching, half-wondering, half-contemptuous look.

Then Sheriff Smalley and T. Joe Cahill went into Tom's cell. He stood up to be strapped, carefully laying down his cigar butt. He would have only ten feet to walk from his cell to the trap. T. Joe pinioned Tom's wrists to his sides.

"Give her another hitch, T. Joe," Horn said.

"Is that better?" T. Joe asked.

"A whole lot better. Thanks."

Horn now wore a hanging-jacket, with two leather handles, one on either side. These were to permit his custodians to steady him on the trap. Jailor Proctor came in with a message from the Irwins.

"They want to sing for you, Tom."

"That'll just be fine," Tom said.

"What'll it be?"

"Oh, anything. Make it 'Keep Your Hand upon the Throttle and Your Eye upon the Rail.' "

"O. K., Tom," said the jailor.

Smalley, Cahill and Proctor stepped to the gallows. Smalley gave the rope a tug. The Chaplain, the Rev. George Rafter, had been at the cell door, and now he came forward to join the group on the upper tier. Horn had sat down on his cot again, trussed up for the kill. Jailor Proctor went to the edge of the platform:

"Gentlemen," he said, "Charlie Irwin and his brother will sing. Mr. Horn says he will be glad to have them do so."

At this word, Horn rose from his cot. He stepped from his cell for the first time in many months. As he came out, he hesitated at the door for he had been misinformed. Standing below him was Deputy Sheriff Snow. He stared hard at Snow, but his hatred seemed to melt when Charlie Irwin called out:

"Hello, Tom."

"Why, hello, Charlie," said Tom, smiling.

The Irwin brothers leaned against the side of the jail wall, and at the gallows' foot. Then they sang:

"Life is like a mountain railroad
With an engineer that's brave;

We must make the run successful,
From the cradle to the grave.
Watch the curves, the fills, the tunnels;
Never falter, never quail;
Keep your hand upon the throttle,
And your eye upon the rail.

"As you roll across the trestle,
Spanning Jordan's swelling tide,
You behold the Union Depot,
Into which your train will glide.
There you'll meet the Superintendent,
God the Father, God the Son,
With the hearty, joyous plaudit,
'Weary pilgrim, welcome home.'

'Put your trust alone in Jesus;
Never falter, never fail.
Keep your hand upon the throttle,
And your eye upon the rail."

Horn now was smiling at his friends. Many of the witnesses were weeping. Horn was nodding approvingly as the song ended. Then Jailor Proctor called out:

"Charles Irwin, step around that stove and come to the gallows."

"Yes, sir," Irwin said, climbing the iron steps. He stepped onto the platform. "Tom, I want to say good-bye. Did you write that letter?" (His declaration of innocence.)

"I did."

"What did you do with it?"

"I gave it to Proctor to give to John (Coble)."

"I want that letter, Tom."

"All right. Proctor has it. He will give it to you. I addressed it to John Coble, Charlie."

"Can I make that letter public, Tom?"

"If you and John think best."

"And there is no doubt that you wrote it and meant every word in it, Tom?"

"I did write it, and I mean it all."

"Tom Horn, I want to ask you if you made that confession published last night?"

"I did not."

"You swear to this?"

"I do."

"Tom, I want to say good-bye. I wish you good luck, old man. I am your friend."

"Thank you, Charlie. Good-bye."

Irwin hurriedly left the platform. He went to the office of Sheriff Smalley, being unable to witness his friend take the plunge.

Jailor Proctor was kneeling beside Horn, putting straps on his thighs. "Are they too tight, Tom?"

"No, they're all right. Make 'em tight; I don't want to kick."

Horn moved from side to side, as though to assist Proctor in adjusting the straps. "Look out, Tom. Not yet," said Proctor, keeping him off the trap-door. "Back this way a few inches."

Horn was inspecting the gallows-beams and the rope in a critical manner. Then he said to Cahill: "Will you steady me a bit, T. Joe? Put your hand on me. I don't want to fall."

He smiled as T. Joe grasped one of the jacket-handles. Then he whispered: "Look at 'em! Did you ever see so many scared sons of bitches in your life?" Then: "Come on, let's get it over with."

Cahill placed the black cap over Tom's head. Sheriff Smalley dropped the noose over the hooded skull. Horn presumably thought T. Joe was working with the loop, as he had requested, for he said:

"Hey, there, T. Joe! A little more smug. There, that's better." He wagged his head, settling his left ear against the hangman's knot. He was on the trap now and the ticking of the Jim Julian waterbox began. The noise seemed to last a long time.

"I hear you've got a new baby over at your house, T. Joe," Horn said.

"That's right, Tom."

"Say, I think your hand is shaking. Are you nervous, T. Joe?"

"Not very."

"Well, don't be . . ."

He fell through the trap. When the body had ceased its jackknife lunges, doctors put their stethoscopes to the hanged man's chest, then pronounced him dead. He had been strung for seven minutes. Officers now cleared the chamber of witnesses. T. Joe Cahill and the coroner's men brought in a creaking basket to receive the cadaver. Jim McCloud began wailing in the cell below the one which Horn lately had left.

"Shut up, you!" a guard said to McCloud. The guard was sawing at the yellow rope. "It's harder than getting a big fish off a hook. Push the basket closer, T. Joe."

As they took off the noose and then the black cap, the coroner said: "His neck ain't broken at all. Strangled."

"That's the way it goes," the guard said, as they carried the basket to a rear window of the jail. "Horn was a bull-necked fellow."

A line of militiamen was on either side of the coroner's wagon outside the window and in a court-yard. The one black horse drew the wagon swiftly away to the morgue, where another crowd was assembled.

Kels Nickell was waiting in the coroner's outer office. He followed T. Joe to a washroom and was wildly excited. A rumor was spreading that *Horn had not been hanged!*

Kels caught T. Joe by the sleeve. "I'll believe you, but nobody else. Did you *really* hang Horn?"

"Of course we did."

"They say he escaped."

"He's right in there. Dead on a slab."

"I want to see him. I've got to."

The officials were willing to permit Kels inside the morgue. They threw their guns into the drawer of an autopsy cabinet, so as not to be armed in the event they would have to manhandle Kels. But the County Attorney objected to letting Kels inside. It was a regrettable decision, for from that day to this, Nickell (and a great many others) believe Horn never was hanged!

The *Post* throughout the trial had published broad hints that "men higher up" were involved in the Horn case. It is improbable that the newspaper had any definite grudge against Horn—unless his accusations of journalistic bias angered the *Post* proprietors. It is likely that the newspaper was after a certain cattleman, or group of cattlemen, hoping through the publicity of the Horn trial to uncover a condition that might give the discoverer of such malpractice a whip-hand over cattle barons and establish greater influence with the citizens of Wyoming. What personal animosities may have entered the *Post's* treatment of the case cannot be discerned now. But after the hanging, there were several attempts by the *Post* to spur officials on to inquiries concerning the cattle industry and its ghosts of slain rustlers. These efforts failed.

Although their attitude had been outspoken and their activities broad, the *Post* partners were tight-lipped as to their motive in prosecuting the Horn case and their search for the "higher ups."

Tom Horn was buried in the university town of Boulder, Colorado, in a little cemetery at the foot of the great hills, where students sometimes sit to read.

16

Cave Of The Winds

THE NOISE-MAKERS, Bonfils and Tammen, dramatized themselves, their newspaper and their community with a carnival whoopla never approached by other publicists. They were exhibitionists of incredible calibre. Their presses stood like impounded elephants behind plate-glass windows, through which gaping citizens might look and marvel.

When the *Post* moved to a new, white-tile structure in Champa Street, Tammen emblazoned a gilt motto across the upper ledge of the press-room show-window:

"O Justice, when expelled from other habitations, make this thy dwelling place."

This golden line stimulated the fancy of Clyde Spencer, the *Republican's* cartoonist—the man who had had trouble with Uncle Sam's whiskers. He speculated on what would happen if Justice were to accept the *Post's* invitation. In his cartoon he depicted the blindfolded lass traipsing into the *Post's* Red Room, with Bon and Tam plunging head-first out the window, and the caption:

TAM—Hey, Fred, who in hell is this?

BON—Damned if I know!

Tammen split his sides over this cartoon, persuaded Bonfils that it really *was* funny and then hired Spencer away from the *Republican.*

In crying its wares, the *Post* overlooked no selling feature, however small. Its lowliest employee was subject-matter for human-interest stories. And when it came to ballyhooing special writers, the *Post* thundered their virtues in a manner elsewhere accorded only to H. G. Wells or Anatole France.

The hustlers who vended the *Post* on street corners were a picturesque gentry, ranging in type from "Birdlegs" Collins, a blind and decrepit Negro pugilist, to the studious, personable Eddie Egan. This boy became successively a Yale University graduate, Rhodes Scholar at Oxford, National Amateur and Olympic heavy-weight champion and companion of Gene Tunney.

These newspaper merchants could out-howl and out-fight a trainload of jaguars. Each had a novel style of commercial address. For example, there was the "Crying Newsboy," a rosy-cheeked little fellow who stood in front of the *Post*, a bundle of papers half-falling from his arms, and bawling as though possessed of the world's largest and most devastating heartbreak. The tears cascaded down his apple-cheeks, and though the townsfolk grew to question the authenticity of his horrendous grief, the Crying Newsboy did a man's-sized business. He one day dried his tears to become manager of a large West Coast newspaper.

Perhaps the foremost character of all *Post* newsies was the poetic O. J. Owens. He was born in Poultney, Vermont, March 7, 1859, and went to Denver in 1879. He began selling newspapers in 1881. In 1895, Owens hawked the first extra editions from the *Post* press after Bonfils and Tammen had assumed charge.

A stranger might have thought Owens a somewhat stupid, inferior person who shouted the day's headlines in rhyme. His customers knew better than that. He possessed a brilliant mind (despite his spluttering verse), and, had it not been for his crippled, almost helpless condition, Owens might have taken a more fancy place in the affairs of life.

He barely could shuffle along. His poor hands were so crippled it took time for him to deal his papers. A customer often had to make change from Owen's pocket, meanwhile listening to rockinghorse rhymes.

It was popularly supposed that Owens had been struck by lightning. Burt Davis, veteran sportsman and tobacconist, who counted Owens' pennies at the close of each business day, said the warped newsboy had suffered an attack of what is now known as infantile paralysis. That misfortune came at a time when he was planning to become a lawyer.

Silver-Dollar Tabor gave Owens permission to stand in front of the Tabor Grand Opera House to cry his wares. Owens and Gene Field would compete in side-walk tourneys, making jingles on any and all subjects, including the indelicate.

This twisted newsboy was acclaimed the best weather prophet in the Rocky Mountain region, and, to the discomfiture of the Federal Government's forecaster, the *Post* for years printed Owens' predictions alongside those of the authorized weather man.

On his fortieth anniversary of selling the *Post,* Owens ignored the headlines of the day and cried out to pedestrians:

> "Through winter snow and summer heat,
> I've sold the *Post* upon the street,
> Filling patrons' hearts with cheers
> For forty long and happy years."

He died in January of 1933, the best-known street-figure in Denver, with the possible exception of a portly old gentleman who wore hundreds of shoe-laces festooned from his arms and neck, and who stood at the corner of Sixteenth and Arapahoe Streets, calling sonorously:

"All kinds of shoe-laces."

The *Post* issued editions, beginning at 10 o'clock each week-day morning and ending with a pink-jacketed 5:30 o'clock edition in the evening. There were numerous extras—a school miss's stubbed toe being enough to start the presses rolling and the newsies shouting: "SOCIETY GIRL STRICKEN."

Each edition found a sidewalk bedlam as the newsboys fought, screamed and all but bit one another in front of the plate-glass press room. Instead of trying to maintain a less homicidal demeanor at its curbstones, the *Post* encouraged the longshoreman fights, the slaughter-house ethics and the gamin hullabaloo as "Punch," the newsboy king, or his firebrand successor, Jake "Humpy" Sobule, distributed the papers to lieutenants. Such uproar attracted attention to the *Post*, and if that were not enough, the partners had a siren on the roof which yowled like a great and pestiferous tomcat whenever Bonfils or Tammen felt like pressing a button.

Inside the want-ad room on the ground floor, with its wall decorations of painted flames and a huge clock among the floor-tiles, customers jostled one another. Premiums were given away with each ten-word ad.

So many were the promotional schemes, stunts, contests, street fairs and publicity inspirations, that I shall postpone detailing them until a later chapter, a glossary devised for the convenience of mob-psychology students.

In keeping with its noise-provoking program, the *Post* sponsored a "Boys' Band." It was intended to rival the prize-winning Cook's Fife and Drum Corps, a local organization of young musicians who looked martial and sounded brave in parades.

The Cook boys wore Civil War Zouave outfits. The *Post*, ever partial to red, garbed its youthful horn-tootlers in British tunics of the George III period. A black-moustached tailor, Sidney Beal, became coach and leader of the band.

Whether it were due to youthful impatience, laxity of lip or a possibility that the *Post* recruits were tone deaf, their drums were jumpy, their woodwinds asthmatic and their brass sour enough to have justified John Philip Sousa in entering the courts of equity. But they made a lot of noise, and that was the chief purpose of the thing.

A seven-foot fellow, Carl Sandall, became the drum-major of this band. He was door man for the Daniels & Fisher Store, a drygoods firm that had reared Denver's tallest structure in imitation of an Italian campanile. Not content with Sandall's natural footage, Tammen outfitted him with a Horse Guards' shako of black bearskin. As big Carl stalked in pageants and at the head of the tuneless boys, brandishing a staff the size of a sewer pipe, he looked like something that had stepped out of *Gulliver's Travels*.

It is not to be inferred that the *Post* was trying to revenge itself on Wagner or Beethoven by maintaining this band. The partners encouraged music in City Park, where taxpayers rode in swanboats or looked upon an illuminated fountain. They praised maestroes in municipal concerts other than their own.

Signor Rafaelo Cavallo had a symphony orchestra, in which young Paul Whiteman played the viola, and it was a meritorious group. The Signor also led the orchestra at the Broadway Theatre, where again young Paul obliged with the viola.

The modern New York impression of Whiteman seems to be that he came from nowhere, from an obscure and untalented home. His own free and easy mannerisms have supported such a theory. The fact is that he was of a well-to-do and cultured family.

His father was Wilberforce J. Whiteman, superintendent of music in the Denver Public Schools, and leader of the fashionable Trinity Church choir. His mother, and his sister, Mrs. Fern Whiteman Smith, were contralto soloists of recognized ability. Paul had a thorough musical background and every other cultural advantage of an artistic family. This upbringing shows in his work, and an analysis of the man would reveal him to be a superlative artist first and a "King of Jazz" afterward.

The elder Whiteman, a slender and wiry gentleman, was popular in all Denver schools, but on the West Side he loomed a hero. In 1905 he incurred the dislike of an especially tough gangster at the Elmwood School. Professor Whiteman had decided that the young thug was a tenor, whereas the affronted young man insisted that he would sing bass or not at all. One word led to another, and when school let out for the day, it was reported that the young man was going to paste Professor Whiteman in the nose. Indeed, the Principal of the school confided to the music master that the boy in question was about the rowdiest citizen in those parts.

"I'd leave by the *side* door if I were you," the Principal advised.

"If you were I," said Professor Whiteman, "you'd leave by the *front* door, and that's how I'm leaving."

When he reached the street, the Professor saw a hundred boys waiting to witness the massacre. As the disgruntled tough stepped up to the Professor, raising a right fist, Paul's father countered with the nicest left hook ever delivered until Jack Dempsey, another Colorado product, began to make pugilistic history. The Professor received an ovation as he walked calmly away to catch a Lawrence Street car. From that time on, a male pupil would even sing falsetto if Professor Whiteman counseled that register.

Paul's father at first was not in favor of his son's methods of conducting an orchestra. Paul had scored his early success and wanted his Pa to see him flip the baton.

When Pa Whiteman had his introductory glimpse of his hefty son wiggling a hip, tapping a toe and doing a bit of modern fly-swatting, he muttered:

"He doesn't know how to conduct!"

A sarcastic drunk one evening asked Paul Whiteman: "How are you doing now?"

"Pretty well," Paul said. "I succeeded in running an orchestra up from a gents' washroom into Carnegie Hall."

The *Post* always spoke at the top of its voice.

As noise begets noise, the whole town took on a louder tone—just as the modern radio has made of each American home a place for conversational shouts. And if a man wanted publicity, all he had to do was perform some act of real or implied loudness.

It was thus that the Rev. Christian F. Reisner, at present New York's go-gettingest pastor, attracted attention through the *Post*. He zoomed into town, finding his brother Methodists rather quiet, tame and resigned to dull sermons in dull surroundings. He was the first pastor—according to my best information and belief—to put an *electric sign* at the portals of a church. That act may now seem commonplace and sensible enough, but in 1907 it was *revolution*.

Grace Church, a mellow-red sandstone structure, had been lingering with Biblical diabetes until the Rev. Chris arrived with evangelical insulin. Soon his church was crowded. He permitted hand clapping and whistling at the close of any portion of the service which had audience-appeal. He launched games, social functions and hinted that God had a sense of humor and wasn't an old meanie with a big, bitter frown and an Assyrian beard. The diehard ritualists resisted this intrusion of life, so they hastened to their own solemn graves, the better to find out first-hand if Jehovah approved of high-jinks in mundane pulpits. The Rev. Chris pulled Grace Church out of its financial hole and went on to New York.

One might expect that the mortician's craft would be the one art to escape a noisy milieu such as that created by the *Post.* Yet the undertakers of the town made quite a graveside din. They advertised profusely. When there was a funeral, the lucky director always made sure there was no mistaking *who* was running the works.

They were a colorful lot, these Denver embalmers. Old Man Walley was the longest-lived and best known. His partner, Bob Rollins, was a machine politician, and the wordy rows between Rollins and Judge Ben B. Lindsey were *something* to hear. The gamecock Judge was as courageous physically as he was mentally, and many the time shook a fist in Bob Rollins' face at political meetings.

Old Man Walley came West in 1859 as a cabinet maker. When the city was founded no thought had been given as to who should bury the dead, until a lynching occurred and the vigilance committee found a coprse on its hands. So the lynchers called on Mr. Walley. He was a very handy man with the tape measure, as well as with hammer and saw. He measured the town's first cadaver and then fashioned a box. He had no ideas concerning an ornate casket, still he was able to construct a form-fitting box, which later was called the "Pinchtoe Model." It got its name because Mr. Walley wished to save as much lumber as possible.

The lynched man was buried in a prairie hole. From that planting grew the town's first cemetery. Later it became Cheesman Park.

Walley conducted funerals thereafter for sixty years, and until a short time before his death walked two miles daily to and from his office at No. 1408 Larimer Street, across from old City Hall. He died at the age of ninety-four years.

One of Denver's undertakers owed his success to having been boycotted in Leadville during the gold-rush days. There had been an epidemic of pneumonia, and when a miner suffered that malady his chances in a high clime were those of a martyr among the Roman lions.

There was a shortage of caskets, the gold-seekers' ambitions having dispensed with the "Pinchtoe Model." As all caskets had to come overland by stage-freight, the undertaker was wondering what he should do about the problem of supply and demand. He struck on a brilliant procedure. He decided that caskets which he recently had buried would be in good condition. So he salvaged them at night and was doing a brisk business until the stage driver became suspicious—he had not been hauling that type of furniture of late. A committee kept watch at the cemetery and caught the undertaker *replenishing* his stock.

He left town at sunrise to become a leading figure in Denver's marble orchards.

The population of Colorado in 1907 was about 700,000, and that of the city of Denver about 175,000. The net daily paid circulation of the *Post* in September of 1907 was 83,000 copies, more than the combined circulation of its three competitors.

It paid to be noisy.

17

A Study In Sawdust

Tammen was forever on the lookout for new and tinkling toys. A boys' band, a coal company, Geronimo's authenticated scalps, the *Great Divide*—his beloved weekly newspaper, now dedicated to farmers and miners—all these were excellent treasures, but Tammen sought a more glamorous field.

So he set his heart upon the thing that any other perennial boy might dream of acquiring. A circus!

Bonfils, brooding over his power and his pennies, didn't sympathize with that ambition at all. Yet he loved Harry and liked to humor him. And Harry *could* talk. He spent eloquent hours in discussing elephants, tigers, clowns and acrobats. Bonfils countered with speeches on Mother Nature. He repeated lessons he had gained in talks with Eugene Grubb, the potato Burbank, and with Enos Mills, the renowned naturalist of Estes Park. Why didn't Tammen use his excess energy in hunting, fishing, studying nature at first hand? Circuses were expensive. Nature was free. Why, there were four hundred and five species of birds in Colorado.

"Birds, hell!" Tammen said. "I want elephants."

Bonfils capitulated, but with the usual understanding: "It has got to *pay* for itself, Harry. Please remember *that.*"

The Tammen circus-toy began as a comparatively small outfit, a dog and pony show. Tammen sought a resounding name for his first tented outfit, so he called it the Floto Dog and Pony Show, after his sports editor, Otto C. Floto. Now, Floto owned no part of this or any of the subsequent circus interests of the *Post,* but Tammen was in love with the name, Floto.

As time went on, Tammen added to his menagerie and increased the spread of canvas. Bonfils became more interested in the enterprise, especially when a chance came to lash out at such competitors as the Ringlings. There were to be circus wars, near-scandal, mergers and all-around activities, with Bonfils and Tammen astounding a group which they dubbed "the circus trust."

To give prestige to their growing circus and to imply that it was a combination of major-league calibre, the partners sought for a well-known name to be coupled with that of Floto. The Sells Brothers' trade-mark was impressive in the world of spangles, so the *Post* proprietors hired a relative of the Sells family to work for them, gave him a qualifying share or so of stock, and thereupon re-christened their show "The *Sells*-Floto Circus." Its billboard posters showed Floto's portrait and *all four Sells Brothers.*

There was a furore. The Ringling interests were indignant on several counts. First, they had bought several circuses, merging them with their own to prevent bothersome competition. Among these shows was the famed Barnum & Bailey Circus, the Adam Forepaugh Shows and the Sells Brothers' enterprise.

Next, it was claimed that the *Post* sent hirelings from city to city, wherever the Ringling shows were billed to appear, to cover up the opposition lithographs, posting Sells-Floto three-sheets over those on the Ringling billboards. The *Post,* in rebuttal, charged the Ringlings with doing that very same thing.

The *Post's* circus also made raids on Ringling's performers, notably the clowns. There was much mystery as to why the joeys were content to work for the Bonfils and Tammen show in preference to larger, more important outfits. Tammen himself supplied the answer:

"You see, clowns are vain fellows, and we couldn't afford to pay them bonuses for deserting other shows. I couldn't see any fun in the clowns we already had—they acted like editorial writers. So I said to Fred:

" 'The one thing we've got to do, money or no money, is provide the best cook tent in the world. All men love their stomachs, rich, poor, white, black, clowns or serious Supreme Court Judges. If we get the best cook tent, we'll get the best clowns.'

"You'd be surprised. We got almost any clown we wanted, letters piling in from Marceline, Johnson, Arthur Borelli, Lon More, John Albion, all the crackerjacks in clowndom. They'd turned down bigger offers to come to us. They said: 'It's a good show, but the cook house is great.' "

Regarding his circus experiences, Tammen said:

"Bonfils and I were foredamned to success in the circus, as in any other undertaking. What made it grow? The truth. A circus is a pack of lies—which is something of a lie itself, because it's only half true, like most paradoxes. So just to be original, we told the truth about our show, advertising it as the 'second largest in the world'. We discovered that the truth was about the only thing that never had been told about a circus, so we hurried up to tell it."

The Ringling Brothers filed a suit in equity against the Sells-Floto Circus as a company, and F. G. Bonfils and H. H. Tammen personally, in October, 1909. The Ringlings asked $2,000,000 damages and a restraining order preventing the Sells-Floto company from using the photographs of William Sells and his family, as well as the name "Sells." This suit was brought in the Federal Court in Denver, Judge Lewis presiding.

The *Post* loosed its editorial guns on the "circus trust" in general and the Ringlings in particular. The following is an example of a *Post* "news" story prior to the preliminary hearing in the Ringlings' suit:

"Probably the real reason this suit is brought is because of the growth of the Sells-Floto Circus, the Sells-Floto being recognized by the Circus Trust as a dangerous competitor. Up to this time a guerrilla warfare has been practised by this outfit by *covering up the bills of the Sells-Floto Circus.*

"For instance, it is customary when showing in any city, and especially small towns, to post bills twenty and thirty miles out in the country and make arrangements with the owners of barns or fences to post Sells-Floto bills, paying therefor either in money or circus tickets. In the course of a little while, one of the trust agents would come along and say to the farmer:

" 'The Sells-Floto Circus is busted up, but Ringling Brothers, or one of the Ringling shows is coming.'

"And so the Sells-Floto tickets are taken up and others given in their place, and a bill is posted, reading:

" 'Ringling Brothers Coming Soon.'

"Their hired men go about the country telling all sorts of stories, and they are aided and abetted by Mr. Thomas M. Patterson (owner of the *Rocky Mountain News*), the idea being that the people in other states, not knowing the Sells-Floto Circus, and the trust having three or four aggregations, it is easy to tear up or destroy the Sells-Floto property.

"But in spite of this guerrilla warfare, which has been kept up continuously for a number of years, the Sells-Floto Shows has prospered and extended its territory this year from coast to coast and from Vancouver to the Gulf. And now that the owners, under the name of 'Sells-Floto', have built up a national reputation, the Ringling Circus trust seeks to take away the name for which the Sells-Floto paid William Sells, after his interest and the right to use his name was paid for more than two years ago.

"At the present time the Barnum & Bailey shows are down in North and South Carolina, Georgia and Tennessee, and so is the Sells-Floto Circus, and the most violent circus war is raging in that section. Down in Texas, where the Ringling Brothers are now showing, they are having trouble with the revenue agent. It seems that they avoid paying the proper license, and have been avoiding it for a number of years, and the revenue agent, up to this time, has attached their circus for some six or seven thousand dollars for taxes unpaid in the year 1906 and 1907, and is also collecting for shortage of taxes in El Paso, Abilene, Dallas, Waxahachie and other places for this year. Their particular and special agent, McCraken, is down there at the present time, protesting and bull-dozing, but Texas is a country where the laws are executed and nobody is afraid of a circus trust.

"This year they began trying to exterminate the Sells-

Floto Circus at El Paso, April 2, telling the people not to patronize Sells-Floto, which played there April 10, and posting bills:

" 'Ringling Brothers Coming Soon.'

"But Ringling did not show there until September 30! Sells-Floto is the only other big show that stands in their way to gain absolute monopoly."

The *Post* labeled the Ringlings as "The Coming Soon Circus Trust." When the Ringlings arrived in Denver, they met an unexpectedly annoying setup. Authorities friendly to the *Post* had declared an extraordinarily large license fee for circuses the size of the Ringling Brothers' contingent. The number of cars a circus used for railway transportation was made the basis of license regulation. The Sells-Floto Shows, however, could qualify for a much smaller license fee, its number of cars barely coming within the prescribed quota for the cheaper tariff.

Thus the Ringlings hurriedly changed their billing and played in a remote suburb and at a loss of patronage. The anti-*Post* Tramway Company did its best to serve the Ringlings in their emergency, which fact did not please Mr. Bonfils very much as he composed "So The People May Know" philippics.

Inconvenienced circus customers raged at the *Post* for shunting the Ringlings into the sticks. To counteract this ire, the *Post* loudly subscribed $1,000 to bring pioneer aviators to Denver, a civic campaign for that purpose having failed. The sky-men came to town, performed nobly and the citizens forgot their rancor concerning pink-lemonade tricks.

Frank Tammen, Harry's younger brother, was manager of the Sells-Floto Circus in its adolescent days. Arthur Bennett, who wrote *Saturday Evening Post* articles under the pen name of Yates, was press agent. His successor was Courtney Ryley Cooper, ace of circus fiction, ex-clown and star reporter of Kansas City and Denver. Coop was about the best literary bet any circus ever had. The advance man for the circus was Frank Cruikshank, former genius of Chicago's White City, and later of Broadway, New York, and the show shops.

A tiger escaped its Sells-Floto cage in May, 1907, while that circus was tenting in Twin Falls, Idaho. J. W. Bell, a blacksmith, and the keepers tried to recapture the animal. It made a dash for the outside, and peculiarly enough, leaped on a pony's back. The terrified little steed began to race in circles, endeavoring to shake off its savage rider, but the tiger clung on with claws and fangs. The attendants closed in and began to club the tiger. Blacksmith Bell fired on it with his pistol. The beast thereupon leaped from the pony's back. Mr. and Mrs. C. E. Rosell of Twin Falls and their four-year-old daughter, Ruth, were in its path of charge.

The tiger knocked the Rosells down, then set upon the girl, closing its fangs on her throat. It then made another dash for the open, but was shot dead. The Rosell child died.

Manager Frank Tammen was arrested, charged with manslaughter, but was acquitted. The circus counsel, David C. Webber, of Denver, introduced evidence to show that the attack on the Rosell child occurred *after the regular performance* and during feeding time, holding that the circus was not responsible for an accident under such circumstances.

The *Post* had this to say:

"Manager Tammen was not in the animal tent at the time the tiger escaped and knew nothing of the accident until some time later. No one held him blameable, but while the owners of the circus were preparing to arrange a settlement, an agent for a certain Denver man approached Rosell and induced him to swear out a warrant for Mr. Tammen's arrest. The agent represented to Rosell that by trumping up a false charge against the circus manager, he would be able to mulct the owners out of a large sum of money. During the trial it was conclusively proved that an accident such as happened last Saturday night, while wholly unavoidable and unexpected, was not liable to happen more than once in a lifetime of a circus."

During the afternoon performance of the Sells-Floto Shows at Riverside, California, April 16, 1908, two gasoline tanks exploded at a filling station three blocks away.

The menagerie tents caught fire from sparks. While Fred Alispaw, the dependable and courageous menagerie superintendent, was preparing to dismantle the tents, an assistant boss canvas man became hysterical. He began shouting for everyone to "get out" and confiding loudly to the throng that the tents were "going to burn." Although Alispaw, one of the coolest men in the animal business, endeavored to assemble the veteran elephants along the picket line to prevent a stampede, four of the great animals broke loose from their leg-chains. The screaming people, and not the danger of fire, seemed to have confounded these big beasts. They knocked down cages and tent-poles and made for the California outdoors.

Alispaw and his men saddled horses and gave chase. They took several elephants along to help capture and calm down the dangerous truants. The four fugitives separated. A posse of townspeople, policemen and firemen, armed with clubs and rifles, also joined the chase, doing more to increase the danger than to lessen it. Alispaw captured one elephant, Frieda, who was mired in a clay bank. She was chained to two of the law-abiding elephants. A second elephant, Old Snyder, leader of the Sells-Floto herd, was cornered in an orange grove, where he had regained some of his usual composure, but was uprooting orange trees, eating the fruit and otherwise making himself as unwelcome as a frost in blossom time.

The two other elephants, Floto and Alice, were spreading fear and destruction. Alice had wrecked a chicken yard, then had gone on to invade a larger poultry farm. She demolished the pens and finally was brought to account, a chicken-coop on her head and her legs entangled in hundreds of feet of wire.

Floto, still thoroughly crazed, was racing along country roads, occasionally feeling the thud of a bullet against his thick hide. Deaconess Gibbs of the Methodist Church was sitting on the front porch of her farm house when she saw this gray animal charging in her direction, trumpeting madly. Instinctively the Deaconess raised her chair in self defense. The elephant shattered that piece of furniture with one swing of his trunk. Then he stepped upon the Deaconess, killing her.

Floto then started back to Riverside, crashing into a horse and wagon that barred his path, killing the horse, then charging the posse. A bullet wound in his trunk increased his frenzy. He went through a barber-shop window, continued out the back way, knocking down the rear wall. Next he crashed into the lobby of a hotel and thence into a music store, carrying the front of the building on his neck. He ended his mad foray by breaking through the wall of a livery stable. It was there that the brave Alispaw fought with Floto for an hour, armed only with an elephant hook.

This unfortunate matter was adjusted for $17,000, there having been no agent of the "certain Denver man" present to "trump up charges."

Whenever the Ringlings ran into trouble, were it from alleged avoidance of license tarriffs or other amusement ills, the *Post* came out with printed hurrahs, and in one instance quoted the Bible:

"Verily, as the prophet said, as ye sow, so shall ye reap."

In April of 1909, Bonfils and Tammen claimed an important recruit in their war against the Ringlings. Their own account of the conflict at this juncture follows:

"Tucson yesterday was the scene of a circus war between organized forces on one side and unlimited resources and a plentiful supply of daring on the other. The fight which is bound to take place is to be engineered by Alf Ringling, representative of the circus syndicate, and John W. Gates, of New York, whose favorite remark, 'Betchu a million', is typical of the man, his character and his disposition.

"The Sells-Floto Circus, owned by Bonfils and Tammen of the *Denver Post,* has an additional backer, John W. Gates, and is to be considered fortunate, indeed.

"The Sells-Floto Shows is one of the traveling circuses in the country outside of the trust or syndicate in which the Ringling Brothers are the principal movers, and the latter have made the boast that they will break and put out of business every circus in the country not belonging to the Ringling syndicate.

"John W. Gates is noted for his nerve, and, hearing of the threat of the Ringlings, he immediately set out to checkmate them.

"His first act was to get for the Sells-Floto Shows, from his personal friend, J. Ogden Armour, the famous Armour prize horses when Ringling was about to close a contract for them. Next followed the Rooney Brothers, one of the Meer Sisters, and the Nelson Family, all prize members of the Ringling organization. The loss of the above valued artists put a bad taste in the mouths of the Ringlings, and their first move, in the line of revenge was to send out a squad of five advertising men to follow the Sells-Floto Circus around, and wherever they advertised, to also advertise that the Ringling show was 'coming soon.'

"Some time ago Tucson was plastered with papers announcing that the Sells-Floto Circus would show in Tucson on the fourteenth of this month. Yesterday a crew of five men, working for the Ringlings, posted bills stating that 'Ringlings will be here soon.'

"This means that the first stage of the circus war is on, and the fight from now on will be a merry one. Ringling Brothers are millionaires, but John Gates is also a rich man; in fact, one of the wealthiest in America, and, rather than be beaten by the rival circus, he will spend every cent he possesses in keeping up the fight."

The Gates threat was only a bluff. Armour was Tammen's closest friend, and it is believed he induced Gates to "stand for a story" that he was allied with the circus. At any rate, Gates faded out of the picture almost as quickly as he had entered it.

Federal Judge Lewis on November 5, 1909, granted a temporary restraining order, forbidding the Sells-Floto Shows to use the four photographs of the Sells Brothers, Ephraim, Allan, Peter and Lewis, on advertising matter. The court, however, upheld the right of the Bonfils-Tammen circus to continue in the use of the name "Sells," granting that it had acquired the name from William Sells.

Later the circus obtained the permanent right to the Sells' name, but had to cease posting the faces of the original brothers on barndoors and fences.

The circus was not a gold mine. We shall return to its sawdust rings in later chapters, with its Buffalo Bill episodes, its parade of champion athletes, Jess Willard, Frank Gotch, Dempsey, and its winter quarters in Denver.

Tammen traveled much with this group, thriving on its steam-organ excitements.

There were legends to the effect that local millionaires were persuaded to buy stock in this circus, or be subjected to editorial attacks by the *Post*. Such supposed holdings were referred to in Denver gossip as "Monkey Stock," but none of the charges ever was brought to court or elsewhere sustained.

Bonfils did not care much for the circus, its people or its fanfare. It was said that he once took a trip with the troupe and suffered a sad experience. The tale represented him as poking about the dressing tents, trying to ascertain if economies might not be exercised in housing the performers. He wandered into the tent of a lady acrobat while she was changing from her tights to civilian frock. Her husband, a lion tamer, heard her scream and saw the nonplussed Bonfils hastening from the dressing tent.

Mistaking Bonfils' parsimonious nature for a lecherous one, the lion tamer began to ply his jungle-whip on Bonfils' head, neck and arms. Bonfils ran off the lot, the lion tamer and his whip in close persuit. Despite the compromising situation, those who knew Bonfils were sure that his intentions were of an economic, not an amorous, character. Nevertheless, he was laid up for some days until his wounds healed.

From that time on he looked distrait when anyone spoke over-long concerning the circus.

18

Mahatma On Horseback

THE *Post* scorned no man's patronage. Its advertising doors bounced open for anyone's admittance, provided there were doubloons in the rapping hand. Its pages burgeoned with occult promises. There were patent-medicine enthusiasms, optimistic messages for the lame and the halt, and a reader might choose from a wide assortment of luxuries, such as potent electric belts, magic ear trumpets, Wizard of Oz philtres for the lovelorn, timely rectifiers of tainted blood streams, or penny-arcade philosophies for teetering minds.

The city, with its multitude of health-seekers, was a camping ground for metaphysical caravans. Crystal gazers with mikado moustaches and knights errant of phrenology spread crazy-quilt banners to the mountain air. Did they but advertise judiciously and often, these necromancers were reasonably immune from raids by stewards of civic welfare.

The columns of the *Post* were slightly confusing to a seeker for spiritual solace. There were go-to-church campaigns, appearing beneath Biblical texts, on the one hand, and voodoo entreaties of broomstick riders on the other. The abracadabra displays in the *Post* were as fascinating as a premature explosion of pin-wheels at a county fair.

The town's Witch of Endor hankerings were indicated as early as 1895, the year when Bonfils and Tammen took over the *Post*. Francis Schlatter, a German cobbler of schoolmaster countenance and zealot's eyes, came to town in July. By August tens of thousands had acclaimed him the "Second Messiah."

This man was in nowise a money-grubber. He lived with great simplicity, underwent devas ting fasts and blessed talismanic handkerchiefs for pilgrims. He performed in the muddy back yard of a North Denver home, where he stood on a plank, "laid on hands" and spoke gently of "the Father's Business." Commercial harpies stationed henchmen in the enormous queue of applicants for the Schlatter blessings and sold such places for large sums. They also went about town vending cheap cambric kerchiefs, which they pretended the North Denver messiah had blessed.

So far as is known, Schlatter never accepted money for his own use, but lived on the charity of a Mr. Fox, in whose back yard he received the faithful. He refused an offer of $5,000 to perform miracles in Chicago. He eventually disappeared from Denver and was presumed to have gone to the prairie, never again to be seen by any who had known his healing touch. The prophet left a note pinned to the pillow of his simple cot:

> "Mr. Fox—
> My mission is finished. Father takes me away.
> Goodbye.
> (Signed) "Francis Schlatter."

The hegira of this Mohammed of the plains received nation-wide notice. His mail had been so heavy that a special postman hauled it in a truck twice daily to the Fox home, and now it was re-routed to the Dead Letter Office.

Schlatter became a part of old men's legends. For years he was reported as having been glimpsed on desert trails, mounted on a white horse, a beatific smile on his face, a hand upraised in apostolic hello—and then he would vanish as though taken to the skies above the Bad Lands, where the mirages are born.

A skelter of bones and a Bible said to have been the evangel's property were found on a desert trail long years after the era of back-yard ministrations.

Although Schlatter had not commercialized his talents, the prophet's success indicated to other healers, psychic high priests and oracles that Denver was "one of those places." Soothsayers, good, bad or indifferent, began converging on the city.

The *Post's* first big advertising account along astral lines was that of Professor Abuhama K. Solomon, self-styled "Tiger Hindoo, Clairvoyant and Palmist." He specialized in "love, courtship and marriage." His advertisements proclaimed him a three-ply product, "Mohammedan, Hindu and Brahman." This dealer in the Vedas weighed more than two hundred pounds and wore a turban of reefed mainsail proportion, with a glass-works Kohinoor the size of an anchor-light. He claimed to have read the hands of Ella Wheeler Wilcox, Mrs. Potter Palmer and Miss Helen Gould.

The "Tiger" had a good season in Denver, but the *Post* passed him by for a more personable, more useful communicant with the Other Shore, Dr. Alexander J. McIvor-Tyndall, exponent of advanced thought, magician and table-rapping king. It was he who drove, blind-folded, through the city streets in search of a hidden needle.

McIvor-Tyndall, handsome, fierce-eyed and with a head of roached, black hair, impressed everyone, including himself. He set up quarters in the Psychic Science Building in California Street and published a magazine called *Swastika.* The *Post* not only printed his ads but publicized him profusely. He sometimes contributed articles to that newspaper, such as the one headed:

"TYNDALL SAYS LOVING MOTHERS
MAKE DRUNKARDS OF SONS"

Whatever his ability to peer into the future, his *Post* story of December 23, 1909, did not prove him exactly stupid in that respect, for he wrote:

"Temperance is the way of health and true success, but prohibition is not temperance. Even if prohibition could stop drinking, I maintain that it would not be the best way to proceed. The method of force always requires force to maintain a position, even if once gained.

"The idea that drink makes drunkards is a wholly erroneous one. There is no power in drink that is able to collar a man on the street and pour itself down his throat. All power is resident in the human mind. The will of man determines what he shall do with his life and his appetite.

"Loving, self-sacrificing, good mothers make drunkards of their sons. Most mothers are ruled by fear. Their very love makes them afraid of possible harm, and it is due to this fear that they make drunkards, or some other kind of weakling, out of their boys. When Little Willie first begins to creep, and from then on, during the rest of his life, the fond mother watches for possible dangers in Willie's way. She would, if she could, take the sun out of the heavens so that it would not burn the dear boy's nose. It never occurs to the mother that she is herself ruining her son by fostering a weak character and eliminating any possibility of choice and discrimination.

"Saloons never ruined any man. The ruin comes through a man's own feeble, uncultivated will, nurtured and fostered by those who love him most. What mother ever blamed herself or the boy? It is always 'the other thing' that is to blame, as it was in the days when the cat was punished and banished as a 'bad cat for scratching the dear baby.'

"When the boy becomes a man, he goes into a saloon. He stays waiting and expecting some one to come and remove it from him. It doesn't occur to him that it is up to him to remove himself from it. That sort of mental training has never been taught him."

Now it occurred to City Editor Joe Ward that Doc McIvor-Tyndall would be just the man to solve murder mysteries. Joe had tried out all local criminologists, professional and amateur—the city was rife with Sherlock Holmes emulators—and summed up with the following remark:

"An ass is always an ass, and a magnifying glass only enlarges the assininity."

Early in June of 1908, Sheriff A. H. Bath of Laramie, Wyoming, was slain near Jelm, in the Woods Landing section of that state. The slayer was not known, but G. A. Summers, alias Ned Davis, a former convict at the Canon

City, Colorado, penitentiary, was in the sheriff's custody on an assault and robbery charge when the crime was committed. Summers had fled.

Citizens of Laramie offered a $500 reward for the capture of the slayer. A posse and bloodhounds searched the countryside with no success.

In the *Post's* city room, Joe Ward waved his long shears and piped:

"Get McIvor-Tyndall on the job."

The *Post* announced that the "Doc" would ride blindfolded "at the head of a posse of determined men." It added:

"The ride may last for ten miles, or it may last for hundreds. How it will end, no one can tell, but Dr. McIvor-Tyndall's success in previous ventures of this character makes him sanguine of success on this occasion.

"To the layman it is hard to explain the mysterious force that he believes will lead him in his chase after the criminal. To the psychic it is easier."

There were interviews with Mahatma Tyndall. For example:

"We believe that all life is a unit. Crime is a disturbance, a variation of the ether waves which surround us all. Those of us who are peculiarly sensitive can get psychic impressions from these waves and follow where they lead us. In this way I hope to trace this man."

Concerning the psychic's *modus operandi,* the *Post* said:

"Those who have seen Dr. Tyndall at work of this kind will be familiar with his methods. He is first blindfolded, mounts his horse, then waits a few instants for the 'impressions'. Once receiving them, he starts, and those who follow him will have a wild ride. The peculiar psychical conditions which produce the impressions and make him susceptible to the vibrations left by the criminal may not be understood, but their workings will be evident."

Dr. Tyndall told interviewers:

"Yes, I must get the 'impression' first. And after that it will be easy. It may take me some little time to get started, but after that the work will be fast. I really think I will have very little difficulty in finding the man, although

I cannot tell exactly what the conditions are until I get on the ground."

Joe Ward's staff observers were able to find out that: "Dr. Tyndall has, during his career, saved one man from the gallows, and in numberless cases restored stolen property to its owners. In Rio Vista, California, after James O'Bryan had been positively identified by the sheriff of the county as a murderer by the name of Lee Harrell, and was sentenced to death, Dr. Tyndall produced positive proof that the man was not Harrell, and also found the guilty man."

And again:

"In a $3,000 jewel robbery in San Francisco, he located the guilty man, induced him to return the jewels to their owners, the Shreeve Brothers, and worked from a considerable distance from the place where the crimes were committed. He has also been interested in the Maybrick case in England, the Luetgart case in Chicago and other celebrated criminal affairs."

The blindfolded seer got on his horse, concentrated, galloped, curvetted and hop-scotched at the head of an awed cavalcade. The *Post* chronicled his progress; every spiritualistic sigh was recorded by Joe Ward's mounted reporters, and even the whinnies of the Mahatma's mustang were construed as solemn evidence of psychic stimuli.

After the ouidja-board hunt had continued for some days, the *Post* wound up the whole business with a story picturing Doc McIvor-Tyndall as "resting" in Cheyenne on June 16, 1908, and after "a series of remarkable psychic demonstrations, in which he engaged in the Woods Landing Country." The story continued as follows:

"Dr. McIvor-Tyndall last Friday went to the scene of the murder and there reduced himself to a sub-conscious state in the hope that he would yet be able to establish telepathic communication with the fugitive. In this he failed, he says, receiving only the faintest of psychic impressions regarding the whereabouts of the murderer, but distinctly receiving impressions of the route he had followed in leaving the scene of the crime.

"Following the telepathic trail, Dr. McIvor-Tyndall

went northward from the place where the killing occurred, leading those with him into the rough Sheep Mountain district until the approach of night caused the experiment to be abandoned for the time being.

"Saturday morning, Dr. McIvor-Tyndall took the train from Laramie to Milbrook Station, and at the latter point was met by a conveyance which carried him to the Porter ranch, where he desired to utilize the horse ridden by Sheriff Bath when he was shot, and afterward ridden by the murderer.

"Blindfolding himself, the psychist mounted his horse, and at once, he relates, was the recipient of psychic impressions which enabled him to describe the course taken by the murderer in leaving the vicinity of the killing. He described in detail the country through which the murderer traveled until he reached Woods Landing, where friends of the man he had killed were yet waiting his return.

"Woods Landing is on the edge of semi-open country, and there, said Dr. Tyndall, the fugitive abandoned his horse, removing the saddle and bridle and hiding them in order that their discovery might not excite suspicion that the horse had been turned loose under extraordinary circumstances.

"Dr. McIvor-Tyndall, still receiving clear-cut impressions, told how the fugitive had struck eastward from Woods Landing, traveling on foot and following all natural cover. The impressions gave him the trail for only a short distance beyond Woods Landing before they failed utterly, and he was unable to obtain further impressions connected with the murderer.

"Dr. McIvor-Tyndall expresses the opinion that the murderer went directly from Woods Landing to the Union Pacific Railroad, and this belief is held by others acquainted with the clues obtained in the search for the fugitive. Furthermore, Dr. McIvor-Tyndall believes that the murderer is now far away, the faintness of the mental impressions he was able to receive indicating this."

When "Doc" McIvor-Tyndall packed his Yogi knapsack and went to New York, the *Post* said in October of 1911:

"Dr. McIvor-Tyndall and the Rev. Christian Reisner are two Denverites who have made good in New York City."

After a boycott on the *Post* by merchants had failed, it was said that the newspaper's income from four large department stores alone was $186 for every hour of the day, seven days a week.

Bonfils' newspaper adversary, the elderly ex-United States Senator Thomas M. Patterson, proprietor of the *Rocky Mountain News* and the *Denver Times,* charged editorially that the *Post* shielded vice and blackmailed merchants. He also published an unflattering caricature of Mr. Bonfils, depicting him as a freebooter.

This situation exploded in the weed-grown lots of East Denver, and almost within the shadow of the State Capitol.

19

Good Morning

Ex-United States Senator Patterson was something of a pedestrian. It was his habit to walk daily, with an Edward Payson Weston stride, to and from his newspaper offices in Welton Street. His punctual appearance on a never-varying route was a signal for neighboring citizens to push away from their breakfast tables and make for the street cars.

The good Senator was somewhat myopic and wore thick-lensed spectacles which accentuated the size of his naturally large gray eyes. If he failed at once to recognize a friend by name, he made up for that lapse by the dignified sincerity of his greetings.

On the morning after Christmas of 1907, the Senator had left home with his son-in-law, Mr. Campbell, the latter catching a street car and the Senator beginning his invigorating mile to the *Rocky Mountain News* and *Denver Times* offices. The one was a morning, the other an evening publication. Both papers had carried Ciceronian attacks on Bonfils and the *Post* two days before Christmas, not to forget the Captain Kidd cartoon of Bonfils. As the Senator was cutting across lots, he heard a greeting by someone behind him.

"Good morning," said that someone.

The Senator was about to turn in reply when a fist descended on the side of his head. Then a second wallop. The Senator fell, stunned, among the weeds. He received yet other fistic tattooings as he lay prone. The dazed and near-sighted publicist as yet was unable to discern who his assailant was, or why the waylaying. Bystanders at length rescued him, lifted him to his feet, and it was then that he recognized Fred G. Bonfils. The *Post* partner was cursing like a twenty-mule-team driver and threatening the Senator with further mayhem.

"If my name ever appears again in your papers," Bonfils was saying, "I'll shoot you down like a dog!"

The groggy Senator replied: "That's all right, Mr. Bonfils. Your name may not be mentioned today or tomorrow, but it will be mentioned in the *News* and the *Times* just as often as conditions arise which seem to require it. A man has but one life to live, and I am going to do my duty at whatever cost."

The ambuscade occurred at 9 o'clock in the morning. That afternoon Senator Patterson swore out a warrant before Justice Thomas Carlon. At the Senator's behest, Bonfils was not served until night, and in the Red Room of the *Post*.

Mr. Bonfils was permitted to telephone his attorney, John T. Bottoms, promising to appear in half an hour before Justice Carlon to give bond. He arrived in court, accompanied by counsel, and was held in $100. The complaint charged:

"Fred G. Bonfils did unlawfully and maliciously commit an assault upon Thomas M. Patterson, wilfully, unlawfully and maliciously beating, striking and wounding the said Patterson in and upon his head, face and other parts of the body of him, the said Thomas M. Patterson."

When the case was called at 10 o'clock on the morning of December 29, Justice Carlon's court was thronged, as were the corridors. Bonfils was arraigned, and when Senator Patterson testified to the editorial incidents leading up to the attack, the crowd cheered.

Much journalistic linen was aired in the two-day session. The alleged tribulations of merchants who dealt with, or refused to deal with, the *Post* cropped up in testimony. These included Edward Monash, former proprietor of the Fair; A. J. Spengel, furniture merchant, president of the Board of Supervisors of the City Council, and former President of the Chamber of Commerce; J. S. Appel, head of a large store, and other business men.

Attorney Harry Silverstein appeared as counsel for the Senator. The testimony in part follows:

Attorney Silverstein—Please state your name and age?

Senator Patterson—Thomas M. Patterson, past sixty-seven years of age.

Q—Where is your residence?

A—On the corner of Eleventh and Penn; it is Number 1075 Penn.

Q—Are you acquainted with the defendant, F. G. Bonfils?

A—I am.

Q—How long have you known him?

A—I don't know. I would imagine twelve years, and more perhaps.

Q—I will ask you to state to the court, Senator, what occurred after you left your house and came downtown?

A—Mr. Campbell, my son-in-law, and I left the house about ten minutes to nine in the morning. Mr. Campbell was fifteen or twenty feet ahead of me. He started to the west side of Logan Avenue, on Eleventh, to take the street car. I, as was my custom, started to walk downtown. I left my home at the corner of Eleventh and Logan. I then turned north on Logan on the east sidewalk of the street; went down the east sidewalk, as I now recall it, to the corner of Thirteenth and crossed diagonally from the corner of Thirteenth to the other corner, and had just turned into a path that goes across lots from near the corner of Thirteenth and Logan that ends on the corner of Fourteenth and Grant.

I recall somebody on a wheel just going by me and almost at the same moment I heard light footsteps and a voice exclaiming: "Good morning." I started to turn my head, but before I could turn it to see who it was, and without seeing anyone, because my head was not turned clear around, I was struck in the face, or on the head rather. It stunned me and I staggered. It was immediately followed by another blow which knocked me down.

I felt somebody on me and blows were rained upon me, and by some means the party who was striking me desisted and I arose and then faced my assailant. I think somebody was restraining the man, whom I now recognized for the first time as Mr. Bonfils. As soon as I was up he commenced denouncing me in the most violent terms.... He kept on calling me vile names, to which I made no response beyond saying: "You came up behind me and struck me like a coward," and he said: "No. I said good morning to you and you said good morning to me." I said: "Nevertheless, you came up behind me and struck me like a coward."

There were one or two gentlemen in front of me and in front of Mr. Bonfils, and I was advised to turn around and go home. I said at once: "No, I have a right to be here," and finally I heard Mr. Bonfils say he would not go as long as I remained, and hearing that, I turned on my heel and went home, brushed off my clothes and cleansed my face from the blood, etc., and then walked downtown.

Q—Senator, were there any bruises, the effects of the blow, or what were the effects of the blows you received?

A—I wear a plate—a rubber plate in my mouth. As soon as I undertook to talk, I found the force of the blows had broken this plate in two in my mouth, and the gums are lacerated and the roof of my mouth is sore; a heavy contusion on my forehead, and in the afternoon it was swollen and slightly cut. This cheek was slightly cut and bruised (pointing). I spent all Thursday afternoon in applying hot lotions to my face to prevent discoloration as much as possible. Yesterday I was at home all day.

The witness then testified as to his usual course of walking to his office, his ownership of the *News* and *Times* and other details. He described his assailant as having been "very white and excited."

Attorney Bottom then began cross-examination, a part of which follows:

Mr. Bottom—Senator, had you and Mr. Bonfils ever had any difficulty before this?

Senator Patterson—I have no recollection of any personal difficulty.

Q—Have you been an enemy of Mr. Bonfils?

A—I don't think that correctly expresses it. I don't consider myself his enemy.

Q—You may answer the question "yes" or "no."

A—I don't consider myself his enemy.

Q—What do you consider an enemy, one to the other?

A—Well, I think it depends entirely upon the feelings

that exist. It would be very hard to describe it. I have very strong views, and have had, about Mr. Bonfils' relations to the public as a journalist, and I haven't hesitated to express them.

Q—Mr. Bonfils is one of the principal owners of the *Denver Post?*

A—I have always so understood it. I think he is the principal owner.

Q—And you have had objections to the manner in which he conducted that newspaper? Is that the idea?

A—Yes, sir. I have had objections, as all journalists should, as to the style of his journalism in dealing with business men, with officials and with matters that affect the safety and morals of the people of Denver.

Q—In what way does that style differ from the style of conducting the *News* and the *Times?*

A—In this: I have had no question in the world but that Mr. Bonfils and Mr. Tammen have used the paper that they control for blackmailing purposes.

There was applause and the court admonished the spectators to maintain quiet. The testimony continued:

Q—When, if ever, did the *Denver Post* or Mr. Bonfils blackmail anybody?

A—I don't desire to mention names.

Attorney Bottom turned to the court. "I insist, your honor, that he mention names."

Plantiff's counsel, Mr. Silverstein, rose to argue it unnecessary to "bring in outside names, and I don't see how it tends to clear up the case in any way."

To this Mr. Bottom, Bonfils' attorney, said:

"Now, may it please the court, it is well known to your honor, I presume, if you have been reading the newspapers, that Senator Patterson and his papers have, not only for weeks or months, but for years, charged that the *Denver Post*, and its proprietors, Mr. Bonfils and Mr. Tammen, have blackmailed people through and with their newspaper. He reiterated that recently, and he has reiterated it since this assault, that he has charged Mr. Bonfils with blackmailing.

"These things have brought matters here that would not be proper, perhaps, in a trial of this kind were it not for our own peculiar statute. This statute is to this effect: that if Mr. Bonfils had pleaded guilty to everything that is charged in this complaint, it would nevertheless be the duty of your honor, or of any jury acting in this court, to hear all the testimony desired to be offered on either side, and then determine what the punishment should be if the man was guilty. That law, your honor, was intended for but one or two or three purposes. Now, what are they? The court or jury shall arrive at the nature of the assault. Mitigating circumstances surround it, and Senator Patterson stated that he never had any difficulty with, and is not an enemy of Mr. Bonfils. If he was an outspoken enemy of Mr. Bonfils, then the matter of the assault would be entirely different from that made by a highwayman or a ruffian, and that it is that your honor may know all of the circumstances that surround this case—whether or not Mr. Patterson is an enemy of Mr. Bonfils, or has so expressed himself, and how is your honor going to determine these things unless you hear the evidence?"

Senator Patterson—Then I have no objection to answering them.

The court admitted the testimony and Senator Patterson began to hang out the journalistic wash.

The Senator—I recall the case of Edward Monash, who ran the Fair. He had been advertising in the *Post*. He concluded that he would advertise simply in the *News*—possibly in the *News* or the *Times*. Almost immediately upon that occurring, the *Post* commenced a series of articles concerning Mr. Monash and the Fair, charging him with violating the child labor law, charging him with cruelties of every kind to the children in his employment, printing the most incendiary articles about Mr. Monash's management of the Fair. It brought Mr. Monash to terms, and Mr. Monash commenced to re-advertise in the *Post*—and immediately every attack of that kind ceased.

I recall the case of Mr. Spengel. Mr. Spengel had been an advertiser in the *Post* and in the *News* and the *Times*. He suddenly ceased advertising in the *Post* and commenced advertising in the *News* and the *Times* exclusively. After a while the attacks commenced on Mr. Spengel. He was charged with violating his oath of office as Supervisor;

with cheating the city in his transactions; cartoons of the most derogatory character were published in the *Post* concerning Mr. Spengel. Mr. Spengel got tired, I suppose, and resumed his advertising in the *Post*. Since then Mr. Spengel has been nothing but a good officer and a splendid citizen. Not another reflection was cast upon him in his official capacity or otherwise.

I recall another circumstance. Mr. J. S. Appel had been advertising in the *Post*. He suddenly concluded that he would advertise in the *News* and the *Times* alone. According to Mr. Appel, he was called up from the *Post* office and notified that he had better send in his ads, but he didn't. For several weeks articles were printed in the *Post* about Mr. Appel, assailing his business, charging him with cheating and maltreating women customers. Everything that could be searched out and brought up and published in the *Post* was charged against Mr. Appel.

The very article that you referred to as being the last that was published, two of the gentlemen interested came in——

Attorney Bottom (interrupting)—Name them.

The Senator (resuming)—came into the office very indignant. It was an article in which he placed a number of our business men upon a common level with professional gamblers, and I was told by one of them that Mr. Bonfils himself had sought to coerce them into letting him have a majority of the Overland Park (a race course to the south of the city) stocks that he might use that place for racing purposes along his particular lines. Those are matters that are plainly noticeable in the conduct of the paper. To my mind they establish to a moral certainty the fact that the paper was being used for blackmailing purposes.

Q—In regard to Edward Monash, did the *Post* publish anything about Edward Monash that was not true?

A—I don't know. I had no knowledge of the method or Mr. Monash's manner of conducting his business. The point I made was that never until Mr. Monash ceased advertising in the *Post* were the attacks made. After he readvertised in the *Post*, the attacks were stopped.

Q—Was not, at the time the *Post* published its statements regarding Mr. Monash's violating the child labor law, was not Mr. Monash violating that law?

Mr. Silverstein objected to the question as immaterial. "We are not trying Mr. Monash for violation of the child labor law," he told the court.

There was a verbal duel between the two counselors lasting for several minutes, Mr. Silverstein holding that Mr. Monash's labor attitudes had nothing to do with inflaming the passions of journalists. Mr. Bottom then retorted hotly:

"If Mr. Bonfils never blackmailed anybody in his life—if Mr. Bonfils was never a perjurer, was never a thief, was never a liar, and never went under aliases, as has been accused in the *News,* then there was all the more reason to give the publisher of such articles a thrashing! I want to hear the answer to this question, and I want to hear Senator Patterson say that he never had any personal knowledge that Mr. Bonfils or the *Post* was ever guilty of anything that he accuses them of, and if he has published these things without any cause, without any reason, then I say to your honor he is in a poor position to stand in a court of justice and try to have a man fined or punished by imprisonment in the county jail for assaulting him."

Attorney Silverstein indicated that the law of libel could be resorted to by Mr. Bonfils if he felt falsely attacked. Attorney Bottom replied that the law of libel did "not apply as final for newspapers."

Senator Patterson—Do you mean by that that the law of libel is not as applicable in favor of a newspaper publisher as for anyone else?

Attorney Bottom—I mean by that, Senator Patterson, that it has been tried in this community, and juries say that newspaper men ought to fight their troubles out between themselves.

The spectators applauded roundly when the Senator replied:

"We never said in the paper that Mr. Bonfils has no right to sue for civil damages. If the things printed about him in the *News* and the *Times* are untrue, he could impoverish me with fines in civil cases and sell out both newspapers!"

The court ruled that both sides develop matters already testified to, but refrain from exceeding the scope of such testimony by bringing in additional specific cases of alleged journalistic irregularities.

Senator Patterson said the attacks on Monash "ruined him and drove him out of business." In regard to the Spengel situation, Attorney Bottom asked:

"Did anyone ever tell you that Spengel had been blackmailed by the *Post?*"

"No, sir," the Senator said. "It was not necessary."

Mr. Bottom—Did you know that the *Post* did stop criticizing him (Spengel) while he was *not* advertising?

The Senator—I am very certain that they did not.

Q—Don't you know? Don't you know that the *Post* criticized him severely with reference to the half-fare rates for children in his bill for the ordinance he introduced for the Tramway Company?

A—I am not certain of everything the *Post* may have said about Mr. Spengel. It may have criticized him, but it also abused and cartooned him.

Q—Now, Senator, what do you know of your own knowledge with reference to the *Post* or Mr. Bonfils having blackmailed J. S. Appel?

A—Well, he didn't succeed with J. S. at that time. Mr. Appel had been advertising with the *Post.* When his contract ran out he stopped. He commenced advertising exclusively in the *News, Times* and *Republican.* Mr. Appel told me he was called up on the 'phone and notified that he had better have his copy in for Sunday ads in the *Post;* that a roasting of some kind had been prepared for him. Mr. Appel didn't respond, and sure enough, the next day, or the day after, some woman had been found who had some complaint to make against Mr. Appel, and it was published. Mr. Appel said that Bonfils sought to induce others to make complaints against him, and they always found ready admission into the columns of the *Post.* Without any hesitancy, as I felt it my duty to protect at least our own customers, I exposed in the paper and challenged the *Post's* conduct in that case as being nefarious journalism.

Q—Don't you know that Mr. Appel was not asked to

put any ads in the *Post,* and at the very time he claimed to you that they were telling him to send his copy that they refused his advertising because he didn't pay his bills?

A—On the contrary, Mr. Appel told me that he was repeatedly solicited by the advertising agent of the *Post,* and he had been so telephoned.

One of the things that aroused the publisher's ire prior to his attack on the Senator was a cartoon of Mr. Bonfils, not very flattering to his handsome face and figure. Regarding this picture and the visit of the "two gentlemen" referred to in earlier testimony, the Senator said:

"Simply for the truth; there was an article printed in the *News* about Mr. Bonfils, in which the picture referred to was associated. On the afternoon of that day there was an article printed in the *Times* concerning the matter. The article printed in the *News,* I did not write. I did not even see it before its publication. Nevertheless, I don't at all seek to waive responsibility. If Mr. Bonfils sees fit to bring a suit, I will justify. The other article—the article in the *Times*—I wrote myself, and that will give you a very fair illustration of my views about that transaction and other matters we have been talking about."

The Senator objected to naming his two callers. The court postponed a ruling on the objection. Attorney Bottom then proceeded.

Q—Senator, in the cross-examination, when asked with reference to what knowledge you had of Mr. Bonfils or his newspaper being guilty of blackmailing, I believe you said it was common talk. What do you mean by common talk?

A—Why, I mean that whenever the name of the *Post* is mentioned, almost invariably it will be denounced as a blackmailing sheet.

Q—Why is that, Senator? Why is it?

A—I don't know, Mr. Bottom. You can judge, perhaps, as well as I can.

Q—I will ask you, Senator, the reason why so many people who talk to you say that the *Post* is a blackmailing paper. Is it because you have, year after year, made such statements in your two newspapers?

A—I suppose that that is a large part of the reason.

Q—You have built up among the readers of your papers, and those who believe that you tell the truth in them, a reputation that the *Post* and its proprietors are blackmailers? You may answer by "yes" or "no."

A—I cannot answer by "yes" or "no." I have told you before that the papers were doubtless partly responsible for it, but I want to add that there have been great numbers of people who had independently accused the proprietors of the *Post* of blackmailing.

Q—Name one of them.

A—No, I won't. I think it's common repute.

Q—Where did you ever hear anybody say of his own knowledge that the *Post* had been guilty of blackmailing anybody?

A—I don't know that I can recall anyone who has so stated on his own knowledge. People who are blackmailed don't like to admit it, I imagine.

Q—Why did your paper say that you were going to drive Mr. Bonfils out of business or into the penitentiary?

The Senator asked to be shown the article in question. He said he had not written it, that his managing editor, Edward Keating (now a labor leader and editor in Washington, D. C.) had written it. He said he was legally responsible for this and other articles, but had not done more than read this particular attack hastily, if at all, before it went to press.

Bonfils was sitting at his counsel's table, and the Senator now turned to him, saying:

"If there is anything in that article that is libelous, you can have me prosecuted for criminal libel or can bring civil suit for damages; for, if they are untrue, you have been sadly libeled, and the man who would libel another in that way deserves punishment under the law."

A paragraph in the article had read:

"Publicity is the most popular weapon to use against a blackmailer. The *News* will continue to use it against Mr. Bonfils, alias Winn, until he is driven out of the business he disgraces or is sent to the penitentiary where he belongs."

Regarding the supposed alias of "Winn," Attorney Bottom asked the Senator who had told him of the false name.

"The newspapers of Kansas City did," replied the Senator.

Q—When did they tell you?

A—Everybody who reads the papers read the aliases.

Q—When?

A—Mr. Bonfils can tell you that.

Attorney Bottom then declared: "Mr. Bonfils tells me that he never went under an alias in Kansas City. He tells me he never went under the name of Winn. Do you know of your own knowledge that he ever went by the name of Winn?"

"Mr. Bonfils has never denied it in the newspaper," the witness said.

"I deny it now for him," said Mr. Bottom.

When court re-convened the following Monday, Justice Carlon ruled that the names of the "two gentlemen callers" at the offices of the *News* be admitted in evidence. The Senator then identified them as Mr. Charles E. Stubbs and a Mr. Wahlgreen.

Mr. Bottom—Mr. Wahlgreen is the gentleman who had charge of the Overland Park racing for two or three seasons, is he not?

The Senator—I don't know about that.

Q—Had charge of it last season?

A—So I understood.

The Senator was asked to give the content of the conversation. It follows in part:

"I cannot give the exact words, but they came to talk about that outrageous publication in the *Post* that related to the Industrial Association and the purchase and ownership of Overland Park. Mr. Stubbs denounced the publication in the *Post* in unmeasured terms."

Q—State the language.

A—I cannot, but as nearly as I can, he used the terms that it was a 'damned blackmailing article', and that it grew out of an effort by Mr. Bonfils to obtain control of the Overland Park property, which was refused him. Mr. Stubbs remarked that he was not familiar with the early part of the transaction, but he turned to Mr. Wahlgreen and said: "You tell Senator Patterson what they were."

Q—You must state what Mr. Wahlgreen said.

A—I will give you the substance of it to the best of my ability. Mr. Wahlgreen went on to explain about his deal for the Overland Park territory; that he had an option from Henry Wolcott, I think it was for the property, and that that option expired just before the commencement or opening of the last races; that he found that if they got the property it would be necessary for him to arrange a syndicate to put up the money that they might get it, and that Mr. Stubbs—I think he mentioned Mr. Stubbs as one of them, and Bennett & Myers (a real-estate firm) and two or three other gentlemen had accepted the proposition and bought the property.

At the conclusion of the races, as it had long been contemplated and had been spoken of for quite a little while in Denver, they determined, if they could, to organize a company that would give to Denver something in the nature of an annual interstate fair, in which livestock, agriculture, manufactures and everything that would go to build up the state would be exhibited annually, and that it would gather into Denver great numbers of people, and that they could conceive of nothing that would tend to promote the interests of Denver and promote its welfare more than that enterprise, successfully carried through. To bring this about, a number of gentlemen of substance and business had organized a corporation to be known as the Colorado Industrial Association; that the capital stock of this company was $300,000, that fifty-one per cent of the stock was issued and paid to the promoting company that had taken up the option for the Overland Park property, and that the property had been conveyed to this Industrial Association for fifty-one per cent of the capital stock; that there was a mortgage of something above $50,000 on the property; that the proceeds of the remaining forty-nine per cent of the capital stock would be used first to lift the mortgage, and then to put up the necessary structures and make other necessary improvements on the Overland Park property for this fair; that before this had been done Mr. Bonfils *demanded fifty-one per cent of the stock*. There was some horseman by the name of Abrahams, or Abrams, whom Mr. Wahlgreen denounced as being disreputable, interesting himself with Mr. Bonfils for this property, and he urged something about the vast amount of money that could be made through horse racing . . .

Mr. Bottom (interrupting)—What was said about that?

The Senator—I cannot give you the exact language, except it was insisted that a vast amount of money could be made by the right kind of management, using Overland Park for horse-racing purposes. Mr. Wahlgreen declined Mr. Bonfils' request, and Mr. Wahlgreen said he had no doubt but that this article, which they regarded as highly defamatory of the business men who went to make up this latter company, was published to coerce them into some action that would give Mr. Bonfils control of this park, or in revenge because they had declined to admit him into the enterprise.

Q—Did Mr. Wahlgreen state to you that he expected to have horse racing—sell books upon the races?

A—Mr. Wahlgreen told me there would be annual races there, but that whatever profits were made would go to this Industrial Association, except ten per cent that would come to him for the management. He went on and said racing is everywhere; it is idle to talk about having agricultural shows and livestock shows without racing; that it is regarded as legitimate all over the world, and that the racing that would occur there would be absolutely under the control of the gentlemen who formed the company, and that it never would or could be anything but fair and honorable racing.

Q—You accepted the statement of Mr. Wahlgreen with reference to Mr. Bonfils and the *Post* as the truth?

A—Yes, sir.

Justice Carlon fined Mr. Bonfils $50 and costs for his assault on Senator Patterson. In passing sentence, he said:

"I deem it my duty to say to the defendant that if, in the future, he is found guilty of a similar offense, the sentence will be far different from that which the court is about to pronounce."

Concerning the attack and the trial, the Boulder, Colorado, *Camera* said in part:

"Senator Patterson is over sixty-five years old and his assailant twenty years younger, though doubtless in a fair fight the editor of the *News* would give a good account of himself. The affair was disgraceful and calls for the severest reprobation of the press everywhere. . . . The truth is that the *Post* is daily a disgrace to journalism. Its policy is for the corruption of the morals of the state. It has raised the black flag of the buccaneer concealed beneath the folds of the American flag."

20

Recessional

JOSEPH SMITH leaned his shotgun against a cottonwood tree and brought a watch from inside his hunting jacket. It was a heavy gold watch on a thick gold chain. Then Smith spoke to the ten-year-old boy who had guided him into the rabbit country near Brighton, Colorado.

"It's getting late, Antone. Maybe we'd better call it a day."

Smith put his watch back, then began to fill an old clay pipe with burley. He half-turned from the boy, as though taking a last look at the hunting ground, which had yielded few jack rabbits on that frosty November second, of the year 1892.

As Smith struck a match over the bowl of his pipe, the boy seized the shotgun. The weapon's kick was heavy against the boy's shoulder. The man's fall was heavy on the frost-bitten ground. His skull was no longer a skull. The boy stood knock-kneed, looking at the slow blood.

Antone Woode rolled the body over and unbuttoned the fleece-lined jacket. The heavy gold chain was stubborn. The big gold watch was warm. Antone shouldered the shotgun and went toward the South Platte River bank to hide. Darkness came on and he halted at a cottonwood grove. He gathered some twigs for a fire, but decided to crouch there in the dark. He took out the watch. Its ticking seemed very loud in the night.

Antone climbed a tree, pulling his gun after him. The branches, leafless and sapped by winter, creaked. It was too cold, perching up there, so he got down and sat against the lee side of the tree trunk. He wished he had brought along the fleece-lined coat. The moon came out, and he saw that he would have to wash his hands. He would knock a hole through the river ice with the gun butt, drink, then wash.

There was a clear dawn. The sun was as big in the east as the big gold watch. Antone wandered among the trees, bearing toward the Arkansas River. His slim legs were tired, his stomach empty. He saw a rabbit's white tail bobbing, but he was afraid to shoot. He had not been afraid to shoot a man. But there were voices and a stirring in the bush. Then two men appeared suddenly as though from the pit of a mine. They seemed as tall as ogres and their breaths came in white snorts.

"Hello, Antone," said one man, pointing a pistol at the boy. "Whose gun you got?"

Antone was very pale and cold. "It's mine."

The second man also had a pistol, but was not pointing it. He wore it in a holster slung from his hip. There was a silver star on his sheep-herder's coat. "It's a mighty fine gun," he said. "Hand it over, Antone. You're coming along with us."

Antone began breathing hard, as though he had been pitching hay. "I don't want to go back to Brighton."

"Why not, Antone?"

"My father sent me on an errand."

"You don't act glad to see us," said the man with the star. "Don't you trust us?"

Little Antone began to cry. The man with the star now had the Smith shotgun. The other man was searching Antone's overall pockets. He held up something. "This watch and chain belong to you, Antone?"

"Let me go," Antone said. "I got an errand to do."

"It's a big watch," said the man. "Look, Sheriff. A nice big one with a gal's picture in it."

The man with the star grasped Antone's skinny shoulder. "Tell us why you killed Joe Smith! The watch?"

"I didn't do it. The gun just went off."

"Quit slobberin' then," the sheriff said. "You'll have plenty of time to think it over. Now move!"

"It's an eighteen-jewel movement, Sheriff," the other man said. "Swiss works."

A jury couldn't agree when the trial of the boy ended in Criminal Court on February 28, 1893. The prosecuting attorney held Antone for a second trial.

"He's a degenerate," said the prosecutor. "No matter how young or how old, he's a born criminal."

If Antone had been less than ten years of age, he would not, under the law, have faced the gallows or a lifetime sentence in prison. But he had been several weeks past ten years old at the time of the slaying. The prosecutor told a second jury that murder was murder, that a murderer was a murderer, and that criminal instinct had nothing to do with age.

The second jury convicted little, skinny Antone.

He didn't appear to know what was going on about him, but rose when a bailiff prodded his shoulder. He knew he once had shot a man, that he had stayed in jail a long time and occasionally had heard men singing and swearing from adjoining cells; that he had had a watch, which had been taken from him, and right now was lying with the big gold chain on a court-room table, with a tag on it, and the marking: "Exhibit B."

"Answer the judge," the bailiff said to Antone.

The judge repeated a question: "Have you anything to say before I pronounce sentence?"

"No, sir," said Antone, "Except I want to go home."

"Your home for the next twenty-five years," the judge said, "will be the State Penitentiary at Canon City."

It was explained to Antone that he was lucky, indeed, that a rope was not placed about his neck by honorable men, the life to be strangled out of him.

A deputy put handcuffs on the boy's wrists, but finally took them off. Antone's wrists were so slender, his hands so long and tapering and thin, that they slipped out of the handcuffs.

"Just don't pull no tricks on *me*," the deputy said. Then they went by train on a long ride to the fortress-like prison in the fortress-like hills of Fremont County.

The warden was interested. "A pretty young one to come here. A killer, eh?"

In keeping with prison tradition, Antone Woode should

have grown up a hardened criminal, a victim of association with the roughest of felons to be found anywhere. Yet, amazingly, this did not occur.

He had had little schooling either in his New Jersey birthplace or subsequently in Colorado, and no one had thought Antone particularly "smart." To begin with, he was frail and somewhat "gawky," and it had been a stupid thing to kill old Joe Smith for a gun and watch.

A college professor, sentenced for forgery to the Canon City cells, discovered latent qualities in this boy. The warden, John Cleghorn, allowed the unfrocked savant—a former mathematical wizard—to tutor young Woode. At the end of the first year of coaching, the professor told the warden:

"He is a prodigy. It sounds unbelievable, but within the year he has mastered a college course in mathematics. Only for want of text books has he not gone into calculus and astronomy. Never have I seen his equal. Furthermore, he has a memory of abnormal type. On two readings, he has memorized Lord Byron's *Childe Harold* in entirety, that being the only work of any literary quality in the library here."

Antone Woode's abilities became known to women's clubs. They sent him books and endeavored to have him pardoned. When the boy was fifteen, he took up the study of music—a violinist of no mean ability having been sentenced to the penitentiary. This virtuoso became Antone's music master. The boy amazed this professor, also, and soon they were playing duets for the Sunday services. Later Antone took to sketching and painting. He had to educate himself in this art, however, as no painters or sculptors were sentenced to the prison during Antone's residence there.

Among the society women who campaigned for Woode's release was Mrs. Madge Reynolds, wife of a wealthy Denver oil man. A letter Antone had written to the Board of Pardons in 1899 had moved Mrs. Reynolds to interview the Governor in Woode's behalf. She promised that she would give young Antone a place in her own household, were he pardoned, and that her husband would see that he obtained work.

It is probable that Woode would have gained his freedom on Mrs. Reynolds' representations, had he not become involved in a prison break the night of January 22, 1900. He was now eighteen years old, a model prisoner, and something of a "celebrity." His undernourished, anemic appearance had vanished. He was a youth of interesting, slightly mystic, manner, and as well educated as many a Master of Arts.

How deeply Woode was involved in the prison break is not clear. But when a group of "lifers" made a rush on Night-Captain Rooney, killing him and taking his keys, Woode was said to have been with them. The leader in the break was the notorious "Kid" Wallace, a desperado.

There was a man-hunt for three days, and judging from the manner in which Woode was condemned by critics of prison reform, one might have thought him the real brains of the prison break. At the end of the third day, Woode and "Kid" Wallace were re-captured. Woode denied in letters to his friend, Mrs. Reynolds, that he had planned the break, or had participated in the slaying of Night-Captain Rooney.

"It is regrettable," he wrote, "that Captain Rooney was killed, but I honestly cannot say I am sorry I sought my freedom. For I longed so to get away from this place."

Mrs. Reynolds began a four-year crusade for the freeing of Antone Woode.

That the paths of Mrs. Reynolds and Bonfils should cross appeared incongruous. Yet cross they did. That the dour, brass-knuckled publicist should melt before the dignified, yet vital warmth of this woman; that he became, for a time, almost humanely pliable, seemed a miracle. Yet the miracle happened.

Bonfils was not a man of amorous exploits. Whether this was so because of lofty moral precepts, fear of emotional entanglements, or in keeping with his obsession of "drawing back" from anything which might tend to master him—such as the ice cream craving—no one can determine. But whatever else might have been charged to him, his chastity was conceded. He carried the burden of physical purity almost to the point of asceticism—a St. Anthony of the money bags.

This seemed all the more remarkable because he was a person of bodily charm and competence, according to experts in such signs and portents. He wore raucous tweeds, it is true, but no bizarre raiment, gambler's tunic or tinhorn chapeau, could detract from the boxer-like body, the military stride, the proud, Corsican lift to the head. In Bonfils' instance, clothes did not *unmake* the man. It is possible that the flamboyant garb somewhat accentuated his charms, the portrait vying with the frame.

Curiously enough, with so many snipers attacking Bonfils on every hand, eager to accept as fact any slur on his character, few scandal-passers placed an unwholesome interpretation on his friendship for Mrs. Reynolds. She, it seemed, had a reputation that precluded adverse comment. Her friendships were open to any examination. She was one of the best beloved women of the city's upper social reaches, partly because of her unsung charities, and partly because she chose to live simply, despite her aristocratic station. She was now in her full-blown thirties, slender, with deep gold hair and large, violet eyes. Bonfils was forty-three.

Not the least odd circumstance concerning the association of Bonfils and Mrs. Reynolds was that her most intimate friends were Bonfils' deadliest enemies. Among them were Senator Thomas M. Patterson, William G. Evans, head of the Tramway Company, and Thomas J. O'Donnell, a leading corporation lawyer. Ordinarily Bonfils could brook no friendship with anyone who even so much as *spoke* to one of his enemies.

The Antone Woode case brought these strongly contrasted persons—Bonfils and Mrs. Reynolds—together.

A hop-scotching Tammen chances on a book of lottery tickets in a print shop; he meets Bonfils, and a newspaper of power and importance is the result. A man-eater practises cannibalism in the far-away past, and the partners are shot down by a firebrand attorney in a more modern day. A boy covets a big gold watch and a shotgun; he blows off the head of a hunter, goes to prison, and a refined woman and a man, described as a Corsair in gingerbread armor, become devoted to each other.

In such strange fields the seeds of tragedy are sown.

They met in the *Post's* Red Room in February of 1903. Mrs. Reynolds had not been admitted at once, for it was Bonfils' custom to keep callers waiting, the better to emphasize his importance, and also to unsettle the nerves of a visitor in case a favor were to be asked.

As Mrs. Reynolds entered the "Bucket of Blood," the czar was having smoke blown in his face by a journalistic serf.

What was her first impression on meeting the arch-enemy of her friends? Perhaps he revealed himself to her as he did to others, an extremely courteous gentleman, beneath whose skin lurked a brooding tiger.

The conversation lasted for more than an hour. Mrs. Reynolds asked Bonfils to launch a crusade for freeing Antone Woode. After the petitioner had gone, he said:

"That woman has the biggest heart in the world."

He stood, looking out the window. Then he announced to his lieutenant that the *Post* would free Antone Woode. On what ground did he reach this decision? Sympathy for an imprisoned boy? A desire to do something for a personable woman? Or opportunity to score a victory over the hated Senator Patterson, and to bestow on this woman a conspicuous favor, whereas her friend, Patterson, had neglected to perform a similar service in the *News?*

The first story that Bonfils dictated in the Woode crusade may contain a clue as to his feelings toward Mrs. Reynolds. In it he grew alliterative:

"Fortunate indeed is Antone Woode to have so eloquent, untiring and forceful a champion to plead in his behalf. Refined, educated, earnest, with a wide circle of influential friends and a liberal supply of money to help the cause along, nobody in Denver is in a position to do so much for the boy-murderer as is Mrs. Reynolds. Her plea is plaintive, her argument amazing, her eloquence effective and her perseverance pervading. With such an attorney, the chains of Prometheus might have been stricken off."

Did Bonfils see himself as Prometheus, chained to a

rock of gold? Without knowing it, perhaps, he was reaching out for a saving human relationship. It seemed he could buy or have anything material in life, yet so many emotional nosegays had withered in his grasp.

There were frequent meetings in the Red Room concerning the Antone Woode crusade. Soon Bonfils and Mrs. Reynolds were riding horseback together, and finally he became a caller at her Logan Street home.

Influential big-wigs of the town visited that cottage. She adhered to no religious sect, but clergymen of various creeds visited her. The more polite politicians, editors, literary folk (forerunners of the modern intelligentsia), corporation directors and attorneys met at the Reynolds cottage. Her home was a place for informal, but discriminating salons. Although Mrs. Reynolds, as a rule, did not show partiality to any of her circle, she made an exception in Bonfils' case and often received him alone. It would have been slightly inconvenient for him to have mingled socially with those whom he daily placed on the journalistic cook stove.

To her intimate friends, Mrs. Reynolds accorded a somewhat unconventional privilege—that of sitting in her kitchen to drink half-pint bottles of champagne and to eat cake which she herself had baked. Her home was a small one, with no servants. She enjoyed doing housework. Mrs. Reynolds could have afforded a sprawling estate with a regiment of lackeys. She preferred the cottage.

It was a day of large kitchens. A rich woman usually had a cookroom big enough for a six-day bicycle race. Mrs. Reynolds, however, had a small kitchen, with what is now called a "breakfast nook." It was decorated in blue and white, and with dainty curtains. The powerful gentlemen of the city regarded it as a club room.

And so the months passed. Bonfils began to call her "Dearest." Their idyll was accepted by nearly everyone as a thing of sentimental beauty, an association rather than an amour. It remained unmarred until the attack on the myopic Senator Patterson. The next day, it was said, Bonfils was forbidden the Reynolds house until such a time as he might "reform." How long he remained away, what remorse he suffered, or how a reconciliation came about—these matters are buried with the dead who once lived through their history.

Although frequenting a house which also extended hospitality to his foes, Bonfils managed to endure that condition. And while he did not pull his editorial punches in attacking these men, or lessen his assaults on other foes, a new light shone in his black eyes and his frown was less menacing. In daily association with his fellow men, he seemed almost tolerant.

What understanding, if any, he had with relatives regarding his unconcealed admiration for Mrs. Reynolds is not a matter of public knowledge—nor, perhaps, one of literary concern. Both Bonfils and Tammen appeared to keep their families in the background, and these relatives did not encroach on the public or professional careers of the partners. The publishers did not parade their womenfolk in print, and, in fact, Tammen once threatened to fire a society editor for mentioning his wife's name in a list of social celebrities.

"Society is the bunk," said Tammen, "and nobody knows it better than you. Just leave us out of it."

In August of 1905, the *Post* predicted that Antone Woode would receive a parole. On September 1, 1905, the Board of Pardons granted a commutation of sentence. A parole would follow as soon as Governor McDonald made out the necessary papers. The parole provided that Antone be apprenticed to Elbert Hubbard's Roycrofter colony at East Aurora, New York. Mrs. Reynolds bought a railroad ticket for Woode and agreed to pay Hubbard a weekly fee of five dollars during Antone's stay at the colony.

Hubbard had promised to appear before the Board of Pardons to make a plea for Antone, but business matters prevented. He telegraphed that he would meet Woode in Chicago, to escort him to the colony. The boy was permitted to take an assumed name, which we shall say was John Goodwin.

"Antone is a good boy," Mrs. Reynolds said. "He is not a degenerate. And he will make his way in the world."

At noon, September 12, 1905, and while prison whistles were blowing, Warden Cleghorn signed the release papers. He handed them to the twenty-two year old Woode.

"Now, Antone," said the warden, "you are free to go."

Antone shook hands with the warden. "Free!" he said. "Free!"

Woode was in his shirtsleeves as the warden signed the release. He did not wait for his coat or hat, but ran outside the prison gate to look at the trees, to touch them. He was crying. The warden's son, Willard, brought Woode's coat and hat to him.

"Just to think I can go where I please," he said to Willard. "I knew this was coming. I thought I had been prepared for it. But I wasn't prepared. It has been a long time—eleven years, five months, three days and a half!"

Something untoward happened to Woode at the Roycrofters' colony. He had arrived under promising auspices, this well-educated youth, proficient in languages, notably German and Spanish, a mathematician of extraordinary ability, a fine violinist and a passable artist. Elbert Hubbard had greeted him with philosophical handshakes and Benjamin Franklin *mots*. But in less than a week after his arrival at Fra Elbertus' man-foundry, Antone wrote a letter to Warden Cleghorn, attacking Hubbard's cult as a commercialized choir of stooges, characterizing his East Aurora paradise as a haven for defeated bores. He demanded a summary transfer from the Hubbard *aegis* to the guardianship of Mrs. Maud Ballington Booth, New York welfare worker and leader of the Volunteers of America.

Woode had composed his beefy letter for private perusal by Warden Cleghorn. It became as public as a tornado. The *Post*, which had campaigned so assiduously for his release, now jumped on Woode with editorial hobnails.

In commenting on the demand for an amended parole, the *Post* carried a headline:

"IS WOODE AFTER AN HEIRESS?"

Before quoting from the story itself, it were well to explain that *any* girl remotely figuring in a romantic bit of news was an "heiress," insofar as the *Post* was concerned. Similarly, there are journalistic fetishes in other editorial climes. A New York reporter seldom encounters a woman who has not been a "Ziegfeld Follies Girl." The late Florenz must have had a stage as large as a World War battlefield.

Another habit of the *Post* was loosely to bestow the adjective, "prominent," on any one who spoke in support of *Post* policies. Contrariwise, the *Post* dubbed as "obscure" certain low fellows whose opinions were in conflict with *Post* views, or *Post* enterprises.

The anti-Woode screed follows in part:

"Why does Antone Woode desire to get to New York? Is it because he wants to have the good influence of Mrs. Maud Ballington Booth, the 'Little Mother' of the Salvation Army (*sic*) about him? Or does he want to get to the large metropolis, that he can tear up the town? Or, possibly does he want to get married? . . . His letter is the talk of the state. It has set the tongue of everyone interested in the young murderer to wagging. Those who opposed him declare in words without mincing: 'We told you so. Woode is a degenerate, and this letter proves our case conclusively.' After being penned up in Canon City for twelve years, Woode shows his ingratitude by asking to be taken from the Roycrofters before he has been there a week.

"Of all the questions that have grown out of the letter, that about the marriage is the one most generally discussed. This is perhaps due to the fact that Warden Cleghorn is said to be one of those who believe that Woode is smitten with a girl. According to a story that is going the rounds of the penitentiary, a New York miss—in fact, an heiress—while touring Colorado last spring, went through the penitentiary. Being of a sentimental turn of mind, it is said she became enamored of the young convict."

This story quoted *prominent* convicts as having said that Woode had spoken of the "heiress" in "endearing terms." The Secretary of the Board of Pardons, C. E. Hagar, was represented as roaring from his official high horse:

"I was talking to a prominent physician of the city. He said that Antone, going under a new name, with his past unknown, would probably attract some young woman, and that a marriage might result in bringing into the world degenerates."

An extraordinary contribution to scientific knowledge, this discovery that an assumed name is the well-spring of degeneracy! Denver was fecund with such Darwinian excitements.

In the days that followed, poor Woode learned the lesson that no one at any time should write letters of private import—not even to one's own mother. Like drowned men, they wash up with the tide.

Not content to spank Antone once and let the matter drop, the *Post* kept up its Rocky Mountain yells, saying that Woode stood a first-class chance of being returned to the penitentiary. The *Post* was clearly indignant that a Colorado ex-convict had inferred that the great Elbert Hubbard was an exhibitionist doing head-stands on a carrousel. When Woode followed up his publicly-private letter with a telegram to Governor McDonald, asking permission to leave Hubbard's monastery, the *Post* got out the bastinado sticks:

"One thing is certain, and that is, the Board of Pardons does not propose that Woode shall dictate to it where he shall go. The board realizes that Mrs. Ballington Booth, the 'Little Mother' of the Volunteers of America, is an estimable woman, capable of taking good care of him, but the members are unanimously of the opinion that Woode has some ulterior motive in mind, and desires to reach New York for some mysterious reason.

"Although the individual members of the board were indignant at the letter, the real straw that broke the camel's back was a telegram received by Governor McDonald yesterday afternoon, in which Woode asks permission to leave East Aurora at once. The members deem the first letter bad enough, but yesterday's telegram they characterize as nothing short of audacious, and pronounce it as adding insult to injury. The members compliment the Governor on his prompt reply, turning down Woode's request."

Woode's longing to go to New York seemed to worry the *Post* no end. For example:

"A report current at the State House this morning was that one reason Woode is so anxious to get to New York is that he desires to go on the stage. It is said that while he was in the penitentiary he frequently spoke of appearing before the footlights when he secured his freedom. In prison, he was an interested reader of the sensational Nan Patterson trial in New York. Time and again he is said to have told his fellow convicts that he considered himself as great as Nan Patterson, and his boast was that the notoriety secured by him is as great as that of the former Floradora girl. When the trial was ended and the reports circulated about her returning to the stage, Woode is reported to have caught the stage fever in virulent form."

In defense of his desire to leave Fra Elbertus to his own devices of furniture making, pamphleteering and motto-moulding, Woode wrote in the *Buffalo Inquirer* in October, 1905:

"In the light of the publication of my personal letter to an intimate friend, and foreign criticism thereupon, I believe it not only right, but a duty, to make an extended explanation concerning the writing and subsequent publication of the matter in question.

"It is with no common satisfaction, but at the same time with a keen sense of diffidence, which I cannot put away or care to deny or conceal, that I approach my subject. In all human probability it is beyond the power of pen or tongue to register even faintly the infinite pathos of a supersensitive soul. I hope but to suggest it—no more. It is from this point of view that I wrote the letter above mentioned (to Warden Cleghorn).

"I am emphatic in my statement that whatever I wrote concerning the Roycrofters or the management of the institution thereof was purely personal. Simply because I said something which some individual, from a sectional point of view, considered a virulent arraignment, ought not to be taken as infallible. Humanity is as prone to err as it is to be carelessly subservient to stronger wills.

"I do not wish for a moment that my so-called 'arraignment' of the Roycrofters be taken as a fact, or that simply because I found conditions other than they were painted to me while I was still in the West—which were that the

Roycrofters were a world-center of love, a veritable co-operative paradise—that the institution was a worthless commercial grab.

"Considered from the viewpoint of a safeguard to moral and physical development, this institution is eminently worthy of all that has been said and written concerning it. . . . Personally, I consider Mr. Hubbard as one of the most remarkable men of our day and age, his writings worthy of comparison with the immortal writings of the Greek and Roman age. His work here, while still in its infancy, scarcely out of the inceptive period, is worthy of the best and highest co-operation. Because I do not deem the means worthy of the end should not preclude the nobleness of his work. I am as yet nobody—he, somebody—my opinions and conclusions are private and personal—his, impersonal and public. Which of the two is correct, time, the infallible rectifier, will evince."

Woode gained permission to leave the Roycrofters in the latter part of October of 1905. In announcing this fact, the *Post* said:

"Antone Woode, the boy murderer from Colorado, is now at liberty to sell peanuts, pins, pineapples or soap outside of the Roycrofters' shop and anywhere within the confines of the Empire State, as long as he observes the terms of his parole."

In June of 1906, Antone Woode was married (under his new name) to the daughter of a New York state judge. She knew of Antone's prison past. The young woman was a musician and elocutionist. On October 15 of the same year he gained a full pardon.

Bonfils suddenly ceased attacking Woode. An abrupt change of tone was apparent in the *Post*'s story announcing the amnesty, with a complimentary reference to Mrs. Reynolds:

"After a little more than a year of liberty on parole from the Colorado pen, Antone Woode has been granted full and unconditional pardon by Governor McDonald. The pardon was issued this morning on receipt of the accompanying recommendation from Warden Cleghorn. (The warden's letter to the Governor was quoted here.)

Ever since the parole was granted, his benefactress, Mrs. Reynolds, has worked unceasingly for a full pardon, with a view of giving the youth, who practically never had a chance to know the world, an opportunity to assert the best in him. The result of her effort was made apparent today."

Although the *Post* had flayed Woode as a person displaying ingratitude, bad taste, and of recalcitrant manners, the pardon story went on to say:

"Good behavior in every instance has stood in his favor."

Describing Woode's career on leaving prison, the *Post* accorded him the courtesy title of "Mr."

"On June 2 of this year, Mr. Woode was married to M—— T—— (the girl's name was stated in full), daughter of Judge T——, one of the most prominent jurists in the state of New York. Up to the present time they have lived in a flat, but this week, a successful business undertaking having favored the young husband, they are moving into a house entirely their own, and are furnishing it completely, although modestly.

"Since his release from the pen, Mr. Woode has installed himself in business. He has a studio where he gives violin lessons to an immense class. He plays constantly in churches, at public functions and social entertainments, and has proved his loyalty to the philanthropic woman who befriended him by repaying her a good share of the funds loaned him.

"Although only twenty-one years of age [NOTE—he was twenty-four], Antone Woode has led one of the most pathetic lives that ever comes to a child."

The *Post* reviewed his prison life and added:

"Having been granted a release from his contract with the Roycrofters, Mr. Woode went to New York, where he received a position as a stenographer with a church calendar company. After he was married to Miss T——, he opened a studio for himself, and has been highly successful.

"Mr. and Mrs. Woode write to Mrs. Reynolds every week. The letters are full of gratitude and love for the kind woman who went out of her way to help a forsaken youth, forgotten by all in his dingy cell in the pen, and only recalled when he accomplished something new in an artistic line or astounded a mob of curious prison visitors."

The public saga of Antone Woode closed with this story, in which the *Post* referred to him, not as "the boy murderer," but as "the young Coloradoan."

How are we to account for Bonfils' erratic attitude during the Woode case? There had been a two-year campaign, boisterously paternal in flavor. Then, without warning, Bonfils reversed his policy and began attacking Woode with saber-teeth. Finally, the assaults abruptly ceased; the *Post* blessed the young man with a sentiment-dripping benediction—then a silence.

The fact that the *Post* changed its mind frequently, long had ceased to be a local wonder. Subscribers were weather-wise to its veering tempests. The *Post* was capable of rapid-fire shifts of allegiance in all causes, great or small. It had a sweetheart in every port—and levied the customary heartaches that follow such philandering.

This fickleness was summed up by Harry Tammen in the case of the Very Reverend Henry Martyn Hart, a venerable Dean of the Episcopalian Cathedral of St. John-in-the-Fields. The *Post* for years had baited the crotchety old Dean, a brilliant though headstrong Britisher of High Church discipline. The paper, without any apparent excuse, one day contained an article so laudatory of him that the Dean glowed all over. The very next week, the *Post* attacked him, gratuitously, charging that the Very Reverend detested the United States so completely that he once had hustled his wife off to England so that their expected child might *not* be born on American soil. Then followed another day of praise, then another attack, with an anecdote concerning the Dean's views on fishing. The article set forth that certain of the Dean's critics accused him of cruelty in feeding sharp hooks to the fish. The Dean, in reply, was quoted:

"Cruelty to fish? Tommyrot! My conscience is clear on that score, for, when I jerk the fish from their native element—the water—into a foreign element—the air—the oxygen makes them *drunk,* and they die happily."

Dizzy from alternate attacks and praises, the Dean drove his gig to the *Post* offices to demand of Tammen why such indignities were imposed on a man of God. Tammen pinched the Dean's arm playfully:

"There, there, old boy! You know how it is. We tickle your —— today and kick your —— tomorrow!"

The astounded Dean pulled his hat down, as though to shield his red ears from such impious philosophies, stomped out of the *Post* building, never to return.

Another example of the *Post's* quick-change propensities was its handling of Judge Ben B. Lindsey. When merchants were imposing an advertising boycott on that newspaper, the *Post* utilized the Judge's child-labor campaign as a ram to bowl over the "employers of pitiful, wan, little girls." Later they gave him the "Dean Hart treatment" of alternate blame and praise.

One of Bonfils' powerful adherents escaped prosecution for allegedly receiving a truck-load of unlawful whiskey at his mansion door. At about the same time, three or four Italians were up for hearing before Judge Lindsey. They had been arrested with several bottles of wine in their possession. Judge Lindsey refused to sentence them, holding that a law which could not reach Bonfils' rich colleague should not descend with vengeance upon the poor. The *Post* barred Lindsey's name from the paper for a time, then attacked him.

Later, when Lindsey took the stump against a certain mayoralty candidate, the *Post* began to cheer the judge. He again resumed the rôle of humanity's greatest benefactor. The man Lindsey was attacking happened also to be an enemy of the *Post*.

Summary change of policy, then, was not of itself the matter to be wondered at in the *Post's* diatribes concerning Antone Woode. The *reason* for the about-face was the challenging factor.

In explanation of Bonfils' weird onslaught, two motivating causes were set forth by those who worked for him. Each was soundly based on his character, and both diagnoses seemed plausible enough, except for one thing: how could he publish the extremely uncalled for slurs, full of hearsay and guesswork, when Woode's constant benefactress was the woman that Bonfils admired?

The reasons advanced were as follows:

1—Bonfils had an almost maniacal hatred for New York City—a rancor sired, perhaps, by his ancient ejection from a Knox Hat Store while giving a gas-mantle sales talk. Woode's demand for a transfer to that iniquitous city was sufficient cause to send Bonfils into a Punch and Judy dither.

2—Bonfils himself had arranged the Elbert Hubbard apprenticeship for Woode. When the young man didn't like the plan, he was committing the unpardonable sin of questioning Bonfils' decree. Bonfils always insisted on doing the thinking for anyone who came within his orbit—possibly excepting Tammen—and, as a master mind, never adjudged his codes to be at fault concerning life, liberty and the pursuit of happiness. When anyone resisted a Bonfils *ukase,* the czar pounced upon and sunk bombastic gaffs into such a gutter-bound ignoramus.

These theorems go part way toward explaining the Bonfils' manner of dealing with his subjects. But in this specific case, there must have been a more potent reason, something less capricious than the usual whimsies born of the Corsican's psychological hangovers. The attacks on Woode not only ceased, as we have seen, but a sugary, backtracking, euphemistic valedictory followed directly on the heels of the blasts. Mitigation of libel certainly was not the motive for the article of praise, for Woode, as an ex-convict, had no reputational recourse to the courts. Furthermore, the critical essays had contained no mention of Mrs. Reynolds, whereas the final article again took up the business of complimenting her.

A biographer, unable to gather all the intimate facts, can come only to the brink of a guess. A novelist, with protagonists such as are presented here, would not hesitate to say that his principal figures had had a significant quarrel; that the quarrel had been precipitated by the man; that the woman, being of a vital, maternal nature, had sub-consciously fixed upon a third party—the young, talented, imprisoned man—with a mental relationship as a substitute for a physical one; that the adoring, middle-aged suitor, his feelings now having mounted to the inevitable urge for complete possession, read a sinister meaning in the younger man's gratitude on being freed, and foresaw continued ministrations by the woman to the boy; that the man became inordinately jealous; that in his rage he sought to destroy the rival, but when that rivalry vanished, his jealousy automatically subsided—the marriage of the rival having provided visible and convincing proof, either of the death of the rivalry, or of its never having existed.

Bonfils' admiration for Mrs. Reynolds appeared reciprocal. By the year 1908, he was evidently trying to analyze for himself just how much he *did* care. He confided in Tammen that "there is something more to life than just piling up money." This was a rather drastic admission for one who had put such store by wealth.

Tammen was too broad-minded to worry much about the social aspects of Bonfils' regard for "Dearest." He did not, however, want his partner to kick up his heels in a way such as might lead him to repent. Bonfils talked with his partner a great deal on this subject. In one conversation he approached the problem in a left-handed way, describing a hypothetical case, wherein a powerful and rich man gave up everything to "wander hand in hand with a beautiful and understanding woman." To which, Tammen replied:

"Well, come out of the ether and think it over first."

On the afternoon of February 21, 1908, Bonfils went riding with Mrs. Reynolds. Whatever the gist of their conversation, he did not see her again that evening or communicate with her. It is believed he had announced some course of action relative to their future.

He came to his office the next morning, Washington's birthday anniversary, and picked up the rival newspaper, Senator Patterson's *News*. Two of the *Post* staff, passing the door of the Red Room, heard a moan. They looked in and saw Bonfils standing there, his knees buckling, his eyes staring. Then he fell to the floor unconscious, the newspaper beside him.

For some time it was not known what had caused this

collapse. Then a headline was seen in the paper which had dropped from Bonfils' grasp:

"MRS. REYNOLDS DIES"

Bonfils did not fully regain consciousness for more than two hours. When he did rouse from his fit-like coma, he seemed baffled, defeated, crushed. He then denied that "Dearest" *could have died.* He behaved as though his enemy, Patterson, had perpetrated some ghastly jest. He mumbled and beat his head with his fists.

All that day he was gripped by a despondency that frightened his aides. He spoke of going to the Reynolds' house, himself to look at the woman who had given life some warm meaning during five years of his middle age. Then he said he couldn't go there. He sent one of the women reporters, a friend of "Dearest," to look at her and to report back. He crouched low in his chair, gazing at the globe on his desk, and awaiting the return of his envoy.

When the woman reporter came back, Bonfils asked almost wearily: "Tell me how she looked."

The reporter, seeking to comfort him, said:

"She looked so peaceful, yet curiously *alive.* Why, her cheeks are as rosy as ever. You wouldn't think she was . . ."

He sprang to his feet, his voice shrill as he interrupted: "She's *not dead!* Don't let them bury her. It's a case of *suspended animation!*"

He spoke of "tests" to make sure of "Dearest's" death. He demanded that the funeral be delayed. Finally he was convinced that she was not alive, that a new embalming process—employing a blood-solvent—accounted for the life-like complexion.

He could not be comforted. He seemed to feel that he was to blame somehow for the woman's death.

There were rumors then, and since then, that Bonfils had announced to the woman that they had better remain apart for a time, to see how matters would work out; that she had ended her own life. The death certificate, however, signed by Dr. O. J. Pfeiffer, attested that she had died of *angina pectoris.*

After her horseback ride, it was said, she had gone

home, had changed from her riding habit to a dressing-gown, then dropped dead.

"Send a blanket of roses," Bonfils directed. "The most beautiful roses." Then he lapsed into a morose silence, his face pale and pinched.

Many persons mourned the death of this woman. She was buried from her little Logan Street home at 2 o'clock, Wednesday afternoon, February 26, 1908. The Rev. Dr. David Utter delivered the funeral address. The story published regarding her death contained phrases significant of the virtues which had appealed so strongly to Bonfils. For example:

"No woman has ever died in Denver whose loss is so deeply mourned by people in all walks of life. Her charities were numberless. But her work was not alone among the unfortunates. She fed and clothed them, and in them rekindled the fires of ambition and hope. She was sunshine in the lives of all who knew her.

"She was an optimist, so full of the joy and good of life that even to talk with her inspired the better natures of those with whom she came in contact. She never saw the wickedness in life, but for every creature she had infinite compassion, and in him found tremendous possibilities of virtue and success. . . .

"Her pity was boundless. It embraced everyone in trouble. She disliked newspaper notoriety, and would not have it when she could prevent it. . . . She had entrée to the most exclusive social circles, but she confided to her friends that the parties she liked best were the ones when she invited all the unfortunates whom she happened to be caring for at the moment to her house for an evening. The old ladies, the boys and the girls, and the aged men came and had the time of their lives. Few other women would have had the tact to make everyone happy in so strange a gathering, but no one ever could be unhappy in her home. . . .

"Her chief characteristics were her infinite sympathy and tenderness and her bubbling joy in living. She loved work for the pleasure of its doing; she loved nature, and in everything she found happiness. Her life was intensely active always, even up to the hour of her death."

Bonfils' name was missing from the list of pallbearers at the Reynolds' funeral. Nor was his car among the carriages which took the chief mourners to Fairmount Cemetery that cold February afternoon. He went there alone, and from a distance watched the men and women assemble near the newly-opened earth.

The pallbearers were the influential men who for years had met in the blue and white kitchen to sip champagne, eat home-made cake and talk over many matters. And these men, with perhaps one exception, were Bonfils' deadliest foes.

He stood apart during the obsequies, for there were few among the mourners whom he could call friends. Even the funeral directors, old man Wally, pioneer undertaker and coffin-maker, and Wally's partner, Bob Rollins, were near-enemies of the publisher, because of Rollins' adherence to political machinery.

Bonfils saw his enemies remove the casket from the hearse and carry it to the grave. He saw the blanket of red roses, which he had sent, being taken from the coffin lid and placed upon the frost-hardened earth.

He saw among the pallbearers Senator Thomas M. Patterson, owner of the *News*, the old man whom he had beaten to the weeds of a vacant lot, and for which act he had suffered temporary banishment from the Reynolds' home. He saw another and hardly less hated enemy, William "Napoleon" Evans, powerful utilities magnate—Evans, also, a pallbearer. Still a third antagonist was a member of this guard of honor, Thomas J. O'Donnell, huge-headed, bull-necked, brilliant corporation attorney.

The spectacle of these enemies, placing the coffin on a lowering device, must have been too much for the brooding Bonfils. For, with the casket now vanishing below the level of the cemetery lawn, and with the Reverend Dr. Utter preparing to administer the last offices, the sprinkling of earth, and the words: "Dust unto dust," an amazing thing happened.

Bonfils suddenly left his place in the background. His

bare head was held high, almost defiantly so. He passed through the ranks of mourners. He stooped to pluck a single flower from his blanket of red roses. He straightened quickly, stepped nearer to the grave, leaned over it, and with a slow movement of the hand, dropped the rose on the lid of the casket.

He did not remain long at the grave, but took one brief look into the earth, then turned away. There were no tears in his black eyes, only a strange, cold light. He walked rapidly to his automobile, glancing neither to his right nor his left. He stepped inside the vehicle. His chauffeur closed the door. They drove off, not waiting for the conclusion of the service.

He had said good-bye in the presence of his enemies.

Certain observers noticed a subdued bitterness in Bonfils' manner for months after Mrs. Reynolds had gone. He seemed to find release in talking of her, always referring to her as "Dearest." He preserved her letters, keeping them where he might consult them frequently. Concerning these, he said:

"Whenever I am in doubt, or in trouble; whenever I want to know the way, I choose at random one of these letters, open it, read it, and there always is a message from her, telling me what to do."

21

The Rover Boys In Missouri

FOURTEEN YEARS had passed since Bonfils' abdication as king of the lottery gypsies. He had moved to Denver, multiplied his original eight hundred thousand dollars and had tasted power. He had seen three Denver newspapers fade in prestige and slump in circulation, while he and his bubbling partner had taken a donkey-press dodger, and from it had built one of the most valuable newspaper properties in America.

For thirteen of these fourteen years, the Messrs. Bonfils and Tammen had withdrawn no moneys from the *Post's* surpluses. All profits had been applied to development of the paper and expansion of its field. It bought mountains of syndicated newspaper features, skimming the cream and throwing away the rest—the purpose being to corner anything an opposition journal was remotely interested in procuring.

The *Post* was client of all available wire services. At this time the only news asset it did not own was an *Associated Press* franchise for its Sunday morning edition. How it obtained that concession will afford a few tinkling paragraphs later in our chronicle.

The leading comic "art" was to be found in the *Post,* Bonfils holding that children today would be *Post* boosters tomorrow.

"Address yourselves to the children," he would say, after the manner of church logicians. "If we can keep them interested in the *Post* until they're twelve, we'll have them for the rest of their lives."

The *Post* was alert in developing local features, such as the comic strips of Maurice Ketten, the weekly essays of "F.W.W.", the fullpage painting each Sunday by Paul Gregg, and a column of daily jests and jingles by Bide Dudley. Paul Thieman, later a Hearst editorialist, Hugh O'Neill, erudite Australian, and George Creel were staff contributors to the *Post's* columns.

In keeping with his adoration of the dime, however, Bonfils encouraged unpaid for, outside writings. Local celebrities, print-infected clergymen, prose-twiddling attorneys and other amateur wayfarers on the road to Parnassus brought the first fruits of their talent to the *Post.* The city was extremely quill-conscious.

In one case, a truly gifted poetess made the mistake of submitting her odes without mentioning remuneration. Her verses had an appeal among thinking people of the town, and Bonfils was very happy about it. After several of her poems had appeared, and no checks were forthcoming, the poetess called on Bonfils.

He was cordial, talked at length about Nature, congratulated Miss—— on her gift for rhythmic assembly of words, and then (for no apparent reason) went into a discussion of the Sermon on the Mount.

The poetess finally interrupted Bonfils. "I am glad you like my verses, and I thought I'd ask why I haven't been paid for them?"

Bonfils seemed genuinely grieved. "You mean you want *money?*"

"I think my work should be paid for. Don't you?"

He shook his head sadly. "I am amazed! Truly amazed! Why, my dear child, Jesus never asked for money."

In 1908, the partners declared their first quarterly dividend of one hundred thousand dollars each. Then they put themselves on a weekly salary of one thousand dollars.

Tammen immediately decided to build a large house. He purchased lots in Humboldt Street, near a park which had been the city's first burial ground, and where old Undertaker Wally had sunk scores of his "Pinchtoe model" coffins. It now was called Cheesman Park, the gift of Walter S. Cheesman, head of the Water Company.

When Mrs. Tammen saw these lots, she said:

"Why, this is where we used to sit before we were married, to watch the sun go down."

"That's the general idea," Tammen said.

The Pullman Car Company sent its best woodworkers to finish the interior of Tammen's new home. Then Tammen and his wife left Denver for a vacation trip, during which they proposed to buy furniture in every large city they visited.

In San Francisco, they inspected an expensive bedroom suite. Tammen was pleased with the outfit and took out his check book. Then he hesitated, looking once again at the boudoir furniture.

"Why the two beds?" he asked.

"They're twin beds," the salesman said. "You know? All couples use them nowadays."

"Is that so?" said Tammen. "Now look here, have you got one great big bed to match this set?"

The salesman was worried. "I'm afraid not, sir. You see, twin beds are the fashionable thing now."

Tammen returned his check book to his pocket. He took his wife's arm. "Come on, honey. We don't want *that* set."

The salesman tagged after him. "But, my dear sir, these twin . . ."

"Look here, partner," Tammen interrupted, "do you think I'd be sucker enough to sleep away from a beautiful woman like this?"

In Tammen's home, there never was any newspaper work or other business transacted—excepting charities.

"I'm not a drunkard in any other respect," Tammen said. "So why should I be a drunkard for work? A home's a place to have a good time in and relax."

It was otherwise at Bonfils' house. He worked there, and worked hard. He brought his newspapers home each night, and, after dinner sat down to check and re-check the various articles, to hunt for mistakes, conjure up campaigns and otherwise prepare for next morning's editorial conference—meetings in which he ranted and fumed, snorted and berated his lieutenants.

In Bonfils' home, lights (for others) were ordered out at a relatively early hour—a curfew law which discouraged sociable gatherings.

Bonfils was studious in his scheming, deliberate in his planning. Tammen was spontaneous and trigger-like in his manifestations of ability. He might now labor far into the night, but at the office he could be depended upon always for timely suggestions, were it a story about Dean Hart's cruelty to fish, or an unconventional headline, such as the one when the explorer, Stanley, died:

"STANLEY GOES TO FIND LIVINGSTONE AGAIN"

Concerning their divergent habits of mind, Tammen said:

"Bon' speaks once after thinking twice. I speak twice before thinking once."

Bonfils and Tammen arrived in Kansas City, Missouri, in 1909, bent on establishing an important branch of their industry in that metropolis. They set up the *Kansas City Post* on the junk-pile remains of a traction-managed newspaper, and until 1922 continued publishing the *Post* from a hole-in-the-wall office at Tenth and Main Streets.

Contrary to popular belief, the *Post's* owners were not afterward expelled from the Missouri stronghold by unappreciative citizens. In 1922, they disposed of their tin-can plant and 190,000 paid circulation (but mighty little else in the way of assets). They sold out for one million two hundred and fifty thousand dollars. It is said the paper had cost them less than *two hundred and fifty thousand dollars,* including the lease on their morgue-like headquarters. The buyer was Walter S. Dickey, multi-millionaire owner of the *Kansas City Journal,* a publicist who was said to have had dreams of a seat in the United States Senate.

Whatever else they did, or did not, obtain from Kansas City, Bon' and Tam' netted a million dollars cash profit from their stay near the Ozarks.

Bonfils never had forgiven Kansas City for the hectorings he had suffered during the lottery days. Nor had he forgotten Colonel William Rockhill Nelson's newspapers, the *Star* and the *Times,* for having heated the tar-barrel and spread the feathers for his exit in 1895.

He now prepared to get even with ancient detractors.

In surveying the moth-eaten plant of the *Kansas City Post,* Tammen said: "Kansas City can consider itself invaded."

The partners then set out to dumbfound the citizens of Missouri and Kansas with a wild hammering of drums and medicine-man huzzas. Tammen began announcing (as usual) that he and Bonfils were "bad babies," and that they were out to kidnap advertisers from Colonel Nelson's conservative newspapers.

The merchants of Kansas City were slow in responding to the circus-poster antics of the Katzenjammer Kids from the Rockies. The *Post* began trying to insult Colonel Nelson. But the latter's papers remained discreetly calm and silent. The *Star* and the *Times,* for many dignified months, did not appear to see the cocoanut-tossings or hear the double-darings of the invaders.

Finally, however, the owner of the Nelson papers took notice of the yammering, yelping, tantalizing *Post* to this extent: the staff members of the *Star* and the *Times* were advised *never,* under any circumstance (including the absent-minded moments of drunkenness), to *read* the *Post.* Instant dismissal was the penalty promised.

This taboo led to a rather peculiar incident involving Lee Taylor Casey, now one of the most able admirals of the Scripps-Howard journalistic fleet. Mr. Casey, at the time of which we write, was a stalwart reporter for the *Kansas City Star.*

When a newspaper obtains a story which a rival journal is unable to get, it is called a "scoop" or "beat." To be scooped is to commit the unpardonable sin. It is much worse than any other known mistake, more poignant than any other sorrow, more embarrassing than losing your pants during graduation exercises at Bryn Mawr. Yet every *good* reporter has been scooped, or beaten, at some time or other in his career—and that may be the thing which makes a reporter good, the watchful and discriminative sense never again to mistake a ball of lightning for a lollipop.

Mr. Casey was scooped by the *Post* one sad day. The story had been a big one, and Mr. Casey feared for the worst. Yet, when he poked his dejected nose into the *Star* city room, no apoplectic editor began screaming the Casey Waterloo or reviewing the Casey family tree. Nor did any brother reporter lift a slyly sympathetic brow in the manner which always makes the scooped victim want to break an Old Crow bottle on a colleague's parietal bones.

There were no howls. Not a word. For, were a *Star* editor to read the *Post*, such perusal was punishable with instant dismissal. And how was an editor to know of the scoop *if he had not read the Post?*

Mr. Casey kept his job, thanked God, and went on to bigger and better accomplishments.

The *Kansas City Post's* plant was not a thing of beauty. The presses belonged in a museum. They grumbled like abused tractors on a heavily mortgaged farm. Baling wire was kept in readiness—like bandages of a first-aid kit—to be wound about fractured parts. The building itself was hardly reminiscent of the Taj Mahal, and the office equipment looked as though it had been bequeathed by a beach comber. Yet there was a spirit of hooray among the palsied rafters, too much excitement to permit of cobwebs.

The first editions of the Bonfils-Tammen *shimbun* jolted the dignified populace of the city. These folk had been accustomed to the *Star's* restrained, one-column headlines; and here was a howling, Dervish-like intruder, its headlines dripping with blood and its body-type stuttering with enough crime news to have given Bluebeard a nervous breakdown, and enough sex sagas to have sent the Marquis de Sade to a hermitage.

Although circulation rose through the first years of the Kansas City invasion, the advertisers were aloof. On one of his frequent visits to Kansas City, Tammen threatened to fire the whole advertising department.

"I'll show you how to sell ads," he shouted. "Just give me a reporter."

Jack Carberry, one of the most animated lads of the city room, was assigned to accompany Tammen.

"Now," said Tammen, "who's the toughest bozo of all the merchants who refuse to advertise with us?"

He was given the name of a leading citizen, whom we shall call Mr. Gissing. That gentleman was an ornament of the town's cathedrals, a stickler for moral discipline, so dignified and religious that a customer almost had to achieve pastoral degrees before being O.K.'d for a charge account.

On their way down the street, Tammen said to Carberry: "Son, I could sell snowshoes in the middle of the Sahara."

They walked into the Gissing department store. As they passed the perfume counter, Tammen sniffed. "Hum! Smells like Jennie Rogers' parlor house used to."

Tammen sent in his name, *via* a footman-faced monsieur in a morning coat. The *Post,* months before, had incurred the intense dislike of Mr. Gissing, so he allowed Tammen and Carberry to cool their heels for half an hour outside his rectory-like offices. This didn't deflate Tammen. He spent his time examining two newspapers which he had brought along. One was that day's issue of the *Post,* with its juicy banner-lines, many of which were in red ink. The other was the *Kansas City Star,* with its conservative make-up and conscientious news stories.

Finally the swallow-tailed messenger re-appeared, glanced uneasily at Tammen's hat, which had remained on his head and showed no disposition of leaving that perch, then, with a mortician-like gesture, ushered the two *Post* representatives inside the Gissing office.

Mr. Gissing, frowning and sitting forward, as though he had some malady of the lower spine, was at his immaculate walnut desk. Tammen walked straightway to that desk.

"Gissing," he said, "I'm Harry Tammen."

The symptoms of sub-spinal inflammation became more pronounced. "Yes, sir. I know you."

"Gissing, I'm over here to sell you some space."

There was a frosty sneer. "I am not interested, Mr. Tammen."

"Not *interested!* Say, you old son of a bitch . . ."

"Mr. Tammen, I am a gentleman. Do not use that kind of language . . ."

Tammen interrupted. "O. K., Gissing. O. K. You're *not* a son of a bitch. But, Gissing, you need us."

Mr. Gissing made some feeble, inarticulate protest. He was as pale as anything in his muslin department.

Tammen was swarming over the desk now, spreading out the two newspapers, side by side. "Now, look here, Gissing." He pointed to the copy of the *Star*. "Just look at that. What does it remind you of, Gissing?"

Mr. Gissing, waving wanly, didn't know. And, if he *had* known, Tammen wouldn't have let him answer anyway. "I'll tell you—it's like your wife, Gissing. Plain, unattractive, no sex appeal. None whatsoever!"

The merchant was startled beyond compare, tongue-tied, and in need of smelling salts.

"But look here, Gissing." Tammen now spread the screaming copy of the *Post* on the polished walnut desk. "What does *this* remind you of?"

Gissing miraculously worked his vocal cords. *"I do not know."*

Thus encouraged, Tammen slapped Gissing's back. "Then *I'll* tell you. It looks like the most beautiful, the most voluptuous whore you ever saw—that's what it is, Gissing!"

Gissing struggled for air. Then he "blew his topper." He jumped up, screaming: "Get out—both of you!"

A platoon of frock-coated employees arrived to form a hollow square about their commander. Mr. Gissing now was seated, teetering in his swivel-chair, as though to shake the devil's pincers from the base of his spine. He was bleating vaguely, his words seeming to belong to some weird language, possibly the patois of the Ogallala Sioux.

In response to a signal from their gargling commander, the troops dissolved the hollow square and went into a rugby formation. The battle of Waterloo was reënacted, with Mr. Gissing in the rôle of Duke of Wellington, Mr. Tammen as Napoleon, and Mr. Carberry portraying Marshal Ney.

When they had reached the street, Tammen said to Carberry: "What do you suppose he got sore at?"

One of the several managing editors of the *Kansas City Post* was Charles Bonfils, Fred's younger brother. Charlie was a spirited fellow. Fred was constantly supervising his brother's affairs, lecturing him as though he had not yet arrived at maturity. To keep Charlie constantly on the job, Fred outfitted a loft for golf practice above the editorial rooms of the *Kansas City Post*. Charlie would practice driving and putting in this loft—the first miniature golf course in America.

Bonfils and Tammen occasionally assigned their favorite feature writers from Denver to Kansas City to bolster up the staff. Otto Floto, sports editor, and Fay King, cartoonist, divided their time between the Bonfils-Tammen papers.

The first time the *Star* mentioned the *Post* was following a political campaign. The *Post* had put over a Democratic slate in the face of *Star* recommendations to the contrary. The latter came out with an attack on the Bonfils-Tammen organ. To celebrate this left-handed recognition by the *Star*, Tammen bought an old goat and put it in the *Post* window at Tenth and Main, with a sign reading: "The *Star's* Goat."

It was the least aromatic of odoriferous beasts, and although the health department was then favorable to the *Post*, it was compelled to yield to complaints by citizens. Tammen removed the animal from the premises.

In trying to outsmart Colonel Nelson, Bonfils had figured without the realization that the *Star* was powerfully entrenched in Kansas agricultural centers as well as in Missouri's large cities. It had been in Kansas that Bonfils long ago had operated his lottery. His legend still remained there to haunt him, and the Post never could get a foothold in that state.

The partners hired the Rev. Dr. Burris Jenkins, well-known lecturer and writer, to edit the *Kansas City Post*. It kept up a whirl of circus didoes to divert the citizens and attract their patronage. Bonfils sometimes called his editors to Denver to witness, first hand, how promotional effects could be achieved.

One of these editors, an ex-*New York World* man, paid

such a visit to Denver and received a lecture from Bonfils.

"You don't know how to display enterprise," Bonfils said.

"Just what do you mean by that?" the editor asked.

"I'll show you," Bonfils said. He pressed a button. Immediately there arose a strident whinny, as though from the stables of hell. He had touched off a siren stationed on the roof of the *Post* building.

"Well, what's it all about?"

"Enterprise," said Bonfils. "Whenever the citizens hear this noise, they know it is the *Post*. It makes them constantly aware of us."

When the editor returned to Kansas City, he was determined to display enterprise. A nationally important engineering project was to be inaugurated by the President of the United States, and the *Star* had announced that it was going to signalize the event by the sounding of a siren.

"Never mind the *Star's* peanut-whistle siren," the *Post* editor wrote in a first-page article, "just wait for the *Post's* signal. The nature of this signal is being kept a secret. But it will be so novel, so astounding, that all Kansas City will know, that when the President presses a button, the *Post* is on the job to relay the news immediately and in a most interesting manner."

The *Post* editor himself did not know what sort of signal he was to give to the city on the appointed day. The nearer the time approached, the more worried he became. Bonfils was telegraphing nightly, asking *what* the signal was to be. In reply to these anxious wires, the editor would answer:

"We are going to display enterprise. It is a grand secret which will delight you. Please be patient."

The day before the President was scheduled to thumb the potentially historic button, the editor was visited by a prospector who sought to sell his life-history to the newspaper.

"I want to give you a true story called: 'The Desert Rat's Adventures'," he said.

The editor was inspired. "To hell with desert rats! Do you think you could set off some dynamite?"

"Whereabouts?"

"On the roof."

"Whose roof?"

"Ours."

The Desert Rat thought it over. "The force of dynamite is tricky," he said, "but mostly it goes downward."

"Can you figure out something to be touched off by electricity?"

When a bit of money was offered, the Desert Rat decided a wire cable could be slung across the roof, and between two poles, like a clothes line, and the dynamite suspended from the cable.

"How much powder should we use?" the editor asked. "We want it to be heard for five miles."

"I guess three or four sticks," the Desert Rat calculated.

"Make it an even two dozen," said the editor.

On the day of Presidential button-pressing, and after numerous editorials had been written, warning the citizens to be prepared for a wonderful demonstration of *Post* initiative, the Desert Rat stood by to work a hand-generator, which would create current necessary to set off the dynamite.

When the news was flashed that the President had pushed the button, any siren which the *Star* had contemplated using was dimmed. For a blast rocked the business section of Kansas City. Windows caved in. The citizens were in a panic. Business men talked of forming posses to lynch the editors of the *Post*.

In accordance with *Post* luck, its own building had not suffered materially from the cataclysm; but all about Tenth and Main were direful reminders of the great explosion.

That afternoon Bonfils got on the long-distance telephone, and for a time the editor could not make out what he was saying. Finally, however, he gathered that Mr. Bonfils wanted to know what in God's name had happened, and what the *Post* management was doing?

"Mr. Bonfils," the editor said, "we were displaying enterprise."

Bonfils hung up. It was as though he had fainted at the

other end of the line. A few hours later, the editor received a telegram.

"Stop displaying enterprise and square everything."

(Signed) "Fred G. Bonfils."

The *Kansas City Star* was one of the first newspapers in the United States to have a radio broadcasting station of its own. It was in a day when listeners had the old-style crystal sets. There was no such thing as strictly allocated wave-lengths for sending apparata. The *Star's* radio program was all the rage in Kansas City and environs. It had conceived the idea of remote control and was broadcasting the Nighthawks—the Coon-Saunders Orchestra—each evening.

This showmanship burned Harry Tammen to the core.

So he, too, put up a station. It was the grandfather of the present-day Columbia KMBC. The equipment was located in Burris Jenkins' office. His reception room was utilized as a broadcasting studio. Carberry was in charge of the programs, in addition to his duties as reporter, oil editor and a sub-manager of the Empress Theater.

The Empress was a branch of the Bonfils-Tammen theatrical interests in Denver, where they now maintained a vaudeville house. The Denver house booked acts from the Sullivan & Considine Circuit. It also was called the Empress and competed with Martin Beck's Orpheum.

To overshadow the *Star's* program, Carberry arranged a night of stellar entertainment at the *Post's* station. He enlisted Phil Baker—his first appearance on the air—and the McCarthy Sisters. The *Star* felt this competition keenly and decided to "top" the *Post* at any cost.

Rosa Ponselle, the Metropolitan diva, was appearing at Convention Hall. Kansas City was strong for opera, as might be inferred from the special train-load of citizens it sent to New York in a more modern era for the debut of Marion Talley. The *Star* procured the services of La Ponselle and announced the first broadcast of a grand opera star—and *how* they announced it! The greatest event in the history of Kansas City; a chance for the people to enjoy this "greatest of American singers."

Now Harry Tammen couldn't stand *that*.

So he concocted a scheme, and Carberry went through with it. Ed Fetting, the station's technical man, turned some dials, and the *Post* was on the *same wave-length* as the *Star*. The *Star* made its announcement:

"And now Miss Ponselle, noted member of the Metropolitan Opera Company of New York City, will sing for you . . ."

Carberry was all set. He had picked up a newsboy, known as "Whitey the Whistler," in the *Post* alley. And as soon as the *Star's* announcer had concluded a flowery introduction of Miss Ponselle, Whitey the Whistler opened up. Could that boy whistle! Not good, but loud. And he whistled, and whistled.

The *Post* would shut Whitey off every time Rosa finished a solo. Then, just after the announcement, they'd turn the shrill newsboy loose again.

That, in Tammen's opinion, was the grandest stunt he ever fostered. He laughed for days and days, and he never saw Carberry thereafter but that he'd pucker his lips and begin to whistle. Then he would roar.

When the *Star* sued Bonfils and Tammen for libel, basing their action on an editorial attacking that newspaper, an examiner was sent to Denver to take Tammen's deposition. The examination was held in the Federal Building. Attorneys for the *Star* sought to show that Bonfils and Tammen, although absent from Kansas City, had directed the writing of the allegedly libelous editorial.

To register that point, the *Star's* attorneys tried to bring out the intense interest each partner took in their two newspapers. The examination of Tammen follows in part:

Q—Mr. Tammen, isn't it true that you and Mr. Bonfils take a very active interest in the details and operations of your newspapers—even to the extent of exchanging lengthy telegrams many times daily on matters of policy when you are not in the same city?

A—That's just half true.

Q—What do you mean, half true?

A—I mean, it's true I send Fred lengthy telegrams many times a day. He doesn't send me so many.

Q—Why do you send these telegrams, Mr. Tammen, if

it is not to keep closely in touch with the affairs of your newspapers—such as the editorial in question?

A—Well, I'll tell you. Fred's so doggone tight that I like to send him those long telegrams, *collect*, just for the fun of thinking how he'll cuss when he has to pay for them.

Later in the examination, the attorney sought to have the record show that Bonfils and Tammen were wealthy men, presumably for effect on a jury in assessing damages.

Q—You and Mr. Bonfils are very rich men, aren't you, Mr. Tammen?

A—I don't know anything about Fred's personal business affairs. I don't know whether he's rich or poor. But I'm a very rich man.

Q—How rich are you, Mr. Tammen?

A—You mean how much money have I? I couldn't even guess.

Q—Then what is the basis for your statement that you're a very rich man?

A—Because I don't owe a person in the world a red cent, and there's not a man in the world I'm afraid to tell to go to hell. Anyone who can say that is a very rich man.

Walter S. Dickey, owner of the *Journal*, approached Bonfils and Tammen with a view of purchasing the *Kansas City Post*. He offered a million and a quarter dollars for the paper—and not even Bonfils' hatred for his enemies, his conviction that he still could whip Colonel Nelson, and that his *Kansas City Post* would dominate, just as his *Denver Post* dominated its field, could stand in the way of that offer.

The *Post* had little more than a circulation and a staff. They did not own their old building, the ramshackle fortress at Tenth and Main Streets. The lease was expiring. The machinery—it was junk, hardly worth moving.

Not only did they receive this one million, two hundred and fifty thousand dollars, but Bonfils was said to have worked in an extra forty thousand dollars profit in an eleventh-hour sortie.

The smart-money boys, Bonfils and Tammen, had put over the deal so that Dickey—not they—had made all the overtures. The papers were about to be signed. Then Tammen got "sick." He went to the Hotel Baltimore. A nurse was called in. Harry grunted and groaned.

Bonfils brought Dickey over for the consummation of the deal. The lawyers were there. Harry, lying on the bed, took pen in hand to sign the documents. Then he hesitated.

"Say," he said, "how about the eighty thousand dollars' worth of white paper in the warehouse?"

"Oh, that's all O. K., Harry," said Bonfils.

"O. K., hell!" said Tammen.

Mr. Dickey intervened. "But, Mr. Tammen, I am paying a good price for this paper. I . . ."

Tammen interrupted savagely. "Say, listen, you bozo! You're getting a bargain. I did not ask you to buy the *Post*. You want it. We don't want to sell it. To hell with you! There is eighty thousand dollars' worth of paper—paper we can use in Denver. The deal is off. All off! Get out, damn you, get out! Get out before I have you kicked out!"

"But Mr. Tammen . . ."

"Get out, I'm telling you! God damn it! Me lying here sick, and you . . ."

The solicitous Bonfils led Mr. Dickey into the hall. "Mr. Dickey, I am sorry. Heartsick. We want to sell the *Post*, and we want to sell it to *you*. But Harry and I have been friends. We have been partners. We have been more than that—brothers for thirty years. I want you to have the paper, Mr. Dickey, but I positively *will not hurt Harry*.

"Now I'll tell you what I *will* do. There is eighty thousand dollars' worth of paper in the warehouse. I do not want my part of it. The forty thousand dollars that represents my interest in it can go by the board. But please, Mr. Dickey, do not let Harry know about this. You pay him the forty thousand dollars and pretend you are paying me my half as well."

Mr. Dickey "went" for this added expense, it is said. Harry and Fred split the forty thousand dollars, twenty thousand dollars each, and Tammen had a great time telling about it.

22

Wine And Balloons

IF ONE were asked for a sample of the sensationalisms which the *Post* utilized in upbuilding its circulation, the Henwood murder case would be a happy choice. This was Denver's outstanding *cause célèbre*. It embraced the premeditated slaying of Sylvester Louis (Tony) Von Phul, aeronaut and wine agent, the unintentional killing of another man, the crippling of a third, and the co-incidental wrecking of Banker John W. Springer's home. The slayer was Harold Frank Henwood. The crime followed a series of quarrels over the zestful Isabelle Patterson Springer, wife of the banker.

The Henwood murder was to Denver what the Thaw case had been to New York. Great wealth, social position, illicit romance were factors in both episodes. In each case the offending satyr succumbed to pistol-fire in a public resort—Stanford White on the roof of Madison Square Garden; Tony Von Phul on the barroom floor of the Brown Palace Hotel. In the Henwood affair, however, a rhapsodic suitor, and not the husband, was the avenger.

The fact that the *Post's* attorney, John T. Bottom, defended the slayer did not lessen the editorial demands for Henwood's summary punishment. No pistol packer could hope for sympathy from newspaper owners, who themselves had been pinked by bullets, and who constantly were receiving target-practice threats from neurotic citizens.

Banker Springer had been a power in Colorado financial and political circles for many years. He was middle-aged, affable, well educated and a member of numerous choosy clubs. He traced his lineage to Landgraf Louis of Germany, born in 1089, a military officer under Emperor Henry IV. The Landgraf's escape from the battlements of Giebichenstein Castle, after two years' imprisonment for a minor political offense, and a prodigious leap of one hundred feet into the River Saale, earned him the name "Springer." On his mother's side, John W. Springer was of the Kentucky Hendersons.

He came to Colorado from Dallas, Texas, in 1896, in the interests of McKinley's presidential campaign. Possessed of considerable fortune, Springer purchased a ranch of ten thousand acres and decided to remain in Colorado. He became one of the foremost cattle and horse growers in America.

In 1902, he assisted in organizing the Continental Trust Company, a banking institution, in Denver. He served as vice president until 1909, then became president. His first wife died in 1904. He afterward married the young and spirited Isabelle Patterson.

Springer's interests often took him to New York City. His vivacious wife always accompanied him. She attracted social attention in the larger city, and among her admirers at the Waldorf-Astoria were several leading artists of the time. Both husband and wife were interested in art, and Springer readily agreed to permit her to sit for portraiture. He commissioned Carl Brabant, a serious and capable artist, to paint his wife's portrait. When Mrs. Springer chose to remain in New York for the social season, and also to continue sitting for the Brabant portrait, Springer acquiesced. He returned alone to Denver.

The Brabant studio was located in the Gramercy Park sector. Mrs. Springer found it a strangely routine workshop, its owner not at all of the carefree stripe limned in rapturous fiction. Nevertheless she became acquainted there with sighing devotees of moneyed flesh. She turned to them for initiation and divertissement.

While sitting one day for Brabant, she confided that a certain artist had invited her to a studio tea.

"What do you think about it?" she asked.

"I wouldn't accept," said Brabant.

"Why not? He seems very nice. Besides, he amuses me. He says I have a figure totally wasted on a woman who cannot, because of her position, inspire some artist's fancy."

"Would you permit that talk in your own drawing-room?"

"That's different," she said. "My drawing-room is not a studio. Artists would not be at home there. *You* might, for you are so dreadfully sedate; but these other chaps would be stifled by the conventions."

Brabant let the matter rest with a warning. Mrs. Springer attended the tea. She made immediate friends in the gay group. There is no evidence to support a belief that she bestowed forbidden tokens during that season on the debonair bounders of New York's Bohemia.

Mrs. Springer at length returned to Denver, resumed her duties as social leader, but sighed for the Christmas-tree glitter of New York. The next year she again was to be found among the pale-faces of Manhattan studios.

She had the spirit of an adventuress and the privy purse of a king's daughter. She experimented with opium smoke, lent her white arm to an occasional hypodermic barb, and began posing in the nude for the counterfeit Titians of the attics.

After a siege at the poppy pipes, and in response to calls from the West, Mrs. Springer declared a moral armistice and went back to Denver. Springer knew nothing of his wife's New York antics. He had trusted her completely, and he himself was deep in business matters. His chief recreation was development of a model farm, where he toyed with steam ploughs and other rural machinery. Among his farm improvements was a lighting system, installed by Harold Frank Henwood, representative of the Blau Liquid Gas Company.

Henwood was a tall, slightly bald man, nearing his forties. He was adjudged handsome and cut a dash in select circles. He had the aplomb of a dancing master and the eloquence of a minor poet. Henwood was interested academically in Arctic exploration, but his lips had a sensual turn, suggesting that his temperament belonged to the tropics.

He became infatuated with Mrs. Springer, but appeared for a time to keep his feelings under double-wraps. She diagnosed his restrained sighs, however, and maddened him with her piquant confidences concerning life and sentiment. Finally he dove overboard into emotional waters.

And now the swashbuckling sportsman, Tony Von Phul, came into the picture. Mrs. Springer had made his acquaintance during her travels. He was a sturdy, aggressive fellow of thirty-four, a wine agent, balloon racer and connoisseur of women. When Von Phul appeared on the Denver scene, it seems Mrs. Springer's flirtation with Henwood died down. He was living too much on his knees.

Although she kept up a pretense of intimate regard for Henwood, it appears she was bestowing her real affections on the parachute-jumping roué. Henwood divined the situation, tried to endure it with a white-plumed complaisance, then became unable to stand the gaff. He called Mrs. Springer to account, accusing her of flitting between bedrooms.

Mrs. Springer told Henwood—truly or falsely—that she was compelled to humor Von Phul, simply to recover from him certain indiscreet letters. She amplified this story afterward, saying that Von Phul was attempting to blackmail her; that he had demanded cash and jewelry. The letters, she explained, had to do with New York indiscretions. She implored Henwood to have faith in her, pointing out that she dared not reveal conditions to her husband.

Partially mollified by this wench-like sorcery, Henwood knighted himself and decided to compel Von Phul to disgorge the letters. Von Phul brushed him aside as though he were a bubble.

"Just quit annoying me," Von Phul said. "If I want to make love to her, what of it? She's fond of me."

Henwood made threats. Von Phul laughed—a graveyard laugh. He was a veteran of emotional wars. He had been involved in many such green-eyed settos. As for attacks, he had had three of his front teeth shot out during a St. Louis escapade. At another time, in Lexington, Kentucky, he was wounded. Too, he had experienced numerous escapes from death, in adventures not involving women, once when his balloon fell eight thousand feet into the Mississippi River. He was holder of the speed record for balloon flights, having sailed a silk bag from St. Louis to Charleston, six hundred and ninety miles, in fifteen hours.

"Just go to hell," Von Phul told Henwood.

When Von Phul favored Mrs. Springer with a rakish account of his rival's deflation, she seemed alarmed. Yet—in that paradoxical manner understood by women better than by men—she continued to fan the bonfire of enmity between the sky-riding Don Juan and the Orlando of the gas tanks.

Springer as yet had no inkling of his wife's extra-marital frolics. Nor did the town gossips know. But an air of the duello hung over the dizzy buckeroos and their warm-blooded Sylvia.

During the afternoon of May 24, 1911, Henwood waited for Von Phul in the barroom of the Brown Palace Hotel. Presumably he wanted a showdown concerning the purported letters, and intended asking Von Phul to withdraw from Mrs. Springer's horizon.

Von Phul entered the bar, pushed Henwood from his path; then, when Henwood persisted, knocked him down.

Henwood rose, then went out to procure a pistol and ammunition at a pawn shop. He wandered about town, thinking over his troubles. He tried to get in touch with Mrs. Springer at her suite in the Brown Palace, but she didn't answer the telephone.

Von Phul that evening attended a performance of the road show "Follies of 1910" at the Broadway Theater. Henwood followed him, perhaps with the thought of slaying his enemy there. The rivals passed each other in the lobby between acts, but no words were exchanged. After the theater, Von Phul and two friends—one of them Manager Rosenbaum of the "Follies"—appeared for refreshment at the Brown Palace bar. Henwood trailed after them.

Several men were standing at the bar as Henwood entered. These included George Edward Copeland, proprietor of the Copeland Sampling Works of Victor, Colorado, and James W. Atkinson, contractor and part owner of the Acacia Hotel of Colorado Springs. Neither of these was known to Von Phul or to Henwood.

Von Phul saw that Henwood was keyed up, for his face was pinched and white and his jaws set. The sportsman looked at him, sneered, then turned to the bar attendant.

"Send some wine up to Mrs. Springer," he said. Then he ordered drinks for himself and his two friends. Henwood edged over towards Von Phul.

As Henwood stood there, apparently too angry to speak, Von Phul addressed him with mock cordiality:

"Oh, hello. I was just going upstairs to make love to your sweetheart."

Henwood was in a frenzy. "You lie! You lie!"

Von Phul said to his companions: "Please excuse me." He then let drive with his fist against Henwood's chin, sending him to the floor. Von Phul returned to the bar, as though nothing extraordinary had occurred.

Henwood, on his knees, was drawing the pistol from his coat pocket. He began firing wildly. The first slug pierced Von Phul's right shoulder as he faced the bar. He sagged forward, weaving like a bear in the minuet. A second shot missed him as he lurched, leaving Copeland in the line of fire. That bullet struck Copeland's right hip, coursing up and through his abdomen and traveling in an arc down his left thigh and leg. He fell, mortally wounded.

Attendants and guests ran from the crowded lobby to the barroom. The bartender was leaping over the counter to disarm Henwood. An elevator man was advancing on Henwood's rear. Henwood continued firing. A third bullet struck Copeland in the groin. A fourth passed through Von Phul's right wrist, ricocheted and caught Atkinson in the left thigh, shattering the hip bone. The fifth and last bullet entered Von Phul's back, near the kidneys, ploughing upward, grazing the heart.

None of the three wounded men had lost consciousness. Henwood was still snapping the hammer of his exhausted weapon as the elevator man pinioned his arms. The bar attendant relieved him of his pistol. Von Phul was reeling through a doorway leading to the smoking room, his right wrist gripped by his left hand to check the flow of blood. Copeland was lying on the floor near the end of the bar. Atkinson was prone at the entrance to the lobby. Guests were attempting to aid the injured men. In the excitement,

Henwood freed himself and walked toward the lobby, pausing however to lean above Atkinson. He was shedding tears as he said:

"I'm dreadfully sorry I shot you by mistake."

Atkinson looked up at Henwood. "You ought to be."

A bell hop followed Henwood through the lobby and to the street, pointing him out to a policeman. An ambulance and detectives arrived. All three wounded men displayed great nerve. Henwood, now in custody, was pale but cool. An ambulance took Copeland and Atkinson to St. Luke's Hospital. Von Phul refused to ride in the ambulance.

"Get me a taxicab," he said. "The rat knew I was unarmed."

Meanwhile the city wondered what motive had prompted the shooting. There had been no hint that Mrs. Springer was implicated in any way. Her husband still was unaware of her duplicity. For a while it was believed that a member of the "Follies" chorus had been the cause.

"Was there a woman involved?" Chief of Police Armstrong asked Henwood.

"Yes," said Henwood, "about eighteen of them."

Henwood gave a midnight statement to reporters at the jail.

"Von Phul was looking for trouble and found it. I never carried a gun in my life until tonight, but I was told that Von Phul always carried weapons and I took steps to protect myself. We had a few words while standing at the bar. He struck me and I protected myself with my pistol.

"I am sorry that innocent persons were injured; I did not intend to hurt anyone but Von Phul, and I have no regret for having injured him, nor will I ever have."

He was asked: "But what is back of it?"

He thought for a time. "I am sorry I cannot tell you fellows just what prompted the shooting, but really, I have forgotten what it was. You may say that he liked the bathing girls in the 'Follies' company; that I disagreed with him."

"Had you ever met Von Phul previous to your first meeting in Denver?"

"I don't think I care to answer that question. I have

said all I care to say, and you can roll your little hoop. Good night."

He walked to his cell.

It is popularly supposed (and perhaps hoped) that a slayer spends the first night after his capture cowering in his cell, hearing spectral voices from within his accusing soul; else, he paces like a harried beast, seeing strange and ghastly shapes, and unable to sleep. Reporters know, however, that a newly made murderer quite often drifts into unbroken slumber—like a workman who has toiled all day with pick and shovel, his digestion good, his snores honest.

Henwood slept.

In the morning he was steady and unperturbed.

"Let me look at the papers," he said. "I want to see how many bouquets the reporters threw at me."

He read of Von Phul's death, which had come during the night. It was as though he were glancing at the obituary of a stranger. He sipped his orange juice as he read that District Attorney McComb and Chief Armstrong had driven to the hospital to take an ante-mortem statement. He ate his cereal as he continued reading:

"The officers knocked at the door; the nurse appeared from Von Phul's room, saying: 'He's dead'."

Copeland died some days later. Henwood was called to trial for the murder of this bystander, instead of for the slaying of the man whom he had set out to kill. The latter case was left in abeyance.

Old Josiah M. Ward, bloodhound of the *Post's* city room, sat at his desk, exercising his long shears. His spectacles rested on the tip of his rum-blossom nose. He was deep in one of those famous *cherchez la femme* moods, a Scotland Yard frown on his brow.

There *surely* had been a woman. But *who?*

Joe had a wealth of mottoes in his journalistic textbook. One was: "Always look to a family friend for dirty work."

To whom had Henwood been an intimate friend? Springer. Who had a beautiful young wife? Springer.

Ward's lamentable experiences with Pepita, in the long ago, had made him particularly keen concerning young and beautiful women. He now squealed with ecstasy. He was sure he had guessed the woman. He called for a trio of reporters, like King Cole screaming for his fiddlers three. He explained nothing—as usual—but merely knocked them silly with his command:

"Get an interview with Isabelle Springer. Ask her why she double-crossed her husband and played the game with Von Phul and Henwood."

"All three for the *one* interview?" asked a reporter.

"Hell, no!" He pointed to each in turn. "You get to Mrs. Springer. You talk to her maid; make love to the maid or maim her, I don't care, but get the story. And you tell Henwood that Mrs. Springer has spilled everything."

The three reporters failed. Mrs. Springer was sequestered in her apartment, with a "Do not disturb" placard on the door. Henwood refused to see reporters. As for the maid—she had been fired three days before the crime.

"Ha!" said Joe. "Ha!"

Ward released the reporters from the assignment. Then he looked across the room at a young society miss—one of those girls whom Bonfils continually was foisting on the irascible City Editor for "training."

"Come here, sister," he said.

He spent some minutes whispering to the young woman. She went away. She returned that afternoon. She had the story! Ward pretended that it was only a trifling scoop, but the *Post* spread Mrs. Springer's interview all over the paper.

The society miss, coached by Ward, had gone to Mrs. Springer's suite with another young woman of her set. She had represented that her family was on the point of hiring a maid. That the maid, on applying, had no written references, but had said she recently had been in Mrs. Springer's service. Would Mrs. Springer kindly vouch for her former maid?

By this ruse the society girls obtained information linking Mrs. Springer definitely to the Henwood and Von Phul feud. The *Post* had its scoop, and the way now was thoroughly paved for Mrs. Springer's descent to hell.

Springer made plans for a divorce, another godsend for Joe Ward's scandal columns. And another boost in *Post* circulation figures.

The Henwood murder trial was before Judge Greeley Whitford, a jurist as sturdy as the old oaken bucket. The hearings were held in the West Side Criminal Court, a small, red building of pressed brick, stuffy and ancient.

Mrs. Springer appeared as witness for the state, and in her domestic extremity threw Henwood on the spears. She characterized him as one who had butted into affairs which had not concerned him.

Henwood, testifying in his own behalf, denied ever having sullied the Springer friendship. He had tried, ill-advisedly, to be gallant, to shield a friend's wife from a blackleg.

After a sensational potpourri of charge and countercharge, hints of perjury and bitter arguments, the case went to the jury. That body found Henwood guilty of murder in the second degree. This verdict carried with it, at the discretion of the court, a penalty of from ten years to a life-term in the Canon City prison.

Judge Whitford denied a motion by Attorney Bottom for a new trial. Then he asked the defendant:

"Have you anything to say before sentence is pronounced upon you?"

To this Henwood replied loudly: "Yes, I have."

Now sentence had not been passed, and for a prisoner to do what Henwood did thereafter seems an astounding thing. He began a half-hour of insult to the very court in whose power lay the choice of giving the prisoner as few as ten years, or as much as the rest of his life.

Henwood was standing. He turned his back to the court and reached for a water pitcher on the attorneys' table. He filled the glass, sipped deliberately, picked up some notes and again faced the bar. He launched into his speech. An excerpt from this amazing oration follows:

"I am not surprised that I am up here for sentence before you, Judge Whitford, after the attitude you have taken toward me since this trial started. I have seen criminals, innocent men and others, afraid of facing you, a prejudiced judge—a biased judge—a man with a mind for just one thing, and that is conviction. A man is guilty before he enters this court-room, in your estimation.

"The woman (Mrs. Springer) was put under morphine before she was placed on the stand—and you know it, and everybody knows it. From my observation, it does not make a bit of difference to you how young a life is, or how old, whether man or woman—your one thought is conviction. It is like kleptomania with you—you have a desire to convict.

"I know that you are going to pass sentence upon me—which is the happiest part of this trial for you. There are any amount of cases in the jail over here just crying to be tried, and the District Attorney knows it, and you know it. But no, you must get mine through with, getting mine before the Springer divorce case.

"Judge Whitford, I cannot call you anything but a prosecuting judge. I wonder if in the past you have ever considered whether or not you are fit to sit and adjudge any man. You should be called a prosecuting judge.

"You took it upon yourself to say that one of my witnesses, this man Garver, had testified falsely. What right have you got to say that?"

Henwood folded his arms.

"I stand ready for your unjust sentence."

Judge Whitford said: "Life."

In the *Post* city room, old Joe Ward was fit for a straitjacket. He had been scooped ten minutes on the Henwood verdict!

There were only two telephones adjoining the old court-room, one near the door and the other up the hall. A nimble reporter for Senator Patterson's *Times* had permitted a *Post* reporter to beat him to the occupancy of the nearer telephone booth—the one near the door. Then, while the jury was filing in to announce its verdict, the *Times* reporter snipped the telephone wires of the *Post* booth.

The *Post* reporter, cut off from his office, had to race across the street to a private dwelling to telephone his report. He lost ten minutes in this manner, and Josiah M. Ward lost his mind—almost.

Henwood waited two years in the county jail while the Supreme Court reviewed his case. He enjoyed every luxury that could be installed in an iron home. His cell windows and the barred front were draped with cretonne curtains. There were pictures on the sheet-metal sidewalls, and shelves containing a library of the newest fiction and books on Arctic exploration. There was also a gas-plate, on which Henwood cooked his own food. The Henwood door was not locked, and he had the "run" of the jail. It was even believed in certain quarters—Harry McCabe, reporter for the *News*, declaring it to be a fact—that Henwood went from the jail, one night each week, to drink and otherwise disport himself at one of the houses of the Red-Light district.

Sympathetic newspaper readers sent him steaks, delicacies, gross-lots of tinned tobacco, hundreds of pipes and innumerable packs of playing cards. These he distributed among his fellow prisoners at the jail.

Henwood grew abusive of McCabe of the *News* one day, and the latter threatened to write concerning jail conditions. Henwood thereafter was kept to his cell.

The Supreme Court ordered a new trial in 1913. Henwood was sanguine of clearing himself of all charges. It had been an accident that he had shot and killed Copeland. Exonerated, he would proceed to beat probable prosecution for the Von Phul killing. He would plead self-defense.

It was with an empty-eyed daze, then, that Henwood received the verdict of the second jury. He leaned forward, as though not sure of his ears. Then his shoulders grew round.

The jury *found him guilty of murder in the first degree!*

And he had *asked* for this second trial!

A request for a third trial was denied. Henwood now

seemed completely crazed, for he was sentenced to be *hanged* the week of November 1, 1913.

The Supreme Court granted a *supersedeas*. The case was again appealed, this time for reversal of the sentence to hang. It was not until July 8, 1914, that the Supreme Court confirmed the decision of the lower court. The higher body decreed that Henwood mount the gallows on October 25, 1914.

Further motions were denied and Henwood sent for a priest. Then Springer unexpectedly came to the fore. He now had divorced Mrs. Springer, and asked that Henwood's sentence be commuted, saying:

"If I had known what I now know, Henwood would not be facing the scaffold, for I would have shot Von Phul down myself."

Governor Elias M. Ammons commuted Henwood's sentence to life imprisonment—over protests of the *Post*—and he stayed in prison for eight years. Then, in 1922, he was paroled. He went to New Mexico, changed his name, and for several months seemed on the road to rehabilitation. Then, with extraordinary lack of balance, he threatened to kill a young woman because she had refused to marry him. This was construed as a violation of parole. Henwood returned to prison in 1923.

He lost all interest in life. His weight fell from one hundred and eighty to one hundred and fifteen pounds. He died in the prison hospital, following removal of his tonsils, September 28, 1929.

Mrs. Springer, divorced, had gone to New York. Although her figure no longer had a vital, slim grace, and her eyes were dull, she sought work as an artist's model. But she was a morphine addict. She could not remain steady while posing, nor work for long at a time. She lapsed into abject poverty. Her gay friends denied having known her in the Bohemian days.

She died, a woman of but thirty-six, in New York City's municipal hospital on Blackwell's Island, a victim of narcotics.

An actress, who once had been a guest at the Springer home, read of her death in a small newspaper notice. She claimed the body and saved it from burial in Potter's Field.

23

Trolley Cars And Ditches

ONLY TWO men, it is said, are left who can—but will not—tell why Bonfils ran afoul of Robert W. Speer, Denver's most notable Mayor. But a hundred thousand men and more can testify that the *Post*, for a dozen scathing years, upbraided Speer, vilifying the man and denouncing his every work.

The *Post* began on amicable terms with the modest-mannered executive, whose genius for city planning transmuted Denver from a frontier town to one of the most efficiently beautiful municipalities in America. Yet two years after Speer had been installed as Mayor, Bonfils began accusing him of graft, malfeasance, misfeasance, nonfeasance, and all the other feasances ever laid at anyone's political door.

The *Post* caricatured him as the puppet of "the interests," today the tool of the Gas Company, tomorrow the catspaw of the Telephone Company—but *always* the stooge of William G. "Napoleon" Evans, Grand Lama of the Tramway and the Water Companies.

Whether Bonfils came to grips with Speer because his school board had declined to buy *Post* coal, or because Speer had refused to award the coveted city and county advertising to the *Post*, we shall let political commentators affirm or deny. Whatever the cause, it was a fight to the finish, including libel suits, contempt of court charges, mud-pie tossings, reform waves and tom-cat furies.

It was patent that Boss Speer's local Democratic party was a "machine," and presumable that friends did not go begging beneath his political eaves. But it is equally apparent that Denver progressed during his administration from a town of jerk-water accents to a city of ample boulevards, broad lawns, playgrounds, fire-proof schools, admirable public buildings, great viaducts, flood-controlled streams, competent fire and police departments and a modern sewage system.

He built a huge auditorium, in which he sponsored free concerts, hiring Madam Schumann-Heink, among others, to sing for thousands of taxpayers who never had heard an operatic voice. He removed all "Keep off the Grass" signs from park lawns.

When the Speer Auditorium was filled to capacity during a Christmas Tree Festival, and the fire warden had ordered the doors closed to the overflow, a man was caught "sneaking" a group of children through a rear door. A police officer collared the offender in the gloom of a rear corridor. He was amazed to find that his prisoner was Boss Speer. The Mayor promised "not to do it again," then went outside and distributed dollars to disappointed children, advising them to "spend it all on candy."

Speer came to Denver from Pennsylvania in 1880. He was a consumptive, and unable to walk. He regained his health, rose from a job as dry-goods salesman to a county clerkship, was postmaster under President Cleveland, then successively became chief of the fire and police board, head of the board of public works, and, in 1904, Mayor. He served two terms before a "reform" movement unseated him. After that furore had subsided like a bromo seltzer, Speer again became Mayor of Denver.

He spent thirty years studying municipal problems. Civic research workers from foreign governments often visited Denver to learn the Speer method. The width of Denver's streets, the care with which they were laid out, testify to his great vision in preparing for automobile traffic, a problem which bedevils so many American cities. Unsightly dumps, a menace to health, became sunken gardens, and if the citizens paid heavily, they at least had something to show for their taxes.

But to the *Post,* Speer was a dozen Boss Tweeds. In every concrete bridge that replaced a palsied wooden trestle of the ox-cart era, Bonfils saw an example of highway robbery. By all accounts, Speer should have left forty-five millions, instead of the forty-five thousands of dollars the widow received at his death.

Not the least of the accusations hurled by the *Post* at this Dick Turpin of the tax rolls was that Speer—as errand boy for Napoleon Evans—sought to foist on the public a sixteen-million dollar Tramway franchise steal. This burning rag was brandished so endlessly that certain citizens wished Speer would go ahead, "steal" the franchise and have done with it.

Traction history is so dreary—save to time-table addicts and devourers of statistics—that one hesitates to present any part of it, however brief. Still, the Bonfils-Tammen brawls with the Tramway Medusa are not without spice.

The street-railway equipment of Denver—in the early part of this century—was of the best. The cars were not of that betrayed, bird-cage quality exemplified by New York trolleys—those kidney-splitting tumbrels that specialize in flat wheels.

A Denver street car was as spick as a Vanderbilt road-coach.

The crews were affable, and a conductor frequently might ask concerning grandpa's latest attack of sciatica, or, were your mother and father living together again. The fare was five cents, but the company (as always happens) wanted an increase. For a nickel, a fellow could take his girl aboard a car at Rocky Mountain Lake and neck all the way to Smith's Lake—a distance of almost six heaven-sent miles.

The only local traction franchise never subjected to argument, pro or con, was that of the horse car at Cherrylin in South Denver. It was a honey! An ageless white gelding with splay feet served as motive power for this chariot. His name was "Gladstone." Every half hour he would drag the ancient vehicle up the long hill, and although Gladstone sported little meat on his venerable haunches, he loved his work. And the reason for such love was this:

At the top of that long hill, the driver would unhitch Gladstone. A drawbridge was let down from the rear platform. Then Gladstone, a capricious light in his pink eye, and without a word of command from anyone, would walk up the plank and onto the platform. He stood there, his head and neck protruding from the port side of the vestibule, his rump, as gaunt as a hatrack, from the starboard side. The driver, casting a casual but appraising glance from the forward platform, to make sure that Gladstone was all set, would release the brake—and the whole shebang *would coast back to the starting point of the journey!*

On the platform, and coming down hill, Gladstone appeared as carefree as any other cheat. But when time came to hobble down the ramp, he seemed as one called upon to walk the plank.

This gangling white eunuch worked twelve hours every day in such manner; but, love his work as he might, he had a supercilious disdain for passengers who sought to curry favor by offering him sugar or stroking his lank withers. Women were the chief offenders in this respect, but Gladstone never was misled by giggling flattery. He did not choose to make himself common.

When the Tramway Company obtained its franchise *via* the polls, Bonfils screamed "Fraud!" In a Vesuvian "So The People May Know" editorial, Bon' spouted lava on everyone concerned with the election. Napoleon Evans sued Bonfils and Tammen for libel in 1912. In commenting on this suit, Bon' charged that Judge Herbert L. Shattuck, before whom the case was to be heard, had obtained his place on the bench through Evans' power. The term "pinhead" was employed in one of Bon's editorials, and the Judge construed it as applying to himself.

Tammen was traveling at this time in Honolulu. Judge Shattuck declared Bonfils in contempt of court. He sentenced the publisher to sixty days in the county jail, imposed a fine of $5,000, and assessed him one-half the costs. Bonfils gave notice of appeal and application for a *supersedeas.* Ten days' stay of execution was granted. The court indicated that Tammen would suffer similar treatment on his return from the hula-hula country.

Although the case had originated in trolley-car problems, it curiously changed into a water fight—Evans having been a Water Company baron as well as a traction boss.

In a stormy court session, Attorney Horace G. Benson, of *Post* counsel, said that Evans had tried to "buy off" Bonfils and Tammen. Attorney Benson added:

"When the *Post* proprietors said, 'No, we won't favor you,' Evans replied: 'We own the Tramway, the Water Company, the Evans estate, the Cheesman estate, the First National Bank; we own the city and we own the courts. We will drive you out of business and put you in jail.' "

Attorney John T. Bottom, also of *Post* counsel, told the court the newspaper was prepared to show that Evans had corrupted legislatures, city councils and other officials. He added that Evans had tampered with judges on the bench and sought to control others.

The *Post* attorneys denied that Bonfils had referred to Judge Shattuck as a "pinhead."

"That term, however," said Attorney Bottom, "*could* be applied to one or two other judges now on the bench."

The article of alleged contempt had contained the comment:

"Judge Shattuck is a man of many personal charms, and, were it not for his long, personal friendship with W. G. Evans, would be an ideal judge to try a (libel) case of this kind."

Attorney Benson said: "The manner in which this case has been filed would make it appear as if Mr. Bonfils and Mr. Tammen were in contempt of Mr. Evans instead of this court. And, if they are being tried for contempt of Mr. Evans, I have no doubt they are guilty."

The court asked Bonfils if he wished to present any statement intended to purge himself of contempt. The publisher made a speech which follows in part:

"I stand before this court today, conscious of my own integrity, honesty and rectitude; conscious of having committed no wrong against any judgment, based upon prejudice, malice or suspicion. I haven't now, and never have had anything but the kindliest feeling and best wishes for this court.

"I wish it were in my power to make your honor the

most righteous judge that ever blessed a struggling people. I wish that your honor might have the divine power to harmonize all mankind with the Golden Rule, and that you might teach the world the beneficence and wisdom of the Master, and that through you and your rulings we might all learn to do unto others as we would have others do unto us; and that your honor's decisions might hasten the final redemption and salvation of all mankind—because I know of no community that so badly needs such a judge."

Bonfils now launched upon a history of his fight with Evans. It began in 1910, he said, and after a conference in which Evans had asked the *Post* proprietors to advocate the city's purchase of the Water Company at an appraised valuation of fourteen million, two hundred thousand dollars; or, if that were not possible, to see if the *Post* would not favor a new franchise. This, Bonfils said, he and Tammen had refused to consider. Bonfils continued:

"Evans replied that he thought we were making a great mistake, and one that we would live to regret. He said he would be glad to pay us a hundred thousand dollars for *Post* support.

"We said: 'Mr. Evans, you clearly misunderstand us and the conditions. We need no money, and to accept such an offer would ruin the *Post*. Successful, profitable papers can never be bought to do a thing that is against the interests of the community. Only papers that are losing money and are unsuccessful are purchasable, and such papers, of course, are never worth anything to any cause, as they have no influence. To plead the cause of the Water Company would mean the ruination of the *Post*, and we would fool nobody if we did, as it would be clear to the most stupid voter that we had sold out the interests of the people to the Water Company."

"He then asked if one hundred and fifty thousand dollars would be any inducement, and we replied again, saying: 'Mr. Evans, you do not understand the situation at all, or you would not make these offers. Your proposition is an absolute impossibility, and even if there were no moral side to this question, from a business standpoint it would be suicidal for us to consider the proposition for a moment.'

"We explained to him that it cost three thousand dollars a day to print the *Post,* and his one hundred thousand, or one hundred and fifty thousand dollars, would only pay for getting out the *Post,* at most, from thirty to fifty days, and that after that we would have ruined our paper. The money that we sold out for would be gone, and we would be disgraced and bankrupt as well. He asked us then not to decide the question finally at that time, but to think it over, and we repeated that, under no consideration, could we accede to his wishes, and left him.

"A few weeks afterward, the Water Company applied for a new franchise. The terms were so unfair and atrocious as to sicken and discourage even the friends of the company. The *Post* immediately began as earnest and as vigorous a fight against this unfair franchise as it was possible to make. And from that day to this, Mr. Evans and his interests have tried to ruin the paper, disgrace Mr. Tammen and myself and send us to prison if possible.

"In July, following the defeat of the Water Company franchise, Mr. Evans bought the *Times* surreptitiously from Mr. Patterson, and Mr. Hugh O'Neill was taken over from the *Post* by Mr. Evans to run the paper for him, with the verbal agreement that Mr. O'Neill should have absolute charge of its editorial policy. Almost immediately Mr. Evans wanted Mr. O'Neill to begin warfare on Mr. Tammen and myself and the *Post,* but Mr. O'Neill, who knew us both intimately and the policy of the *Post,* told Mr. Evans that he was absolutely wrong in his estimate of the *Post* and its owners, and he would only hurt himself and the *Times* by such an attack.

"In January, 1912, Mr. Evans made it so uncomfortable for Mr. O'Neill that he resigned, and Mr. Speer [NOTE—A reform administration, embodying a commission form of government, had displaced Mayor Speer] was put in by Mr. Evans as the owner of the paper, preparatory to the late city campaign, and then started a campaign of personal abuse against the *Post* and its proprietors, that for coarseness, untruthfulness, vulgarity, willful and persistent criminal libel, has never been equaled in any civilized community, even by the most despicable sheets.

"Nothing was too indecent, absurd, foul and degrading to be laid at the doors of the *Post* and its owners. Conscious of our own decency and manhood, we remained silent (*sic*) with patience and forbearance for weeks under the unjustifiable, disgraceful and premeditated attack from Mr. Evans and his paper. But the attacks kept right on, and peace seeming absolutely impossible, in an article, 'So The People May Know,' printed May 12, the *Post* called attention to the kind of campaign Mr. Evans and Mr. Speer and the *Times* were waging, and suggested that the public was not interested in the opinion of the *Times* as to whether the *Post* or its owners and Mr. Arnold [Henry J. Arnold, reform candidate, and later Mayor] were saints or sinners, and asking that the campaign be kept free from personal abuse.

"This seemed to make Mr. Evans and the *Times* more vicious, indecent and foul-mouthed than ever, and they kept up the warfare with increasing bitterness, untruthfulness and libel, until we were forced for the protection of our property, our own good names and for the defense of our wives and children, to reply to Mr. Evans as editor and owner of the *Times;* and almost immediately upon the publishing of this open letter to Mr. Evans, both Mr. Tammen and myself were arrested for criminal libel, upon the affidavits of Mr. Evans and Mr. Speer, thus showing a premeditated plan of action with the criminal libel charge in view."

After a considerable further review of the case, including the failure of bailiffs to find Tammen at the depot as he "started on his vacation" for Honolulu, Bonfils concluded:

"The finding of this court is a small matter to me personally. I am but a unit in this great population, an unimportant cog in the great whirling wheels of this busy world. I ask no greater rights or privileges than the humblest man who has ever stood before this court, and I am entitled to no more consideration, but in whatever I have done, I have been guided by what I thought was for the best interests of all—the greatest good for the greatest number. I have tried in a humble way to smooth out the rough places for those that follow after, to make the world a little better for having passed this way, and, if through ignorance I have trangressed any of the laws of this state, I shall not plead *that* as an excuse, but stand ready with mental tranquillity and repose—unafraid and unashamed—to receive the judgment of this court."

As an example of Bon's tranquillity and repose, the day after Judge Shattuck had sentenced him, he began pawing up the earth in the manner of a bull in a Madrid arena. He wrote on July 27, 1912:

"Yesterday I was sentenced to sixty days in the common jail, and to pay a fine of five thousand dollars and one-half the expenses of the trial, and to stand committed to jail until the fine and the costs were paid. This is equivalent to a sentence in jail for two thousand, five hundred and sixty-five days, or exactly seven years and five days, allowing two dollars per day on the fine for remaining in jail; and this sentence was imposed on me for alleged constructive contempt of court by Judge Shattuck.

"In this contempt trial, I was denied by Judge Shattuck, first, the right of trial by jury; second, I was denied the right to show that the articles alleged to be in contempt were true. I wasn't allowed to introduce any evidence of any kind; third, Judge Shattuck decided to act as both judge and jury, and gave me the severest sentence ever given in a Colorado court for contempt, and there are only one or two more severe in the United States. The Judge failed to say in his sentence in what respect the newspaper article complained of was contemptuous. In an effort to add to my humiliation, at the close of the sentence, Judge Shattuck beckoned me to approach the court, and held out a little Bible, but I *declined to take any part in a sacrilegious burlesque of that kind.*

"And, friends, all of this happened in Denver, Colorado, in the twentieth century of the world civilization, and not in darkest Russia."

In a "So The People May Know" follow-up, Bonfils wrote in part:

"Mr. Bonfils and Mr. Tammen claim to be just ordinary, plain individuals, such as you will find by the thousands all over this country, having no greater ambitions, no greater claims, nor seeking greater rights than the average, law-abiding, peaceful man. And in this threatened destruction and imprisonment for fighting the people's rights, we remain unafraid and resolute.

"Two thousand years ago, did not the courts consent to the crucifixion of the Greatest Man who ever lived, because He tried to teach the doctrine of the Golden Rule?

"And didn't the authorities force Socrates to drink the deadly hemlock, because he taught the equality of mankind?

"And didn't Columbus, who gave a new world to mankind, languish in prison for debt?

"And wasn't the immortal Lincoln assassinated because he freed five million slaves?"

The article—a long, characteristic epistle to the Denver Ephesians by St. Bon'—closed with this walloper:

"The people will take away a power that despots of Russia would hesitate to assume, because, when the time comes that men shall be denied by our judges the right of trial by jury, the right of submitting evidence in justification of their conduct, Liberty becomes a farce and a jest, and the white silken gown of Justice has been torn from her, and she had become garbed in the scarlet robe of a harlot."

Harry Tammen returned to town with Hawaiian posies about his neck and a chip on his shoulder. He did not pause to ally himself with the Messiah, Socrates, Columbus or Lincoln. Nor did he describe the new sartorial robes of a honky-tonk Justice.

He marched into Judge Shattuck's court, listened for a time to the charges, then, when sentence was about to be passed on him for his part in the contempt, jumped to his feet, spun the attorney's stand out of his way, shook his fist at the Judge and bawled:

"Look here, Judge Shattuck, you know and I know that this is nothing but a cat and dog fight. You can put me in that jail for twenty years, if you want, but *I'll get you yet!*"

As Tammen sat down, amid a horrified silence, the Judge, his face white and his mouth open, stared for nearly thirty seconds. Then he said huskily:

"Case dismissed."

Nor did Bonfils go to jail after his rip-snorting partner had made this speech, in itself one of the most bare-faced acts of contempt on record.

24

With Creel In Darkest Africa

BONFILS WANTED to be known as the "friend of Presidents." He had met Theodore Roosevelt during Teddy's Rocky Mountain hunt for grizzly bears. And now, what with Roosevelt in Africa, bowling over harte-beests and gnus, Bon' suffered twinges of the wanderlust. He conceived the grand gesture of crossing the seas to renew presidential acquaintance beneath mosquito netting in the Dark Continent.

Once decided, Bonfils began to speak fluently of Kalulu cataracts, Ujiji jungles and Chumbiri lairs. For the time he forgot his knowledge of Colorado's four hundred and five species of birds, in favor of Congo buzzards and the herons of the Nile. Filled with enthusiasms for pith helmets and quinine, he displayed the qualities of a new Stanley en route to find a modern Livingstone.

He chose George Creel as traveling companion on the African jaunt. Creel was among the *Post's* better editorialists, an animated crusader, who afterward served President Woodrow Wilson as director of war-time publicity. Creel had found Denver a congenial place, both for personal happiness and the free exercise of his talents. He was preparing to wed that distinguished lady of the stage, Miss Blanche Bates, and was full of hearty ideals. He was an admirer of Ben B. Lindsey's Juvenile Court and participated with the Judge in drives against municipal corruption. He wrote with a militant, red-hot quill concerning labor wars and commercialized vice. Creel was a firebrand, a tireless, colorful worker.

Bonfils and his companion set out on friendly terms, but the publicist was not distinguished for his Cook's Tour charms. Anyone journeying with him had to bear down on the dimes. He grudgingly had provided Creel with an advance of one thousand dollars, half in cash and half in traveler's checks, for incidental expenses. Hardly had the train left Denver's Union Depot than Creel announced that he had lost—or been robbed of—the thousand-dollar ante.

This monetary setback was almost enough to make Bonfils leap through the Pullman window and give himself to the wheels. It certainly cast a shadow on his anticipation of gripping Teddy's virile hand in fellowship, of hearing once again those heartening cries of "Bully," and "De-e-ee-lighted."

Bonfils telegraphed his office to stop payment on the traveler's checks, and brooded all the way to Paris about the missing five hundred dollars in cash.

There were several other rifts during the trip, nearly all of which were inspired by Bonfils' devotion to the United States mint. While they were crossing France in a *wagon lits,* Bonfils neglected to provide an extra berth for Creel. The latter did not propose to sleep with his boss, come what might, and so protested.

A compromise was effected, whereby the Friend of Presidents bought an "upper" for his lieutenant. It was a hot night, the roadbed bumpy, and the *Post's* editorial ace felt his principles at stake. Below him Bonfils was enjoying the luxury of the better bed.

Creel simulated a nightmarish spasm of horrendous intensity. He leaped, with sinewy ferocity, on Bonfils' chest, bellowing in his ears—the shocked publicist struggling as though beset by tremens. Creel then apologized, explaining that he usually suffered such embarrassing fits if confined to an upper berth.

"Claustrophobia," Creel said. "I have to be where I can look out a window."

Shutting his eyes to expense, Bonfils procured a lower berth for his companion. It was an emergency.

The sojourners reached Cario, Egypt, there marking time till Roosevelt emerged from the jungle. Bonfils stood—in imitation of his celebrated ancestor, Napoleon—looking up at the Sphinx. He didn't think it comparable in size or beauty to the natural sculptures of the Garden of the Gods in Colorado. Nor could he stomach the claim that Pharaoh Cheops had worked a hundred thousand men during a period of twenty years in building the great pyramid. A sheer waste of money.

Word came that Roosevelt would powwow with Bonfils at Khartoum. Bonfils chartered a boat, after much haggling, and started up the Nile—not neglecting to assess Creel *half* the cost of the voyage. When Creel resisted this levy, Bonfils was so pained he could not enjoy the progress through the Valley of the Kings. And on a side-trip, land-excursion, Bonfils' lectures on earning and saving money echoed through the hypostyle halls of Karnak's mighty temple.

When Bonfils and Roosevelt finally met at Khartoum, no historians were present to record their converse, but it is said they spoke mainly concerning Mother Nature and her thoughtfulness in making so many things fit to be shot. Soon thereafter, Bon' and Creel turned homeward. Arrived at Denver, the only thing the publicist had brought as a souvenir of his pilgrimage was the word, "Boma." He had heard it while listening to Roosevelt's account of the hunt. Boma, Teddy had explained, was a species of thorny vine used by natives for building stockades.

"This Boma thorn," Bon' said, "is a mighty fine thing. It safeguards a man's property—his cattle and family. It keeps out the wild beasts."

Bonfils set up a holding concern for his business interests—and called it "The Boma Investment Company."

"The Boma Company," he said, "will keep out the human wild beasts."

Creel took up his editorial duties where he had left off, assailing Mayor Speer and the local machine. But his heart was not in his work. Bonfils kept asking concerning the money lost when they had started on the African journey. Finally Creel left the *Post,* moving his professional belongings to Welton Street, where he became flag officer of Senator Thomas M. Patterson's *Rocky Mountain News.*

Bonfils was infuriated at this defection. He applied the

terms "ingrate" and "unbalanced hot head" to Creel. He never forgave him; nor did he ever forgive anyone who quitted his service in a patently independent manner. An example of this vindictive quality of mind was his treatment of Hugh O'Neill.

O'Neill, a many-sided writer, had left Bonfils to become editor of Napoleon Evans' *Times.* When Bonfils *did* take O'Neill back to the fold, it was with an understanding that he reveal everything which had occurred in editorial conferences in the Evans' sanctum—which O'Neill proceeded to do.

Bonfils was paying O'Neill ninety dollars a week—a fancy salary for that time—and when O'Neill received an offer of thrice that amount to join Frank Munsey's magazine force, Bonfils pooh-poohed the matter. O'Neill, however, accepted the Munsey invitation, and Bonfils was inconsolable.

In the East, O'Neill refused to attack Theodore Roosevelt editorially, as requested by Munsey. His position became untenable. He telegraphed Bonfils, asking if he might return to his old job with the *Post.* Bonfils sent him an enthusiastic "yes, my son" wire, and Hugh came home.

When O'Neill received his next pay envelope, he found that, instead of a ninety-dollar emolument, he was being paid only thirty-five dollars a week! That was Bonfils' idea of punishment.

Bonfils practically ignored him when they met in the hall or the city room. An old cricket injury had impaired O'Neill's sight. Realizing he had but a few years of light left, he pressed himself to complete his law studies. This eye strain hastened blindness. When word came that he had passed the bar examination, O'Neill no longer could see. He tried bravely to represent such clients as sympathetic judges assigned him, but could not go on. He had courageously quit the *Post,* unwilling to stay there as an editorial millstone. His spirit was broken. He died in bitter poverty.

In Creel, Bonfils realized he had no dignified Hugh O'Neill to punish. He would have to wait until the fiery writer left himself open. Creel was a temperate man, of fine moral caliber, so Bonfils had no chance to attack him on that score. But Bonfils was an excellent waiter.

Proud of his new and tempestuous crusader, Senator Patterson allowed Creel great leeway. He encouraged Creel's continued attacks on Speer, a campaign for the abolition of commercialized vice, and a new city charter calling for commission form of government.

Henry J. Arnold, a former real-estate promoter and assessor in the Speer cabinet, was forced out of office, and now headed the reform ticket. Creel supported Arnold energetically. And when the commission government assumed power, Mayor Arnold appointed Creel to the important post as head of the Fire and Police Board.

In the reform campaign, the *Post* and the *News* had found themselves for once on the same side of the fence. Now that the *Post's* great enemy, Speer, had been dethroned, Bonfils had time to deal with lesser foes. He began to bait Commissioner Creel. The latter's zealous activities in cleaning up the city lent themselves to ridicule by the *Post*.

One of Creel's first orders deprived policemen of their guns and clubs. He cited that London was one of the best-patrolled cities in the world and pointed out that the "Bobbies" carried no weapons. When Denver thugs learned that their foes, the coppers, carried no persuaders, they began to kick up their heels and shellac a few of the less muscular flatties of the force. This, the anti-Creel critics declared, undermined the morale of the gendarmes.

The Commissioner then set out to deal with Denver's Red Light District—Market Street. When he took office, his survey revealed that approximately seven hundred prostitutes were sitting for company in the Market Street cribs and parlor houses. In conjunction with the Ministerial Alliance, Creel considered ways and means to discourage this trade.

Politicians and others who profited from crib-rentals and bribes resented Creel's assault on the vice stronghold of the city. However, he rounded up the lax ladies, compelled them to submit to medical examination and thus discouraged all but three hundred and fifty scarlet sisters from continuing publicly in business.

Creel planned to establish a "farm," on which the non-diseased survivors of the medical tests would be sequestered. No liquor or player pianos (those classic luxuries of the bawdy house) would be permitted. No minors would be allowed within the enclosure. This plan was devised to bridge a gap between an immediate closing down of the Market Street zone and ultimate city-wide suppression of the traffic. Due to official tardiness in preparing quarters for segregated women, Creel's love-farm plan fell through. Nevertheless, he prepared to seal all Tenderloin doors.

Creel estimated that twelve thousand Denver citizens were infected annually by "social diseases." He used these figures in arguing his case for abolition of licensed prostitution. Critics expressed a fear that the girls, once driven from their known addresses in Market Street, would scatter through the residential portion of the city. Others resisted Creel's plan, saying the district was needful because of "the physiological necessity of the male."

"The first argument," said Creel, "can be met by the sequestration of disease in hospitals, to prevent the spread of contagion—and the subsequent wiping out of a plague spot. The problem then will simply be one of law enforcement. The other argument is a lie. Ask any competent physician, and he will tell you that the 'physiological necessity of the male' is a fake, and has always been a fake. It is simply the case that we have allowed the absurd claim, perpetuated the double standard of morals, and refused to impose the same restraints upon boys that we impose upon girls."

Creel closed the Red Light District. It has never reopened since 1912.

Many famous landmarks passed out with the Creel order, among them the old Alacazar Burlesque Theater. Here was a robust show, with beefy chorus ladies and raucous comedians. Between acts (in fact, during them) the chorus ladies would rustle beer customers among an audience which sat at tables. Men waiters served the drinks.

It was here that Eddie Foy and Raymond Hitchcock lingered until four o'clock one morning. Hitchcock, a teetotaler, nevertheless found a certain enjoyment in watching others sip beer. Foy, who liked beer, finally fell asleep at the table.

The waiter was going off duty, but didn't want to miss a tip. So he touched Foy on the shoulder, waking him.

"Is there anything else I can do for you, sir?" the waiter asked.

Foy looked at him wonderingly. "No. There's nothing else on earth you can do for me. If you could, you wouldn't be a waiter."

In criticizing Creel, the *Post* referred to the money advanced him by Bonfils at the start of the invasion of Africa. Creel read into the article an inference that he had "stolen" it. When he denied that he had, the *Post* had this to say:

"As for the thousand dollars which Creel says he has been charged with stealing from the *Post*, the *Post* has never made any such accusation. When Creel was on the staff of the *Post*, Mr. Bonfils took him along on his trip to meet Roosevelt at Khartoum, after the African hunting expedition. To pay Creel's personal expenses on that trip, the *Post* gave him one thousand dollars. Just before the train left Denver, Creel said that he had been robbed of either all of the one thousand dollars or of five hundred dollars. The money was never recovered, and the *Post* forgot all about it."

Creel answered in the *News* of February 3, 1913, saying in part:

"What could be more delicious? Note the delicacy with which it is intimated that I lied when I said that I had been 'robbed of either all of the one thousand dollars, or of five hundred dollars.'

"Mark the *Post's* royal disregard of money, insinuated in that noble phrase: 'The money was never returned, and the *Post* forgot all about it.'

"What is one thousand dollars to the *Post?* A bagatelle! a trifle! Does it not offer million-dollar rewards for the first humming bird that flies over Pike's Peak? for the first trip to Mars? and for the discovery of the man that struck Billy Patterson?

"What is one thousand dollars to a man who was a barkeeper before the invention of cash registers, or to one who was a pioneer in the lottery business? Laughable! Excruciatingly ridiculous! Why, they cannot remember whether the amount was one thousand dollars or five hundred dollars. It was never recovered, and they forgot all about it.

"There is an impudence so sublime that it is almost enchanting! Who does not know that money is the heart's blood of these men? *Why, until every cent of the lost money was repaid them, they never knew rest, nor did I know peace.*

"I was writing for the *Denver Post* when Bonfils determined to make the African trip, and he regularly assigned me as an employee to go on the trip with him as a writer, to keep him well in the public eye and to write for the *Post*. The one thousand dollars which he advanced for expense money was half in bills and half in traveler's checks. Bonfils stopped payment on the checks, and he had that five hundred dollars returned to him in Paris.

"I have often wished that a stenographer could have been present at that final settlement after my return. In addition to the five hundred dollars that I had lost, a certain portion of my expense account was disallowed, my agreed salary was cut materially, on the theory that I should pay for my own meals while out on an assignment, and Bonfils deliberately assessed me one hundred dollars as my share of a steamer that he chartered to make a trip up the Nile.

"Having thus safely and securely brought me into their debt, they welshed on a written agreement to raise my salary. Their books will show that from May, 1910, to midsummer, 1910, I drudged for them, paying off to the last cent this manufactured indebtedness, week by week.

"This is how 'the money never was recovered,' and how 'the *Post* forgot all about it.' "

Mayor Arnold, listening to anti-Creel counsels, sought a painless way in which to oust his police commissioner. When Fire Commissioner Thomas F. McGrew and Creel quarreled, Arnold requested that both men resign. McGrew complied with the request. Creel refused to get out. Arnold thereupon suspended him on charges of rather vague and flimsy foundation. A list of the specific accusations amounted, in sum total, to allegations that Creel had caused dissension in the ranks of the department of safety, and that he had sworn at his colleague, McGrew, accusing the latter of bending the elbow.

Under the law, the Mayor was sole judge of the sufficiency of a subordinate's defense. Consequently, when Creel had a hearing in the Mayor's own office, the proceedings were so cut and dried as to seem farcical.

The Mayor permitted the trial to last for an hour, then, in a rather self-conscious way sustained his own kangaroo charges and dismissed Creel on February 16, 1913.

The hearing marked the last important public appearance of the aging Senator Patterson. He made a speech in behalf of Creel, to which the Mayor listened with the air of a bored husband.

Evans had turned his *Times* back to Patterson, but the old gentleman was tiring of the grind. And on October 23, 1913, he sold the *News* and the *Times* to John C. Shaffer of Chicago. Shaffer also bought, in a tri-party deal, the *Denver Republican.* He merged the latter with the *News* and then—of which, more anon—began to receive torrential abuse from the Post, such as was hard for a gentleman from Chicago, a patron of the opera, to understand.

25

The Unsinkable Mrs. Brown

MRS. MARGARET TOBIN BROWN encountered the hoots of her Western sisters. But she hoisted herself by the bootstraps of heroism into huge *Denver Post* headlines.

Molly Brown was as naïvely colorful as she was brave. She mistook her own enormous zest for a symptom of artistic ability, her ingenuous thirst for human relationship as evidence of social grace. She was received abroad by titled big-wigs because of her lack of worm-eaten sophistication. That selfsame lack barred her from the portals of a Denver society that was as hidebound as it was provincial.

This vital Amazon lived a novel of Eulenspiegel dimensions. Her father was old Shaemus Tobin. Molly liked to fancy her sire an Irish peer, but he was in fact a tin-roof Celt of the Missouri River bottoms. Old Shaemus was a man more ready of song than of cash, red-haired and tempestuous.

A cyclone occasioned Molly's birth two months before the laws of nature warranted such an event. The mother, father and two sons had skurried into a cellar while the twister tucked their shanty under its arm and raced like a monstrous half-back over a gigantic field.

Old Shaemus fashioned a crude incubator for the seven-months baby, then collected a new supply of scantlings and tin cans for another shanty. The mother died and Shaemus borrowed a goat as Molly's wet nurse.

Molly's premature arrival on earth was in key with her aggressive temperament, but the frailty of the tiny infant in no way augured a maturity of power and red-headed vigor. She grew up in the river bottoms near Hannibal, Missouri, hated house work—particularly that of a shanty—and spent all her days hunting in winter and fishing in summer.

When she was twelve years old, Molly became acquainted with Mark Twain. Mr. Clemens, too, had been fishing. He at once saw her for what she was, a female *Huckleberry Finn.* He admired her flaming red pigtails, her almost fierce blue eyes, and invited her to fish from his rowboat. She delightedly gave her up home-made raft to angle from the bow of the author's punt.

Mr. Clemens found that Molly didn't have the most remote idea that she was a girl. She could whistle like a calliope, and before Mr. Clemens could gather his celebrated wits together, she had disrobed completely and dived overboard, with an absence of mock modesty that characterized her entire life. She engaged in porpoise-like maneuvers, laughing and shouting and blowing water, but came to grief. Her head got stuck in the mud, and Mr. Clemens pulled her out, half drowned.

She looked like some weird clay model as he began scraping mud from her eyes. He helped her on with her garments, and from that day, Mark Twain was Molly's god.

When Molly was fifteen, she concluded that the shanties of Hannibal held no promise of adventure. She and her brothers packed a single carpet bag and ran away from home. They traveled by stagecoach to Colorado, arriving in the gold camp of Leadville.

She did not know how to cook—nor did she wish to learn that art—but went to work as a "pot-walloper" in the cabins of miners. She washed their dishes, rearranged the bedding on their bunks, and sometimes acted as nurse for sour-dough prospectors. She and her brothers pitched a discarded tent at the end of State Street, a noisy avenue of honkeytonks, saloons with long bars and gambling hells.

The rigors of the mining camp only strengthened the body and courage of this illiterate hoyden. Three weeks after her arrival she met and married John J. Brown, called

"Leadville Johnny" by intimates at the Saddle Rock Saloon in Harrison Street.

Leadville Johnny was thirty-seven years old, as homely as a hippopotamus—although not so fat—unlettered, open-fisted, and had red hair. He seldom was in funds, but when luck infrequently came his way was foremost among the belly-up-to-the-bar boys. Homely or not, he had a way with the dance-hall girls.

In less than two months after his marriage to fifteen-year-old Molly, Leadville Johnny struck pay dirt. He was offered three hundred thousand dollars *cash* for his claim. He accepted, imposing but one condition.

"Pay me off in thousand-dollar bills," he said. "I want to take it home and toss it into the lap of the prettiest gal in this here camp."

He came bellowing into the cabin, did a bear dance with his young wife, then gave her the money, all of it. He found it necessary to explain at length just how much money three hundred thousand dollars was—a genuine fortune! Her mind did not go beyond a silver dollar at most.

"I wanted you to see it; to hold it," he said. "That's why I didn't put it in a safe. But you got to hide it, even if it *is* all yours."

"Where?" asked Molly.

"You figure that out, honey. It's yours. I'm goin' down to celebrate at the Saddle Rock."

He kissed her and was gone to receive the back-slappings of Saddle Rock pals. In an hour he had forgotten that he was a rich man; he was having such a good time of it. He stayed at the saloon until early morning and was brought home by two of his intimates. He was sober enough to make two requests. One was that the "boys" would not disturb his pretty young wife; the other that they fetch some kindling and start a fire.

"I'm freezin plumb to death," said Leadville Johnny.

The boys put him on a bunk, then made a fire. Molly rousing from deep sleep, had an uneasy feeling. She sniffed as the new fire sent wisps of smoke through crevices of the stove. She felt the mounting heat. Then she screamed. She got up, while her husband's pals retreated hastily from the cabin. She scorched her fingers on the stove lids. She couldn't find a lifter and used a steel-pronged fork instead. She almost set herself and the cabin on fire. She delved among the burning sticks, but it was too late. Of all places, she had hidden the money in the stove, and now her fortune had gone up the flue, three hundred thousand dollars floating in the Leadville morning sky.

Johnny rallied somewhat and announced that he was freezing to death. Then he wanted to know if his wife was freezing, too. If so, she should come sleep beside him. For half an hour she wept, yammered and howled in his ear. When it did penetrate his haze that the money had been burned, he sat up and said:

"Don't you worry a bit, honey, I'll get more. Lots more." Then he reiterated the fact—or fancy—that he was freezing plumb to death.

Molly began to shower kisses on Leadville Johnny's red head, his face and lips, it appears that she had not been screaming and wailing because of the lost fortune, but from fear that her husband would be angry.

When Johnny sobered up next morning, he actually *laughed* about the loss. "It just goes to show how much I think of you," he said. "There's plenty more."

"Lots of men would be mad," she said.

Leadville Johnny slapped his chest grandly. "Mad? I'll show you how mad I am. As soon as I get a drink into me, I'll go right out and get a bigger and better claim. Where'd you put that bottle, honey?"

Fantastic as it may seem, Leadville Johnny went out that very afternoon and located "The Little Johnny," one of the greatest producers of gold in Colorado history. It is estimated that he took twenty million dollars from this bonanza.

"Nope," he said to the men who had bought his other property, "I won't sell this one."

"There's another three hundred thousand if you do," his bidders said.

"Nope, let's have a drink instead."

"Why won't you sell?"

He slapped his chest. "I don't trust chimneys. It's safer in the ground."

The meaning of money began to dawn on Molly. It was the commencement, critics said, of her progress from Leadville to lorgnettes. The Browns moved "up the hill," where mine owners and bankers had mansions. Leadville Johnny went the limit in building a house for his bride. As a climactic touch, he laid concrete floors in every room of the house, and embedded silver dollars, edge to edge, in the cement surfaces!

Leadville now was not big enough to hold Molly. She had heard of Denver society, of the gay balls and salons.

"Denver it is, then," said Johnny. "Just name the thing you want, and Big Johnny (slapping his chest) and Little Johnny (pointing in the direction of his claim) will get it for you."

The Browns built a mansion in Pennsylvania Avenue, Denver's Capitol Hill, where the *élite* resided. Leadville Johnny contemplated paving this place with *gold pieces,* but was dissuaded. He compromised by having two huge lions made by a cemetery sculptor. The lions were placed flanking the doorway.

The new mansion was a "show place", where rubberneck—"Seeing Denver"—buses paused and tourists stared while a spieler narrated the drama of the Little Johnny. Inside its stone halls, conniving spongers and fake grand dukes partook of the Brown bounty. But so inexhaustible were the Little Johnny's veins that the attacks of these leeches were hardly felt.

The town's preening dowagers would have none of this redheaded upstart from the hills. Not one of them—their own husbands but once removed from the pick-handle and the stope—was kind enough to advise Molly in her social adolescence. Still in her teens, unschooled and impetuous, how was she to know the emptiness of display?

She hired the largest orchestras, gave the costliest balls, drove the finest horses, but met with snobbery. She often attended, uninvited, the social functions of her neighbors. Indeed, she became such a nuisance as a "gate crasher" that the ladies decided to crush her.

As part of a cat-like hoax, Molly was solicited to write a dissertation on Denver society. This she did, laboring at a desk inlaid with gold from the Little Johnny shaft. Her husband admitted his inability to judge literary works, but said he guessed she knew what she was doing.

"As for me," said Leadville Johnny, "I'd rather be back this minute at the Saddle Rock."

Molly's "article" appeared in a magazine owned and edited by Polly Pry. The effort was published, word for word, as written by Mrs. J. J. Brown. She was very proud of it until the whole of the city's upper crust began heaving with merriment. The new author's misspellings, fantastic verbiage and artless philosophies were there for all to see.

At last conscious of her ignorance, and shamed by her social shortcomings, Molly left town. Johnny said he guessed he'd stay home.

"I never knowed how to spell and never claimed to," he said, "and as far as society is concerned, I ain't aimin' that low. Good-bye, honey, and don't forget the name of our bank. It's all yours."

Denver saw nothing of Mrs. Brown for nearly eight years, and heard little. It was something of a sensation, then, when she returned to the city, gowned in Parisian creations. More, the word spread that Molly had two French maids, with whom she conversed fluently in their native language. Indeed, during seven and a half years in European capitals, she had become proficient in five languages—she who had left town unable to spell in English!

There were other incredible surprises for the hometowners. Molly had made friends with the Divine Sarah Bernhardt, had received stage lessons, and even contemplated playing the Bernhardt rôle in *L'Aiglon*. She had received instruction in painting and singing and had appeared with some success in a charity concert in London and had sung aboard an ocean liner on the voyage from Southampton to New York City.

The hardest blow to her critics, however, was the fact that celebrities and titled foreigners made the Brown home their headquarters while visiting Denver.

But despite her education in the polite arts, Molly Brown's real nature was manifest at all times. She permitted herself the luxury of forthright speech, and, if in the mood, used slang and cursed like a pit boss. Her detractors, still unable to stomach her social ambitions, described her as "eccentric."

"Sure I'm eccentric," she said. "But I have a heart as big as a ham."

When Leadville Johnny refused to "gad about" in Europe and elsewhere, they separated. But he never shut her off from his great purse. He still loved and wanted her to have a good time. All he desired for himself was privacy and the privilege of sitting with his shoes off in the parlor.

Mrs. Brown acquired a seventy-room house and estate near New York City. She entertained the Astors and other Eastern notables—all of which agonized her Denver scoffers.

In April of 1912, the home town which had refused flatly to receive Molly as a social equal, passionately acclaimed her as its very own celebrity. The *S.S. Titanic* had gone down, and Molly had been its heroine.

Suddenly her virtues were sung in nearly every paragraph of a front-page layout in the *Post*. She became known as "The Unsinkable Mrs. Brown." The New York press called her "The Lady Margaret of the *Titanic*."

Now that Mrs. Brown had received the accolade in alien fields, her townsmen's praises resounded like songs in a beer stube.

The tardy cheers for Mrs. Brown were in keeping with the psychology of the provinces. Similarly, Eugene Field had been tolerated as an amiable prankster, a thistle-down jingler and something of a sot during his Denver interlude. Then, his fame having been certified abroad, and death having corroborated his genius, Denver was the first of cities to rear a monument to his memory.

Perhaps it was an instinctive feeling for another free and generous soul that led Mrs. Brown to purchase Field's old Denver home and set it aside, a shrine for children.

Mrs. Brown was thirty-nine years old when she left Liverpool for New York on the *Titanic's* maiden voyage. Instead of a girlish slimness, she now was ruggedly and generously fleshed. Nevertheless, she still bubbled with a seldom-varying vitality.

She sang in the ship's concert and was popular with the traveling notables despite her growing eccentricities. She amused some and terrified others with pistol-feats, one of which consisted of tossing five oranges or grapefruits over the rail and puncturing each one before it reached the surface of the sea.

Although she spent great sums on clothes, she no longer paid attention to their detail or how she wore them. And, when she traveled, comfort, and not a desire to appear *chic*, was her primary consideration.

So, when Molly decided to take a few turns of the deck before retiring, she came from her cabin prepared for battle with the night sea air. She had on extra-heavy woolies, with bloomers bought in Switzerland (her favorite kind), two jersey petticoats, a plaid cashmere dress down to the heels of her English calfskin boots, a sportsman's cap, tied on with a woolen scarf, knotted in toothache style beneath her chin, golf stockings presented by a seventy-year-old admirer, the Duke Charlot of France, a muff of Russian sables, in which she absent-mindedly had left her Colt's automatic pistol—and over these frost-defying garments she wore a sixty-thousand-dollar chinchilla opera cloak!

If anyone was prepared for Arctic gales, Mrs. Brown was that person. She was not, however, prepared for a collision with an iceberg.

In fact, she was on the point of sending a deck steward below with her cumbersome pistol when the crash came.

In the history of that tragedy, her name appears as one who knew no fear. She did much to calm the women and children. Perhaps she was overzealous, for it is recorded that she refused to enter a lifeboat until all other women and their young ones had been cared for, and that crew members literally had to throw her into a boat.

Once in the boat, however, she didn't wait for approval—she seized command. There were only five men aboard, and about twenty women and children.

"Start rowing," she told the men, "and head the bow into the sea."

Keeping an eye on the rowers, she began removing her clothes. Her chinchilla coat she treated as though it were a blanket worth a few dollars. She used it to cover three small and shivering children. One by one she divested herself of heroic woolens. She "rationed" her garments to the women who were the oldest or most frail. It was said she presented a fantastic sight in the light of flares, half standing among the terrified passengers, stripped down to her corset, the beloved Swiss bloomers, the Duke of Charlot's golf stockings and her stout shoes.

One of the rowers seemed on the verge of collapse. "My heart," he said.

"God damn your heart!" said The Unsinkable Mrs. Brown. "Work those oars."

She herself now took an oar and began to row. She chose a position in the bow, where she could watch her crew. Her pistol was lashed to her waist with a rope.

The heart-troubled rower now gasped and almost lost his oar. "My heart," he said. "It's getting worse!"

The Unsinkable one roared: "Keep rowing or I'll blow your guts out and throw you overboard! Take your choice."

The man—who really *did* have a fatty condition of the heart—kept rowing. Mrs. Brown sprouted big blisters on her hands. But she didn't quit. Then her palms began to bleed. She cut strips from her Swiss bloomers and taped her hands. She kept rowing. And swearing.

At times, when the morale of her passengers was at its lowest, she would sing.

"The God damned critics say I can't sing," she howled. "Well, just listen to this . . ."

And she sang from various operas.

"We'll have an Italian opera now," she said at one time. "Just let anyone say it's no good."

She kept rowing.

And so did the others. They knew she *would* throw anyone overboard who dared quit, exhaustion or no exhaustion.

She told stories. She gave a history of the Little Johnny. She told of the time she hid three hundred thousand dollars in a camp stove, and how it went up the flue.

"How much is three hundred thousand dollars?" she asked. "I'll tell you. It's nothing. Some of you people—the guy here with the heart trouble that I'm curing with oars—are rich. I'm rich. What in hell of it? What are your riches or mine doing for us this minute? And you can't wear the Social Register for water wings, can you? Keep rowing, you sons of bitches, or I'll toss you all overboard!"

When they were picked up at sea, and everyone was praising Mrs. Brown, she was asked:

"How did you manage it?"

"Just typical Brown luck," she replied. "I'm unsinkable."

And ever afterward she was known as "The Unsinkable Mrs. Brown."

Perhaps because it is the thing most lacking, heroism lifts anyone above caste. *Still,* the Denver social tabbies would not admit Mrs. Brown to their select functions. But now she no longer cared. She went in for thrills.

She took world tours and explored far places, always meeting adventure half way. Once she almost perished in a monsoon in the China seas. At another time she was in a hotel fire in Florida. But the Unsinkable one was Unburnable as well. She rescued four women and three children from that fire.

In France she was given a Legion of Honor ribbon, with the rank of chevalier, in recognition of her charities in general and her work in establishing a museum for the relics of Sarah Bernhardt in particular.

She now was legally separated from old Leadville Johnny. But still he had not tied the purse strings. Molly could go where she wanted and do what she wanted. It was his way. As for him, he stayed in the parlor with his shoes off, or bent the elbow a bit with old-time pals. The Little Johnny continued to pour out gold as from a cornucopia.

Although her husband was a mine owner, Mrs. Brown always took the side of labor, and sent food, clothing and money to the families of strikers.

During the World War she contributed heavily for the welfare of soldiers and for the hospitalization of wounded warriors of the Allied arms. If she had been hooted by a handful of social snobs in her home town, she now received the prayers of thousands of soldiers. The Allied nations awarded her all the medals it was possible for a civilian woman to receive. She was recipient of personal congratulations and the thanks of kings and princes.

After the war she took another of her world tours. When reporters met her in New York, she said:

"I'm getting to be more of a lady every day. In Honolulu I learned to play the uke. In Siam I mastered the native dances. In Switzerland I learned how to yodel. Want to hear me?"

And she astonished the customs guards by breaking into Alpine melody.

Rumors were circulated that the aged Duke of Charlot was planning to marry her—old Leadville Johnny having died in his stocking feet—and Mrs. Brown confirmed the report. Forty-eight hours later she declared the romance ended.

"Me marry *that* old geezer?" she said. "Never! Give me every time the rugged men of the West. The men of Europe—why, in France they're only perfumed and unbathed gallants; in England, only brandy-soaked British gents. Pooh! Pooh! And a bottle of rum."

In keeping with his character, Leadville Johnny, a multi-millionaire, *left no will*. There was an unpretty fight now. The Unsinkable Mrs. Brown was left floating with little financial ballast. Her eccentricities were cited; her charities construed as loose business affairs. She was awarded the life-income of one hundred thousand dollars annually.

"Just to think," she said with a gay smile, "and I burned up three times that much in one bonfire."

Mrs. Margaret Tobin Brown died in October, 1932. Apoplexy was the cause. She had been singing in her town apartment at the Barbizon Club, in East Sixty-third Street, New York City, then became dizzy and faint.

She was buried at Hempstead, Long Island, in surroundings that she loved almost as well as she had loved her Colorado hills.

26

Gavotte Of The Peeping Toms

THE CORPORATION-LOGGED *Republican,* in a last bid to keep afloat, hired Josiah M. Ward away from the *Post* as general manager and editor-in-chief. He didn't want to change journalistic horses, but when Bonfils refused him a salary boost, Josiah huffed out of the "Temple of Justice," squealing that he never would return. He bought himself a new pair of scissors, and on April 20, 1912, the *Republican* announced:

"By securing the services of Mr. Ward, the *Republican* is but carrying out its well-known policy of having the brainiest, most capable men and women possible to secure, for the handling of live news of the city and the state, the nation and the world.

"Because of Mr. Ward's long experience and training, his energy and initiative and originality, the owners and publishers of the *Republican* anticipate that he will clinch even more strongly than at present the *Republican's* standing as the leading newspaper between St. Louis and the Pacific Coast. . . . New departures are contemplated, both as to style and features for the home-folk."

The announcement contained a hint, however, that Ward would not be permitted too free a rein, saying:

"The *Republican's* business policy is constructive, not destructive; its news policy will be bright and snappy, but not 'yellow'."

Despite his high-powered agility as a news hawk, Josiah M. Ward never could function brilliantly under censorship.

He needed elbow room. Suggestions of purity and high resolve cramped his style.

He gathered together a brilliant staff and immediately began to pump life into the prudish columns of the ancient journal. Its owners grew uneasy. Josiah was a man of spice, and although circulation rose, the publishers winced when Ward began serving his favorite dishes—juicy scandals and peppery murder yarns. It was as though the chef at the old Waldorf suddenly had introduced an aphrodisiac into the spinach of a staid and circumspect dowager, impelling her to caper against her will with the elevator boy.

Ward brought no red headlines to the venerable *Republican,* but he did inject a sparkling Burgundy into the veins of the paper, with here and there a suggestion of pornography—photographs of feminine slayers, with heavy accent on knees, calves and ankles.

Josiah was immediately unhappy in his new offices; the prison-gray stones of the building seemed to weigh upon his soul. Constant criticism from higher up bothered him. The miserable equipment, typewriters which should have been in the Smithsonian Institution, presses that suggested the Inquisition's torture chamber, a shortage of typefaces, decrepit linotype machines that spilled molten lead to an ancient floor, which in turn was so full of cracks that it relayed the metal flux to the city room below—all these things sent Josiah into retreat at Tortoni's elegant bar, to bolster his soul with whiskey. Sometimes he would hide out at the Navarre, a gaming house of storied age—not to gamble, but to sit in the tall, leather-backed stalls of the dining salon, sipping his drinks.

In the old Navarre, Josiah found memories of departed grandeur. There were dim oil paintings, most of which had been scarred by roisterers of the nineties. There was the masterpiece, "A Parlor Match"—a pioneer had thrown a bottle through its canvas. It had been stitched up noticeably, like an early-day appendectomy incision. Another noted painting, "The Queen of the Harem," had been vandalized—some prankish patron had put a lighted cigarette through the painted mouth of the nude "Queen." Also looking down on Josiah, as he brooded over past triumphs, was a framed menu, a bill-of-fare which had greeted hungry stage-coach travelers, and on its cover a picture of an Indian squaw with a papoose feeding at the breast, and the caption: "Hot meals served at all hours."

"The world is going to the dogs," said Josiah. "But I can't bark with the rest of the pack."

When he complained about the rickety plant, Josiah received promises from the multi-millionaire, Crawford Hill, principal owner of the *Republican*, that a complete renovation would be ordered, beginning with the press room. The leading printing expert of Chicago was summoned to make a survey of the antique Hoe presses, two spavined hulks that crouched like sick dinosaurs in a filthy basement.

The expert arrived, spent two days in stewing over the ancient steel, then came upstairs, a deflated vacancy in his eye. He sat down in Josiah's grimy office.

"I'm leaving for Chicago tonight," he said.

"How's that?" asked Josiah.

"I'm sixty years old," the expert said. "I've spent forty years, man and boy, on all kinds of presses."

"Well?"

"And I'm leaving you, flat. These damned presses are *before my time!*"

On the night of October 23, 1913, the *Republican* staff had put the mail edition to bed. There was a breathing spell until time to assemble the home edition, the one designed for local subscribers. Josiah had gone to Tortoni's to swig a nightcap. His cabby waited outside with a blowy chestnut horse to take Joe home, once the editor had received and glanced through the final edition, which ritualistically was delivered to him at Tortoni's.

The staff, meanwhile, as was its nightly custom, waited for a report from an office boy, who traditionally went at this hour to the roof. He would stay there until able to announce "something of interest" to the world-weary editors and their gallant reporters.

The *Republican*—luckily—was situated next to a rooming house, in which lassies, expelled from Market Street by the Creel anti-vice order, engaged furtively in their profession. Men and vice crusades being what they are, there were many such rooming houses in existence.

The tar-and-pebble roof of the *Republican* building was accessible solely by way of a stationary wooden ladder leading to a hatchway. Forty feet to the south of this small hatch, there rose a two-foot fire-wall, a stone rampart separating the *Republican* roof from that of the adjoining rooming house. On the rooming-house roof was a shallow pyramid of glass, a skylight canopy over the boudoir of "Cincinnati" Louise, a brunette of mercenary practices.

Although none of the staff ever had been introduced formally to "Cincy," they felt that they knew her well—and, in fact, they did. In a manner of speaking, they had watched over her night after night. All her secrets were theirs, including the time an assistant district attorney gave her a black eye when she happened to mention money.

On the eventful night of October 23, 1913, an evening in which frost had gathered on the pumpkins, Marty, the office boy, descended hurriedly from his vigil on the roof. He reported good luck. The scriveners postponed their visit to the saloon across the alley and advanced in a body to the ladder. Long practice and strict observance of seniority rights allowed a speedy, silent and efficient ascent to the tar-and-pebble roof.

Once on the building's crest, these amiable fellows removed their shoes and stalked with the slinking caution of Utes toward the firewall, across it and to the illuminated skylight of Cincy's boudoir. Beneath the panes of this glass canopy, Cincy was entertaining a burly gentleman known to the staff as "Stuttering" Clem, a soldier of fortune, a veteran of several Mexican revolutions. Clem's holster and Bisby model revolver could be seen on a chair, if one cared for such details.

Pressure on the skylight was so extreme that an editor—a man noted in newspaper circles for excellence in spelling names of foreign towns—rammed his knee through a pane of glass. The cave-in was a signal for immediate retreat. The editor, his leg a fountain of gore, extricated it from the jaws of the trap, and curiously enough outdistanced all his colleagues to the hatchway haven.

Alas! this editor, so skillful in spelling, was awkward at descending ladders. He had a somewhat obese middle, and became wedged in the aperture. His underlings, disregarding his dignity and power, unmindful of his repute as a self-made atlas, began kicking his shoulders to get him out of their way. Fortunately their feet were shoeless; nor did anyone take time to hunt for boots. It was every man for himself, and as the fugitives converged on the all-too-narrow hatchway, the congestion resembled that of Times Square subway kiosk during rush hour.

The routed troops had difficulty in getting their wounded editor's belly loose from its vise. The moon came out to spread an accusing glow, and the great speller looked as though he were wearing the roof for a huge skirt. Then another misfortune befell. The burly soldier of fortune, having shaken the glass shards from his astounded back, had put a settee on the bed, then had mounted the settee to thrust a brawny arm through the frame where the skylight pane had been. And in Stuttering Clem's hand was the Bisby model revolver. Bullets began to plop against the fire-wall.

The fugitives escaped injury, however, finally prying their editor from the hatch, as the soldier of fortune descended from the firing tower to get a fresh supply of cartridges. They did not dare wait to recover their shoes, but descended and re-formed their ranks at the base of the ladder. It was time to resume the more prosaic duties of assembling the home edition. Pale and winded, they started through the hall and toward the city room. And, as they entered that sanctum, a new menace confronted them.

Standing there, beside the forsaken city desk, was the chief owner of the *Republican*, Mr. Crawford Hill, and on his right a bald, ministerial gentleman in a black, swallow-tail coat. Mr. Hill rarely visited his property. He belonged to, and frequented, the exclusive clubs, leaving the *Republican* to stagger along as best it might. His advent to the city room, therefore, was of ominous import.

There was a wondering expression on Mr. Hill's dignified, handsome face. He twiddled his sandy moustache as though to ask why the office had been depopulated, and why these pale and shoeless men now were coming in like refugees. Mr. Hill looked fixedly at the array of sox. He seemed astonished at the plight of the bleeding editor. The latter explained:

"We thought we heard a cry for help and went to investigate."

Whatever his doubts, Mr. Hill had other business of moment. He gestured in an almost tired way to his ministerial companion. "Gentlemen, this is Mr. John C. Shaffer, owner of the *Chicago Evening Post.* Mr. Shaffer today purchased the *Rocky Mountain News* and the *Denver Times* from Senator Patterson. He also, and at the same time, purchased the *Denver Republican* from me and my colleagues. Gentlemen, Mr. Shaffer is combining the *Republican* with the *News.* The *Republican* is suspending publication. Good night."

With a parting glance at shoeless feet, Mr. Hill handed a typewritten announcement to the bloody best speller of foreign-town names. Mr. Hill's statement appeared in the home edition, proclaiming the sale and suspension of the *Republican.*

Despite all dangers from gunfire, the men went to the roof to reclaim their shoes. Then they got out the last edition—indeed the last one for the old *Republican.* In keeping with tradition, these newspaper men gave their final performance to the best of their ability. The lacerated executive was just as careful with his orthography in editing a cable report involving a riot in an obscure Russian town as he ever had been.

Then the doleful castoffs retired to the Press Club. It had been an evening which Fate had gone the limit. They had been shot at while indulging in a little game of I-spy, had regained safety only to find themselves shoeless, and, facing an owner who probably never in his coddled life had enjoyed the thrills of roof-gaping, had found themselves without jobs. They drank until morning, asking themselves that old, old question:

"Now, why in hell did I ever go into the newspaper business?"

John C. Shaffer invaded the city with a vast bankroll, a love for tried and true hymns, and a silk hat. He was publisher of the *Chicago Evening Post*. He also owned a sprinkling of lesser journals in the State of Indiana. He called his properties "The Shaffer Group" and dreamed of establishing a journalistic chain which would make of him a perfumed Hearst.

Shaffer felt that the people of Denver would support a new champion if there were money and courage enough to fight the *Post*. He surveyed his field correctly to a certain degree. Probably no man till then had had Shaffer's opportunity for competing solidly with the Champa Street Lion.

Thousands of citizens cowered beneath the *Post's* despotism, too timid to risk their fortunes by incurring its hatred. These bewildered persons were secretly willing to give allegiance to a new champion, were that champion interesting, capable and steadfast.

The town was now wearying of *Post* politics. The citizens too late saw the folly of reform government, which, in essence, had resulted from the *Post's* unrelenting war with Bob Speer. The old machine system, with all its graft and nepotism, was preferable to the wasteful, ineffectual shadow boxing of the newcomers. Speer at least had built boulevards, established parks, playgrounds, and reared public structures. The new regime had erected but one edifice of an ornamental or useful sort—a little wooden pagoda at the corner near the Capitol, where a dozen persons might sit while waiting for street cars.

The *News*, it is true, had participated with the *Post* in bringing to life the reform administration. But the *Post* had been the chief factor in its conception and birth. It controlled officials, as witness the ousting by Mayor Arnold of Creel, the *News'* adopted son.

Shaffer's survey indicated that a purchase of the two morning newspapers, their subsequent merger, and the introduction of new and more virile policies would find favor with subscribers. This he undertook to do, with a forceful and brilliant managing editor, William Forman, former sports editor of Shaffer's *Chicago Evening Post*, and

William L. Chenery, spirited but sane editorial writer. Chenery afterward became editor of *Collier's Weekly*.

Shaffer's business manager was Dave Towne, who subsequently acted as Hearst's general treasurer.

The invader's crew was effective, his finances unlimited and his field enthusiastically hospitable. Furthermore, he was known as a patron of the arts, a term that used to fill provincial hearers with hay-shakers' awe. Shaffer contributed to the support of the Chicago Grand Opera Company, and when he announced that that melodious body would appear for one week at the Denver Auditorium, the city felt that it had acquired quite a personage in this Medici from the Gold Coast.

Mary Garden, Ruffo, Tetrazzini and the baton-wielding Campanini gave the ermine-wearers of the town a grand chance to walk the dog up operatic aisles. Not only did Mr. Shaffer—now dubbed John "Clean" Shaffer by the grumbling *Post*—bring classic arias to the Queen City of the West, but helped promote a sprawling tabernacle for Evangelist Billy Sunday. That athletic soul-saver romped into town, howled that hell's asbestos gates were ajar, and bespoke the evils of liquor so loudly that Denver soon thereafter went bone dry. Shaffer could be seen and heard in the front row of Homer Rodeheaver's tabernacle choir, singing a noble bass to "Brighten the Corner Where You Are."

This publicist was an asset, no doubt of that.

Other events augured success for a newspaper which might choose to be independent of powerful interests. The most grievous of Colorado's labor wars now was beginning —the Trinidad coal strike.

Governor Elias M. Ammons, a God-fearing farmer, a thin-voiced wisp of a man, had sent state troops to the strike zone to protect interests of big coal operators. These included the Colorado Fuel & Iron Company, a Rockefeller subsidiary, the Rocky Mountain Fuel Company, the Victor and other coal corporations.

Ammons, sitting low in an overstuffed chair of a hotel suite, and in secret session with the operators, hesitated to declare martial law in the coal fields. The operators danced beside his chair like cannibals about a missionary *en casserole,* shook fists in his face, browbeat him, cursed him and otherwise behaved in a manner which frightened the frail People's Choice. The Governor agreed to send troops under the command of Adjutant General Chase, a fleshy and aging oculist.

Managing Editor Forman leaped at the strike news. Without coloring it in any way, he presented the facts as obtained by Harvey Duell, a capable journalist who afterward became editor of *Liberty Magazine* and the *New York Daily News.* The strike reports were interpreted editorially by Chenery. And circulation began to boom. At last the people of Denver were getting unbiased news and temperate comment.

But Mr. Shaffer seemed to lack staying qualities. He backed down from telling too much about the actions of imported gunmen in the ranks of the State Militia. There was the infamous "battle" of Ludlow, when gunmen fired on a camp, although it was known beyond the vestige of doubt that the striking miners' wives and children were huddled there, unarmed and unprotected. It was a massacre.

The *Post's* worries now were over. Mr. Shaffer sent his progressive editor, Forman, back to Chicago, and replaced him with a fat and harmless fellow. Mr. Shaffer seemed sad every time the *Post* shot arrows at him. He didn't like the epithet, "John 'Clean' Shaffer." He played into the *Post's* hands.

Until now the *News* had used red headlines—although not so prodigally as did the *Post.* Denver, therefore, was accustomed to boisterous presentation of the news. Shaffer decided to abolish the crimson banners from his columns. This conservatism cost him circulation. And now, instead of spending money to fortify his position, he sought to pinch the pennies. He was said to have had a conference with the *Post* owners looking toward broad reductions in salaries. Tammen blocked a proposed cut in wages.

In his purchase of the *Republican,* Shaffer had found himself the owner of *two* morning franchises of the *Associated Press,* whereas he needed but one. The *Post* had a franchise for its afternoon editions, but none permitting access to *A. P.* reports for the Sunday morning paper. Tammen now asked Shaffer to release the defunct *Republicans* Sunday morning press privileges. Shaffer refused to do this.

Tammen thereupon printed coupons in the *Post,* together with daily editorials, claiming that thousands of citizens had petitioned the *Post* to give Denver a "live and worth-while morning paper." If anyone wanted to add his voice to the movement, "sign the accompanying coupon."

Shaffer's men at first laughed at the threat. They did not believe the *Post,* lacking an *Associated Press* morning telegraph service, would undertake publication of a daily morning edition. Still, Shaffer was not so sure. The *Post* partners were aggressive foes.

One day Tammen called Shaffer by telephone.

"I'd like to come over, John."

"Very well, come over," Mr. Shaffer replied.

Half an hour later, Tammen climbed the stairway of the Welton Street publishing house. He went immediately to Shaffer's office. They talked for a time about "calling off this constant attack on each other," then Tammen said:

"How about selling us the Sunday *A. P.* franchise?"

"Oh, no," said Shaffer. "Be sensible."

"That's all I wanted to know," said Tammen, rising. "Well, then, we are going to start a morning paper next Monday."

Shaffer chuckled. "You are joking."

Tammen walked to the window. "Come here, kid. I want you to see something I brought over."

Mr. Shaffer joined him and looked down at the street. He saw a truck, filled to overflowing with bundles of paper. "What is it?"

"John, that truck is full of coupons we have received from Denver people, each one duly signed, imploring us to start a morning *Post.* Joking, eh?"

Mr. Shaffer grew thoughtful. "If you *really* intend to start a morning paper, I, of course, can't afford to have you enter a field which I myself narrowed by scrapping the *Republican.*"

"We can't afford it either, kid, but we'll do it."

Mr. Shaffer motioned to a chair. "Don't go yet. Now here's what I'll do. I'll *give* you the Sunday *A. P.* rights, if you'll call off your plan for a morning *Post*."

Tammen rose. "Shake, kid. Now you're showing some sense."

Thus the *Post* won its coveted Sunday morning franchise without paying Shaffer a penny for a property estimated variously to have been worth from fifty to one hundred thousand dollars.

Furthermore, Bonfils and Tammen never had had the smallest intention of starting an opposition morning newspaper. And, as for the truck-load of coupons, Tammen winked slyly.

"A fake. Nearly all the papers were phoney. All it cost us was the time of some clerks to cut up paper the size and shape of our printed coupons. Of course, there was the wear and tear on the scissors to be considered."

27

The Royal Family

NEAR THE close of the Sells-Floto circus matinee, April 27, 1912, in Salinas, California, the herd of elephants became wildly excited. Trainer Fred Alispaw and his aides were endeavoring to pull a burlap sack over the head and eyes of the laboring Princess Alice. Snyder, her eight-thousand-pound husband, was swaying from side to side, straining at his leg-chains and trumpeting urgently. The entire menagerie seemed uneasy, as if aware of an unusual event.

Finally, at five-thirty o'clock in the evening, and after much travail, Princess Alice gave birth to the first calf elephant ever bred and delivered on American soil.

The nearness of humans enraged the mother. She lowered her great head, flexed her wrinkled knees and then rolled upon her baby, with intent to kill. Alispaw tried to reassure her, to implore Alice to rise of her own free will. Her answer was a lashing blow of the trunk, knocking the trainer a full twenty feet across the sawdust floor.

She had shaken off the burlap blinder, glimpsing the one-hundred-and-eighty-pound son afresh and renewing the murderous attack. Attendants used wooden bars as levers, to pry her heaving belly from the squealing newcomer. They applied elephant hooks to her defiant trunk and brought other persuasive instruments into play. Finally, the hysterical mother reared to fight off the men, and now they dragged the bewildered, half-suffocated calf to safety. He stood as though vastly puzzled by a world which greeted strangers with such sudden, ugly brutality.

With elephantine wisdom, he showed no desire to return

to the vicinity of the gray hulk, within the ribs of which he had been carried for nearly two years.

He looked somewhat like a fatigued ant-eater, was pink all over and had a trunk six inches long. Black hairs bristled from his blushing hide. He was twenty-four inches high and thirty-six from tip to tail and had an orphans'-home personality.

Alispaw named the calf in honor of Manager Fred H. Hutchinson of the Sells-Floto Shows, calling it "Baby Hutch." It became the most valuable elephant in the world. An extra fee was charged for viewing Baby Hutch, who earned a thousand dollars a day for his owners, Bonfils and Tammen. No baby elephant had survived infancy in America, consequently zoölogists everywhere were concerned with Baby Hutch's career.

The Salinas Chamber of Commerce presented mother and son with two elaborate sets of gold-embroidered trappings. Harry Tammen, as proud as a father, assigned a special car for transportation of the pair, with two attendants constantly on guard. The mother finally become reconciled to her son, but refused to feed him at the breast.

The circus exhibited this tiny fellow on a plush-covered table. In street parades, Princess Alice pushed ahead of her a gilded chariot, in which Baby Hutch rode. On the gleaming sides of the car was the matronly boast: "This is my baby."

Two months after the accouchement, in Pendleton, Oregon, Baby Hutch fell ill. Veterinarians thought pneumonia had developed. They were unable to alleviate the pain. The baby died in the afternoon, shortly after Princess Alice had come from the show ring. She knew at once what was happening. She had known all along, for inside the big top, half an hour before her act was done, she had experienced a premonition of tragedy. She had seemed nervous and had behaved badly. And now, her baby dead, she set up a weird, mournful rumbling.

Alice refused to eat. Her keeper and twenty other men had a hard time restraining her in the picket line. They double-chained her, hobbling her legs, for they feared she might break loose, as had happened during that stampede of April 16, 1908, in Riverside.

The Salinas coroner performed an autopsy on Baby Hutch's body. This revealed a putty-like substance in the calf's large intestine, indicating digestive faults. Alispaw sent the body to Denver for mounting by a taxidermist.

Tammen was desirous of replacing the infant, urging Alispaw to encourage another mating beween Snyder and Princess Alice. And from here on, the story might be that of the Romanoffs of Russia, as told in terms of elephants. A dark cloud seemed to foreshadow the coming of a pachydermic czarevitch, and the life of these creatures supplied a parallel of exile and executioner's bullets. The circus itself was a glamorous court, a background for elephantine sorrow, and the multitude of men and women who looked on might seem, not one, but a hundred thousand Rasputins.

Snyder himself was a great, brooding creature, with a reputation for being "bad." Zora, Fred Alispaw's wife, the foremost of women animal trainers, warmly defended Snyder. She had cleared him of the charge of killing Deaconess Gibbs in the Riverside debacle. It had been another member of the stampeding herd. Yet Snyder had a bad name. The term "killer" was applied loosely to him. He seemed an ill-starred Nicholas II, although more majestic and personable, in that he received the blame for many things.

Princess Alice did not recover quickly from her loss. Her bereavement imbued a distaste for motherhood. She appeared to wish Snyder's company in nothing more than a platonic way. Alispaw had a shrewd, sixth sense that made him the most valuable of American menagerie men. He believed that climate had an enormous influence on the honeymoons of elephants. The great creatures apparently were most strongly inspired during warm, drizzling rains, which imparted a marshy quality to the soil, with alluring mud wallows.

With this theory, Alispaw chose not to wait for a return to winter quarters in Denver, a city where a dry climate was the rule. So, in the Middle West, and after there had begun a succession of August rains, the trainer segregated the pair.

He noted that Alice now seemed less cold to Snyder's advances. Certain misunderstandings, however, arose between them. Alice became irritable, capricious and highly nervous.

There had been incessant rains. A thick floor of swampy mud was available. Alice was in a condition indicative of susceptibility. Small vents at her temples were open—a sign of pachydermic fecundity. Snyder implored, and, after a few ineffectual meetings, relations were established successfully.

The rains stopped and Alispaw remarked that the mating ceased abruptly, once the skies had cleared. Snyder still was attentive, but his Princess paid little heed to him. Her manner was of a polite sort. She was congenial enough during the circus performance, but elsewhere she countenanced him with a bored complaisance. She transferred her affections to her keepers, became unusually quiet and reflective. Alispaw concluded that she was pregnant.

When the circus returned to winter quarters, Alice treated Snyder with increasing restraint. He, at first, seemed a devoted husband, an understanding spouse, with a swain-like desire to pet Alice and to humor her whims. Still, when she persisted in thinking of things other than conjugal embraces, Snyder took to pouting. On the whole, however, he deported himself with that resigned compassion assumed by blue-grass gentlemen toward ladies in a delicate condition.

When the circus went on the road in the spring of 1913, Alice was docile and pensive against a background of calliopes and spangles. Although beginning to grow heavy with calf, she performed smoothly in Zora's act, with its pyramids of great bulls and cows, its grotesque, Brobdingnagian waltzes, with brassards of bells on clay-colored legs, tub-walking and colossal prancings. She seemed glad to be busy. Snyder, leader of the herd, watched her jealously, and seemed happy only when she was holding his tail with her trunk in the march to the center ring. As for Alice, she hardly noticed her husband except in the routine meetings of the arena or while parading on the street. Snyder reacted to this indifference wth a sort of subdued moaning and locomotive sighing that might be termed lover's asthma.

The August following her pregnancy—a full year having elapsed—Alice and her court entered another rainy season in the Middle West. Snyder grew obviously uneasy as he felt the soft mud underfoot. Assuredly he was remembering last year's idyll. He paid ardent attention to Alice, but she was exceedingly cold to the advances of her four-ton Romeo. He had been patient, understanding and resigned during the cold, dry months. Now he grew embittered, sulked and muttered. Then, when he persisted, and Alice snubbed him openly, slapping him in public with her trunk, he positively screamed. He raised his trunk as though to return the blow, but with a self-conquering moan repressed the cad's instinct that had flared momentarily. He turned sadly and began to pick up hay, twisting it into great brooms to fan himself. He kept aloof and snorted like a jilted prize-fighter.

In winter quarters again, Alispaw observed that the Princess' breast was filling. It was now December of 1913. Alice had lost her calm, peaceful manner. She became temperamental. She would rock from side to side, then stop suddenly to gaze at the barn walls, as though trying to penetrate the hidden meaning of life. Then she would begin rocking again. She seemed particularly upset whenever Snyder came within yards of her, and her small, pensive eyes held a Reno stare.

And now she began to experience weird cravings, comparable to those of many another lady "that way." Unlike her human sufferers, however, those expectant mothers who awaken at midnight to lust for oranges, ice cream, champagne or boiled potatoes, Alice preferred a rather bizarre fodder. And her trainer was ready for her. He had noticed, during the time she had been carrying Baby Hutch, that she occasionally *ate earth*. A genius for anticpating animal problems, Alispaw had laid in a large supply of humus and loam, in which were bullrush and cattail roots. He now placed that delicate offering before the expectant mother. She almost purred her gratitude. Every other day, Alispaw brought a barrow of this soft earth to her, and she *ate fifteen pounds of it at a time!*

The sympathetic keeper exercised Alice in the approved manner, taking her for peaceful walks, but never permitting her to exert herself, to lift heavy objects, push or pull wagons or climb steep ramps. He sat with her by the hour, talking soothingly, as though to impress on her that having a baby elephant was a most normal and natural function.

Alispaw restricted the Princess' diet to non-fattening foods. He made her as comfortable as possible. He examined her teeth and manicured her big toes, scrubbed her and pampered her. Indeed, in Snyder's eyes, Alispaw might have seemed a Rasputin toadying to a czarina.

There was, obviously, great interest among animal experts concerning the forthcoming event. Alispaw counted back to the time he first had suspected life to have stirred within the Princess' distended loins. He computed that the baby would arrive some time in May of 1914. He thereupon notified leading zoölogists and other scientists when to come West for the birth. He decided against taking Alice on tour in the spring, suffering her to remain behind in winter quarters, the better to receive scientific care in a quiet and more desirable environment.

Well remembering her attempts to crush Baby Hutch, the thorough Alispaw moved to forestall a repetition of that act. He anchored two steel posts in the cement pavement of the elephant barn. They were placed fourteen feet apart. They were eight feet high and were rigged with heavy chains with contrivances to be snapped on Alice's halter and throat chains, that she might not lower her head to attack the baby. The posts and chains also were designed to restrain Alice while the calf was put to the breast.

The obstetrical chamber was now ready for the gargantuan delivery.

Alispaw's presence was needed with the circus, once it took to the road, so he solicited help from the Central Park Zoo of New York. Keeper George Brown, who had assisted Alispaw in the delivery of Baby Hutch, came to Denver to watch over the Princess.

Alice seemed to know that the circus was preparing for the road. When the bustle began and the newly painted wagons rolled from the winter sheds, the pregnant Princess grew restless. The extra chains placed on her churn-shaped feet added to her alarm. And now that the animal house was deserted and Alice saw her brother and sister elephants marching out, two by two, she set up a dismal grumbling. She strained at her shackles, lay against them and swung her trunk in pendulous resentment as keepers closed and barred the tall doors.

Alispaw had intended accompanying the show-train to Albuquerque, where the circus season was to open. When he saw what was happening to Alice, he decided to remain behind for another day. He was afraid she might harm herself. He watched her struggle in her chains for eighteen frantic hours. She would not be calmed. She now was twenty months "gone." Her baby was not due for upward of six weeks. But Alispaw had reason to fear a premature birth, and so informed Harry Tammen.

Intensely alarmed, Tammen notified all the medical men, veterinarians and scientists he could think of, and hastened to winter quarters. Alispaw simply had to join the show, and, leaving explicit instructions, went on to Alburquerque.

Tammen stayed at the barn and behaved like an expectant father. He paced the cement floor, kept the doctors awake and seemed personally injured when Alice didn't want any of his sympathy.

On the night of March 28, 1914, Princess Alice was unmistakably in labor, with great, recurrent pains, followed by intervals of relative quiet. Dr. T. Mitchell Burns, Colorado's foremost obstetrician, had answered Tammen's summons. Other prominent medicos arrived at the barn, all of them as meddling as midwives. Mayor James M. Perkins, himself a physician of long practice, and the widely known surgeons, Drs. C. B. Ingraham and F. W. Kinney, the latter a stomach specialist, were gathered there. State Veterinarian W. W. Yard and City Veterinarian M. J. Dunleavy were present to manage the accouchement.

The six-thousand-pound Princess was in a frenzy. She bellowed and her great chains clanked. Rumbling continuously, she labored throughout the night, then, at ten forty-five oclock on the morning of the twenty-eighth, the czarevitch was born. He was pink and had black hairs all over his body. And now the keepers were in for a battle. They had been unable to snap the throat chains to the wildly sagacious Alice's halter. There was a savage trumpeting. The doctors, who had been swabbing the baby with bolts of finest cheesecloth, scattered to the far sides of the barn. Alice, lowering her head, let down her body to crush the life out of her child.

Keepers fought hard to rescue the baby elephant from the trunk-lashings and the whale-like rollings of the infuriated mother. They succeeded at last, but for some minutes thereafter the calf showed no signs of life. The obstetricians massaged the calf's pink hide with oil and poured some whiskey down its throat. They slapped its pig-like rump with barrel staves. Finally it began breathing and then started squeaking like a huge mouse.

Although Alispaw had left explicit orders to "get the calf out of the mother's sight as soon as possible," the doctors thought they knew better. They theorized that the offspring should be kept within sight and scent of the Princess for twenty-four hours, holding that she would become accustomed to the infant and permit it to feed at the breast.

The keepers slackened Alice's chains somewhat. The doctors withdrew from the barn, leaving a keeper inside long enough to shove the bull-calf by slow degrees toward the now sullen, swaying mother. Then the keeper, too, went outside.

No sooner had the little fellow raised his tiny funnel to explore Alice's milk-distended breast, than she screamed with rage. She worked the calf beneath her, and, crouching suddenly, proceeded to crush him. Keepers threw open the barn doors and raced inside. Then began a fierce struggle with the three-ton mother, who was sprawled upon her one-hundred-and-sixty-pound son.

When at length the emergency squad had dragged the calf from its mother, it was as senseless as a pancake. Whiskey and herculean massage revived it. Twenty-four hours now had elapsed without the calf having tasted anything but defeat—and a morsel of Green River whiskey. It was evident that Alice didn't want anything to do with her heir, and it was equally sure that the child wanted no part of such a violent parent. Each time the Princess trumpeted, the calf would assume a hang-dog expression and cower against the wall of a shed to which it had been removed.

The new calf was christened "Tambon," a combination of the nicknames of Tammen and Bonfils. Two of the medical witnesses were noted dieticians. They now conferred as to what food should be given the infant. They found, from a cursory analysis of a specimen (procured at some risk from Alice's udder) that elephant milk was richer and considerably sweeter than that of a domestic cow. They decided to approximate the natural supply with a mixture of whole milk from a thoroughbred Jersey, and a fortifying amount of sweetened condensed milk.

Tammen ordered agents to scour the agricultural vicinity for a pure-bred Jersey. Such an animal was discovered, one that had freshened two or three days before Tambon's birth. The agents purchased it and placed it in Tambon's nursery, an indoor corral.

The dieticians now procured a second specimen of Alice's milk and tested it exhaustively. They found that it contained seventeen per cent more fat and three and a half times more sugar than the milk of an average cow. So they modified the diet, adding unsalted butter and sugar of milk to the Jersey's product and doing away with the condensed milk. This pap was at first given Tambon from a large bottle, but as soon as an apparatus could be devised, the fluid was kept in an automatic, electrically warmed feeder, at which Tambon could nurse whenever he felt the urge—which was often.

He consumed fifteen quarts daily. His night-feeding was at intervals of four hours. In the daytime his practice was to romp for an hour, then nurse for fifteen minutes, then sleep for two hours. He would reawaken, look about him with sly alarm, perhaps wondering if he were to be crushed by a mountainous body, then begin romping again. A keeper placed him on the scales each morning; he gained from eleven to fourteen pounds a day.

Tambon showed an especial fondness for his wet nurse, the Jersey cow, snuggling up to her, once she had recovered from her first suspicion that her adopted child was a "ringer." He seemed to feel safe in her company, yet never forgot the manhandling he had received from his mother—indicating by his actions that reports concerning an elephant's memory were in no way exaggerated. Whenever Alice's rumbles came through the walls of the elephant barn, Tambon would cease frisking, run to the side of his cow-guardian, squint and shiver.

Meanwhile, Princess Alice had fasted for several days. Finally she began gloomily to eat an occasional flake of hay, like a convalescent taking a shredded-wheat biscuit. Within ten days she had regained a normal appetite, but whenever her calf bawled, she would pause in her eating to make cynical noises through her trunk.

Tammen now decided that he would permit the townsfolk to view Tambon on Sundays, between the hours of ten in the morning and five in the evening—and for a fee of ten cents a person, of course. Tambon was eight days old when the first crowds came to see him in response to a slightly erroneous advertisement in the *Post* promising that they would see "a baby elephant *nursing* from its mother."

Alice, relieved her chains, was not permitted outside the barn.

By April 9, Tambon weighed 254 pounds, had been broken of a habit of sucking his trunk, and gave evidence of cutting teeth. Two sore lumps appeared on his lower jaw. His keepers found a large chunk of raw rubber, on which he chewed in a philosophical fashion, drooling the while.

It was estimated that Tambon would be kept to the bottle until he had attained four months, then he would learn to drink from a bucket, and finally, at seven months, would have at the hay.

He seemed very healthy, aside from gum-worries, and Tammen was anxious to send the calf and its mother on the road as soon as medical advisers deemed it the safe thing to do. Then, on the night of April 26, 1914, Tambon flatly refused to take his milk. He lost interest in play. He lay upon his side for several hours. The medical men arrived to diagnose the case. They listened through stethoscopes to his heartbeats and sounded his lungs. Thinking his kidneys might be affected, they applied hot blankets to Tambon's back. He appeared to have some sort of colic.

Then, after twenty-seven and one half days of life, little Tambon quit breathing. Alice, nearby in her barn, seemed uneasy, but did not mourn. An autopsy revealed an intestinal condition—with a deposit of putty-like solids—similar to that which had obtained in Baby Hutch's case.

Tammen, as disconsolate as if a child had gone, sat beside the dead elephant. He choked up when taxidermists claimed the body. He had been so fond of this elephant that, when it had been stuffed, he placed it in a big glass case and set it outside his office at the *Post*.

Alice now displayed a sullen disgust in regard to male elephants. Nor did she like *small* elephants. Perhaps they reminded her too keenly of calves. When the Sells-Floto Shows acquired two four-year-old bull elephants, "Kas" and "Mo," Alice tried to maim them. She made the newcomers very unhappy on all occasions.

The Princess found an "escape" in work. She went through her circus routine more expertly, more seriously than ever before.

Alispaw, however, did not despair of rearing a baby elephant in captivity. Despite the now frigid nature of the Princess, Snyder succeeded in winning her over temporarily during the seductive Middle-Western spring rains. And, on April 15, 1916, Alice again went into the obstetrical throes.

She bore a third bull calf, called "Little Miracle." She repeated her unmotherly didoes, tried to kill her son, refused him food and treated him as an outcast. This third elephant to be born in America lived until September, then died of the same unidentified malady which had taken his brothers.

Incredibly enough, Alice became pregnant for a *fourth* time, establishing a record never equaled before or since by any other cow-elephant in captivity; and now she was simply tired of the whole business. If ever an animal had cause to advocate birth control, Princess Alice was that creature. Her work in the show ring now became clumsy and lax. Her husband, the ever-brooding Snyder, was violently unbearable to her sight. After all, he had been responsible for those long months, during which she had had to carry about an extra one hundred and fifty to one hundred and eighty pounds of morning sickness.

In 1917, it was decided that Alice no longer was fit for show purposes. Tammen offered her to the school children of Salt Lake City, who were asked to contribute dimes toward the purchase. And so she was exiled to the Utah Siberia, and there gave birth in 1918 to the aforementioned fourth son. But it, too, went the way of her other offspring, and Alice, when last heard of, was leading a conventual life, her elephantine brain preoccupied with the fundamental problems of birth and death.

Snyder, feeling himself an utter failure in life, grew morose and vengeful. He had been sad enough, God knows, with Alice, what with her blowing hot and cold on his emotions, accepting him one day and snubbing him the next—not to mention the public slaps. But now, without her, he was lost. His mind wavered. He became "bad" in fact as well as in name.

His keepers decreed he must die. Executioners, firing steel-jacketed bullets, felled him in his chains. Before he sagged, however, he gave a last, regal trumpet call, and died as bravely as any czar.

Tammen always had liked Snyder, despite the constant love-sick lunacies of the big bull. And after Snyder was slain, Tammen sent one of the great tusks to Japan to be carved into small ivory elephants.

Two weeks before Christmas of 1922, Mr. and Mrs. Tammen received a package from the Orient. Tammen opened it with his usual enthusiasm for trinkets. He spread one hundred ivory elephants on a table, and then, as he touched them, a curious thing happened. Each one split cleanly in halves as he handled the figures. He never was able to explain this strange falling apart of the little elephants, although he presumed the dry climate and some peculiarity in the grain of Snyder's tusk had something to do with the matter.

He called for glue and mended the figures. He sent them as Christmas gifts to his friends. He enclosed with each cracked elephant a card, printed as follows:

GREETINGS 1922

> Once I owned an elephant. The largest one in the world, and the best performer. He went to Elephants' Heaven, and I inherited one of his tusks. A good man carved it up into little elephants. They are not perfect, because he was not a perfect elephant. Folks said he was a bad one—I thought he was wonderful, because *he was my elephant.* Am sending you one of these little elephants, with love and good cheer, and wishing you a Merry Christmas and a Happy New Year.

Years afterward, and as though to complete the tragic, Romanoff-like cycle, a mob of street-car strike sympathizers became enraged at the *Post's* labor editorials. One night they descended on the paper, bent on wrecking its plant. The first thing they seized upon was the stuffed figure of Prince Tambon. They shattered the plate-glass case, hurled the elephant to the floor, broke its small, uplifted trunk and ripped pieces of hide from the plaster mold.

28

Mustangs And Sacred Cows

BONFILS AND TAMMEN were bargain hunters. They had developed a circus from a dog-and-pony show, and now they coveted Buffalo Bill's famous Wild West concern. They set out to ensnare it in 1913, at a time when Colonel Cody was harassed by debt, entangled in a partnership with Major Gordon W. Lillie ("Pawnee Bill") and confused by those sad and disillusioning truths that descend of a sudden with the snows of old age.

The celebrated Colonel's biographers have made him a Sir Galahad of the plains, an Indian fighter and scout superior to the rugged, taciturn Kit Carson. His critics have gone to the other extreme, portraying him as a bellowing faker, a butcher of buffaloes, a glutton for rum and romance. The man himself is lost between legend and calumny. All agreed, obviously, that demigod or satyr, Buffalo Bill was an institution.

Perhaps the handsomest American of all time and a symbol of adventure, he was envied by men, beloved and spoiled by women and emulated by growing boys. He sat his white stallion like a dream prince, and every time he shook his long curls, the lassies of many towns suffered the amatory jitters. In one respect, the Colonel's tresses were a godsend to mothers who persisted in the most villainous hoax ever perpetrated on the lads of America—the infliction of the Lord Fauntleroy coiffure on screaming boys.

"You look like Buffalo Bill," these addled mothers would plead. It was a telling argument, heaven forgive! For Bill was the Pied Piper on the trails.

Half English and obstinate, half Irish and sentimental, the Colonel drove through all obstacles to the fortunes which he never could hold. Indiscreet, prodigal, as temperamental as a diva, pompous yet somehow naïve, vain but generous, bigger than big today and littler than little tomorrow, Cody lived with the world at his feet and died with it on his shoulders. He was subject to suspicious whims and distorted perspectives, yet the sharpers who swindled him the oftenest he trusted the most. And sometimes he repaid quiet devotion with thundercloud doubts and ruthless attack. That of his wife, Louisa, for example.

Mama Cody for forty years had remained in the background, mending the scout's buckskins, speaking well of him to her circle at North Platte, Nebraska, and becoming known as a virtuous, temperate and patient wife. In 1901 Pahaska left home, not to return until 1905 with a set of charges as loony as any ever brought in a prairie divorce action. In a secret hearing, with only a reporter for the *Post* as spectator, Cody claimed that Louisa drank to excess, used foul language and *had tried to poison him!* This was too much for a judge or a neighborhood to believe. Colonel Cody didn't succeed.

Contrasted to that brainstorm was Cody's long association with Johnny Baker. When Johnny was seven years old he caught the hero-worshipping fever in earnest. He lived with his parents in North Platte, and whenever Cody rode into town, Johnny was waiting in the background. He would stay out of sight until the Colonel tied his horse; then Johnny would go to the hitching post, untie the reins and stand holding them until his hero reappeared. Sometimes Cody would be gone for hours.

Touched by this adoration, Cody became friends with Johnny and, when the boy was ten years old, invited him to join his Wild West Show. He was permitted to adopt the lad and taught him to ride, shoot and rope. Cody billed him as the "Boy Wonder" with the target-shooting act of Annie Oakley, whose ability to put bullet-holes in calling cards led to her name being used as a slang term for theatrical passes.

Cody took his foster son all over the world, Johnny sitting on the box of the Deadwood Stage Coach when it carried Queen Victoria and several princes during a command performance at Windsor Castle. And Johnny never stopped hero-worshipping Buffalo Bill. After Cody's body was buried atop Lookout Mountain, Johnny and his wife established a museum there, called "Pahaska Lodge," lived in it and displayed relics commemorating the Colonel's deeds. And when Johnny Baker died in 1931 at the age of sixty-two, his last words were to his wife:

"I want you to stay at Pahaska after I'm gone. I want you to keep alive the memory of the Colonel."

Cody's record on the frontier was that of an intelligent, brave and capable scout. The five-cent Buffalo Bill novels of Ned Buntline and the paeans of press agents led the public to believe that Cody had slain thousands of venonmous hostiles and almost single-handed had won the West.

Foremost among his exploits was the victory over Chief Yellow Hand at Warbonnet Creek in the summer of 1876. Circus Homers described Yellow Hand as having been an aboriginal Goliath. Cody's Denver critics sought to belittle the whole affair, Scout Wiggins, grizzled disciple of Kit Carson, holding that Yellow Hand was in the last stages of tuberculosis at the time, or, as sportsmen say, a set-up for Cody.

The Colonel himself usually refrained from speaking of the Yellow Hand fracas. Pressed by Bonfils to give the details, he declared that Yellow Hand had been anything but an invalid.

"I was chief of scouts for General Merritt," the Colonel said. "I had been guiding the Fifth Cavalry to cut off a band of Sioux and Cheyennes under Yellow Hand and Crazy Horse on their way from the Red Cloud Agency. The news of the Custer Massacre had reached us only a few days before, and we sought to prevent Yellow Hand and his band from reinforcing Sitting Bull, author of the Custer debacle.

"Custer had been my friend. I had served with him at Pilot Knob and elsewhere during the Civil War. I wanted revenge for my friend. I reported that we were across the line of march of the redskins, and General Merritt ordered his men to lie in wait. I went out to scout and happened on an immigrant train. I escorted them to our lines, but in so doing revealed our presence to the enemy. It could not be avoided.

"Immediate plans were made for an engagement. The Indians were on the other side of the valley. Warbonnet Creek separated us from them. I was in advance of our troops. Suddenly I saw an Indian ride forward, shouting and waving a white cloth. My interpreter, Little Baptiste, translated the shouts:

" 'He says that Yellow Hand is the greatest of their warriors and would like to fight Pahaska (Buffalo Bill), as he is the bravest of palefaces.'

"I said: 'That suits me fine,' and put a spur to my horse. The herald retired, and now a huge and fiercely painted chief—Yellow Hand—came charging from the Indian lines. There was a great whoop from his warriors and a cheer from my comrades.

"When we had come within about fifty yards of each other, we simultaneously raised our rifles and fired. My horse stepped a forefoot into a gopher hole and turned a complete somersault. I fell and rolled, losing my rifle. I was on my feet immediately, however, and shouted to let my comrades know I had been uninjured. I ran forward, drawing my bowie knife. I noticed now that Yellow Hand also was on foot. My shot had killed his horse, and the Chief had been thrown and had lost his rifle. He was brandishing a tomahawk.

"As we closed in, I crouched to get inside the arc of his blow. He had his left hand extended to ward off any thrust I might make. Instead of stabbing at him, however, I feinted and dashed inside the arc of his blow, just as his tomahawk began its descent. I grappled with him, caught his right wrist, twisted him about, and then sank my bowie between his ribs, beneath the left armpit. He fell, his mouth open and his eyes staring."

Mr. Bonfils suspected further details. "Didn't you scalp him?"

The Colonel was somewhat embarrassed. "I regret to say that I did. I was carried away by the thought of Custer. I jerked off Yellow Hand's big bonnet, grasped his war-lock and cut it off."

"It was a bloodthirsty thing to do," said Bonfils.

"Yes," said the Colonel. "And that is why I seldom refer to the duel. I sent the scalp to my wife, and when she saw what it was she fainted."

Colonel Cody told Bonfils this story in January of 1913. It prefaced a plea for a loan. Buffalo Bill for some time had been in a circus partnership with Pawnee Bill (Major Lillie) under the title: "Buffalo Bill's Wild West and Pawnee Bill's Far East Show." Of late they had encountered a deficit. They owed for lithographs and other matters.

Bonfils and Tammen lent the show twenty thousand dollars, payable in six months at six per cent interest. It was the beginning of much trouble, of litigation, and perhaps hastened the end of Buffalo Bill's glamorous career.

In lending that enterprise the twenty thousand, it was understood that Cody would split company with his partner, Pawnee Bill, and throw his name and his Wild West interests into the Sells-Floto camp. An anouncement in the *Post* of February 5, 1913, containing "shots" at the rival Ringlings, follows:

"The most important deal ever consummated in American amusement enterprises was closed in Denver a few days ago, when Colonel W. F. Cody (Buffalo Bill) put his name to a contract with the proprietors of the Sells-Floto Circus, the gist of which is that these two big shows consolidate for the season of 1914 and thereafter.

"The Pawnee Bill interests now associated with Colonel Cody's Wild West Show are not included in this agreement—the idea being that the Sells-Floto Circus shall continue in its entirety, and the 'Buffalo Bill Exposition of Frontier Days and the Passing of the West,' with the historic incidents associated with them, shall also be preserved, added to and given with the circus performance.

"This means, not only from a showman's, but a layman's standpoint, the strongest combination ever formed in the history of American amusements, if not in the world. . . .

"This departure marks the culmination of many years' hard work by the Sells-Floto people. The circus is now in its twelfth year, and during that time has been made the target for more opposition and battling in various ways than any organization that ever went on the road. Everyone conversant with the history of the show business remembers the fights with the Ringling Brothers, Barnum & Bailey Circus and other attractions which combined in what is commonly known as 'the circus trust'. For many seasons the smaller circuses went up and down the country losing money in many thousands, but ever since they instituted the cut-rate price, their rivals have been more than willing to make way for them and to prefer their room rather than their company. The present combination would seem to make the Sells-Floto Shows the monarchs of the amusement field."

And now began a legalistic stampede destined to wreck the Cody-Lillie organization. The *Post's* owners got what they were after—Buffalo Bill, in name and in person. The confused Colonel gallantly kept at the business of shooting smoke balls from the saddle of his white horse, McKinley, to keep himself from the poorhouse and his charger from the boneyard. His spirit was broken in the manner which usually seems to please one's critics immensely, but his pride remained with him to the grave.

No sooner had the Cody-Lillie show pitched its sallow tents in Denver, July 21, 1913, than the debtors' lariats began to fly. The United States Printing and Lithographing Company brought attachment proceedings. The General counsel of that concern, Adolph Marks, issued a statement eulogizing Buffalo Bill, but saying an attachment was necessary to protect upward of sixty thousand dollars due the company for lithographic matter, programs and date bills for the season of 1913. Major Lillie was described as the god of the machine in Buffalo Bill's financial downfall.

Although Attorney Marks was general counsel, it is of interest to note that the lithographic firm's Denver attorneys were John T. Bottom and Charles H. Redmond, legal advisers of the *Post*. Mr. Marks took pains to say that the *Post's* owners had been very generous and thoughtful throughout the situation; that they had refrained from interfering with the printing company's legal processes, despite the Lillie-Cody default on the twenty-thousand-dollar loan due Bonfils and Tammen.

Major Lillie was described as having gained control of most, if not all, of Cody's real estate and in a manner not entirely laudable. When asked to release the Cody ranch at North Platte, valued at one hundred thousand dollars, and the Cody Hotel at Cody, Wyoming, so that the Colonel might transfer those assets to the printing company, Lillie refused.

The *Post* lambasted Lillie for declining to "give up." Bonfils and Tammen themselves attached the show for the twenty-thousand-dollar debit. Attorneys Bottom and Redmond represented the Sells-Floto owners in this proceeding. The complaint charged Major Lillie with obtaining money under false pretenses, although it was Cody who really had promoted the loan.

The next step was a petition in involuntary bankruptcy in the United States Court of the Colorado district. Two small creditors appeared as petitioners. One of them cited a claim of three hundred and forty dollars for merchandise. The other alleged he was owed thirty-six dollars for stock feed.

Meanwhile, in the East, Pawnee Bill was not idle. He had a separate action in involuntary bankruptcy filed in the federal courts of New Jersey. Fearing that Lillie would succeed in his counter-move, and that a receiver friendly to him would be appointed in New Jersey, the *Post* loosed reams of propaganda against Pawnee Bill. The newspaper charged that he had deliberately stranded the men with his show, once it had been attached in Denver and its properties levied by the sheriff. The *Post* accused him of "selling" the show the day after it fell into the Denver sheriff's hands. A certain Thomas A. Smith was represented as the "purchaser," and was said by the *Post* to have plastered his name on seats, wagons and other paraphernalia of the defunct show, to create the inference of ownership.

This action, the *Post* maintained, was calculated to

checkmate a public sale in Denver of the Wild West and Far East property. United States Judge Robert E. Lewis overruled efforts by Lillie and Smith to delay or prevent the auction, which was held on August 21, 1913.

Two days before the sale, Attorney John T. Bottom appeared as counsel for Buffalo Bill in a personal action against his old partner, Major Lillie. The Colonel, somewhat bewildered, and hardly knowing whom to believe or what to do, charged that Lillie had failed to account for receipts aggregating five hundred dollars for every exhibition during a part of 1912 and all of 1913. This action was brought in Denver's District Court.

The Cody complaint, charging fraud, asked that the Lillie partnership be dissolved and Lillie compelled to account for all moneys received and expended by him as general manager of the show, and that the transfer, or pretended transfer, of the property from the co-partnership to the corporation, in which Smith appeared, be annulled.

The *Post* said that Attorney Bottom would introduce testimony to show that Buffalo Bill had been made "the victim of misplaced confidence."

The public sale itself was colorful. Mirrored wagons, steel cages, bronchos, ring horses, oxen, mustangs, sacred cattle from India and several camels were on the list. The show train, too, was up for disposal.

Lillie's attorneys made an eleventh-hour effort to check the sale, but failed. Someone had distributed dodgers among the spectators, with the heading:

"Warning! Let the buyer beware!"

The body-type of the circulars set forth an allegation that the property was fully mortgaged by Thomas H. Smith to the Federal Title and Trust Company, and that if anyone bought in, suits for damages might result.

The auctioneer shouted: "The City and County of Denver stands behind this sale."

He then waved his gavel from the old ballyhoo wagon and disposed of the mustangs and sacred cows. But the sale was set aside afterward because of a technical error in its conduct.

There was a re-sale on September 15, attended by many leading show magnates and circus equipment dealers.

To enumerate all the litigations arising from the Cody-Lillie split-up, would be endless. Suffice that the hand of the *Post* was felt, if not seen, in much of the court-storm. Bonfils and Tammen had set out to get what they wanted, for little more than the "loss" of their twenty-thousand-dollar loan. They hardly accomplished that aim, for Major Lillie managed to escape with his buckskin pants, his horse, and a few thousand dollars. Not so the gallant Colonel. His creditors realized about fifty cents on the dollar from the sale, but from now on Cody was virtually a chattel of the Sells-Floto Circus.

His fight to retain his pride was a much harder battle for Cody than had been his struggle against Chief Yellow Hand. His heart was cutting up, and uremic poisoning was spreading through his system. He tried baths at one of Colorado's spas, but knew that death was not far off. He didn't seem to care much about death, one way or another, except that he wanted to meet his end in Wyoming.

He was disappointed in this. Shortly after his family had breakfasted on January 10, 1917, at the home of Cody's sister, Mrs. L. E. Decker of Denver, he called out that he was going to die. An hour later he was unable to speak. He resorted to sign language he had learned among the Indians. He died in the early afternoon at the age of seventy-two years.

When the *Post* began speculating on what sort of memorial should be placed at Cody's tomb at Lookout Mountain, the *Boulder Camera* of Boulder, Colorado, had the following comment:

"Why not let the *Denver Post* proprietors determine the kind of shaft to erect over Buffalo Bill? He was their meat. It was they who brought him down after a gallant career, by breaking his proud heart. Why should not the shaft be crowned with a miniature 'Red Room', bearing this device:

" 'Abandon hope, all ye who enter here.' "

29

Gentlemen-In-Waiting

THERE WERE two stalwart men of unwavering loyalty who seldom left Bonfils side. They were Volney T. Hoggatt and Michael Delaney. Their lives were his for the asking, but immediate remuneration for such fealty was a thing never mentioned. It would have hurt Bonfils' feelings.

Hoggatt, a citizen of six feet four inches from heel to crown, had met Bonfils during the Oklahoma land-rush days. Something of a rover, and a man of extremely good nature, he had been walking the new streets of Guthrie—streets which Bonfils had had something to do with laying out—when he encountered the descendant of Napoleon.

Mr. Hoggatt at that time was fresh from the law college of Ames. He thought the new town needed a mayor. So he became a candidate. When he saw Bonfils, he shouted:

"Just a minute stranger!"

"I'm in a hurry," Bonfils said.

"It's a human failing," Hoggatt said. "But you're not in too much of a hurry to pledge your vote. I'm running for mayor."

"What are your qualifications?"

"None at all, except I'm young," said the huge Volney. "Look at me. Look how young I am. There are too many old men running the world. That's why it's such a sad place. You're young and I'm young. How about that vote?"

"I'll think it over."

Hoggatt became mayor, then roamed some more. He went to Alaska and then Goldfield, Nevada, chumming with Tex Rickard and Wilson Mizner. His first meeting with Mizner occurred when Volney arrived in an Alaskan mining camp street, which was lined on both sides with saloons and gambling hells. He asked Mizner:

"Could you direct me to a saloon?"

Such an introduction, of course, endeared Hoggatt to the Voltaire of the Klondike. They hobnobbed nightly at Rickard's Great Northern Saloon and gaming house, and afterward foregathered at Goldfield. It was from there that Hoggatt, remembering his Guthrie friend, wandered to Denver. From that time on he became Bonfils' constant companion, his man Friday and court jester.

According to his creed that a friend should be paid by taxpayers rather than by the *Post*, Bonfils sought to obtain a public office for Volney. After six months of prodigious effort, Bonfils prevailed on Governor Elias M. Ammons to name Volney as registrar of the State Land Board. As a matter of record, the good-natured and personable Volney was an excellent registrar. Ammons, however, failed to re-appoint him to that office when time came for an encore.

This neglect infuriated Bonfils. He immediately launched an attack on Ammons, who became—as was frequently the case with a Colorado Governor who flouted the Bonfils' power—"the worst executive this state has ever had."

Ammons' successors in the gubernatorial chair, Carlson and Gunther, became "worst Governors," largely because they wouldn't take care of Bonfils' cronies.

Everyone connected with the *Post* was happy to have Volney on hand to behave as jester. Whenever Bonfils suffered majestic spleens, the athletic Hoggatt would turn a flip-flop without touching his hand to the floor—a trick which he could perform until his seventieth year. The spectacle of the huge fellow kicking up his heels, with watches, keys and fountain pen spilling to the floor, ofttimes calmed the Bonfils soul. Volney also would shoot his false teeth to the fore, grimace and bark, and was founder of "The Ornery and Worthless Men's Club of America."

Hoggatt frequently watched beside Bonfils' bed while the publicist slept. The latter was said to have been subject to horrendous nightmares. It was thought that the unremoved bullet of Lawyer Anderson, fired during the Packer maneater case, had something to do with these seizures. Critics, however, would have had it that a guilty conscience was the real reason.

On the golf links, during hunting or fishing trips, or in the parsimonious junkets to watering places, Volney was to be seen, devotedly towering over Bonfils. He came to the office with Bonfils and waited all day to accompany him home. In later years, Bonfils and Tammen made Volney editor of the *Great Divide,* the *Post's* weekly brother.

When "Trixie," one of Bon's favorite poodle dogs, died, the master was disconsolate. To comfort Bonfils, Volney said: "I know it's hard, Fred. Terribly hard to lose Trixie; but, Fred, don't worry—you still have *me!*"

Bonfils' second factotum, Mike Delaney, fulfilled a more burly rôle than did Hoggatt—that of official body-guard. This dour, middle-aged bravo once had served as Denver's chief of police, and, as such, had been the bane of criminals. During that regime, organized vice had prospered, but organized crimes of violence practically had disappeared.

Delaney knew all the sinister "boys," and it was said that whenever a visiting Raffles arrived in the city, Delaney's office was the first port of call—it was a mark of good breeding. Chief Delaney would come directly to the point:

"I know you and you know me. See? Now don't pull off anything here that your mother would be ashamed of. Just lay low. The minute a job is pulled off while you're here, you'll be thrown into the can, guilty or not, the hell beat out of you, and maybe you'll go up for a long rap in the Canon City bandhouse. See? If you get so hard up you got to crack a crib in some other berg, that's *your* business, not *mine.* It's each berg for itself. See? Just don't work in *my town.*"

The Delaney system worked admirably—for Denver.

Although the toughest sluggers of the underworld had feared him, ex-Chief Delaney trembled whenever the Bonfils' baritone lost its customary musical timbre, and the roars of indignation or rage reverberated through the *Post* building. Mike watched all who came to the publicist's door. He followed Bonfils everywhere, eager to prove his loyalty with fist or gun.

Bonfils had greater difficulty in finding Mike a job than in placing Hoggatt The politicians were wary of Delaney, for he was regarded as an important cog in the Bonfils' espionage system. Furthermore, he was believed to have double-crossed Mayor Speer, under whom he had served as chief, when Bonfils sought to bring Speer's record before a grand jury. The position of body-guard had been Mike's reward.

Speer again was mayor, resuming office shortly before America entered the World War. The *Post* redoubled its attacks on this ancient enemy, and at the same time fought Governor Gunther. Speer and Gunther replied in kind. Bonfils couldn't expect to find state or city berths for his two associates, so, in Delaney's case, he elected to foist the former chief on sportsmen of the town. Failing to grab off the racetrack monopoly at Overland Park, Bonfils crusaded *against* racing. It was said he had intended putting Delaney at the track on an easy job.

Bonfils now arbitrarily "cut Mike in" on the local prize-fighting promoters. Bonfils gave publicity space in the *Post* to fistic carnivals, wherepon the sponsors of such affairs were compelled to share profits with Delaney—the silent partner. Oddly, if a boxing contest incurred a *loss*, Delaney never was called upon to share *that*.

Prize-fighting was not legal in Colorado, but there were twenty-round "boxing contests," operated on a club basis. During the War, the *Post* was anxious to have such a bout between Jess Willard, the then heavyweight champion of the world, and the challenger, Fred Fulton, the pallid plasterer. Not only would such a titular match provide profits for Body-guard Delaney, but it would publicize Willard, then a feature of the Sells-Floto Circus.

It was the practice of that circus to hire leading athletes as added attractions. Willard and Gotch were such assets.

Carpentier and Jack Dempsey appeared under the Sells-Floto big top at a later date. Jack Curley, the famous sports promoter, and a lifelong friend of Otto Floto, usually managed the athletic end of the circus.

When the *Post* proposed the Willard-Fulton battle, Governor Gunther said:

"If they want to fight, let them go to France."

This rebuff but strengthened Bonfils' belief that Gunther was "the worst Governor Colorado ever had." Bon' was mollified somewhat, however, by the record crowds that turned out when Willard and Gotch first visited Denver. Bonfils heard that the advance sale was the heaviest in Sells-Floto history, so he went with his body-guards to the showground. There Curley confirmed all that Bon' had heard about the sale.

The party stood at the lee side of the ticket-wagon, Bonfils listening to the clink of the coin with that happy but profound expression that is to be seen on the face of a scientist while the atom is being bombarded. Then, of a sudden, he paled. Thinking Bon' was going to collapse, Curley and Delaney took hold of the master.

"What's wrong?" asked Mr. Curley.

Bonfils spoke with emotion. "Out of the way! Everybody! Hide quick!"

Curley glanced about him, thinking a holdup was materializing. But the only excitement was that of the usual circus throng with its peanut-munching enthusiasm. Nevertheless, the party hid behind the ticket-wagon.

"What in the world?" Curley asked.

Bonfils, still very uneasy, muttered: "They're coming! Don't let them *see* us!"

Curley now peeked from his hiding place. He saw a couple approaching the ticket-wagon. They were Bonfils' daughter, May, and her husband, Clyde V. Berryman.

Curley looked wonderingly at the crouching publicist, who was whispering hoarsely: *"Make them pay!"*

The most enlivening affair participated in by Bodyguard Delaney was the time his fifty-two-year-old employer ran afoul of the sixty-year-old Thomas J. O'Donnell. The latter was a corporation lawyer, a fellow capable of mental or physical action, and a long-time foe of Bonfils. In the old days he had been one of the intimate group in Mrs. Reynolds' salon. He, too, had been an honorary pallbearer when Bonfils had walked before his enemies to throw a rose into Mrs. Reynolds' grave.

O'Donnell was a member of, and counsel for, the Taxpayers' Association. That body had recommended that the city purchase the water plant owned by Napoleon Evans. Bonfils had been opposing the sale, had been castigating O'Donnell, and finally had instituted a taxpayer's suit, asking for an injunction to prevent the purchase of the plant.

As counsel for the taxpayers' body, O'Donnell outlined his reasons for the plant purchase and referred to the *Post* as "the Black Hand." Rumors spread that Bonfils was going to give a thrashing to the robust, aggressive O'Donnell. A few days before the hearing on the injunction was to be had, O'Donnell received warning of the proposed attack.

One February morning, Lawyer O'Donnell reached the court house to find Bonfils, Tammen and Mike Delaney waiting at the curb. Bonfils spoke to O'Donnell, who laughed and started for the court-house steps. Delaney passed him and went hurriedly up the stone steps and through a swinging door leading to the court-house corridors. Bonfils came close to O'Donnell and entered an animated conversation. They walked up the steps together, Tammen following.

O'Donnell carried a heavy walking stick in his right hand. Under his left arm he had a fat brief case. He was wearing spectacles. He had a revolver in his hip pocket, as also had Bonfils.

As they entered the court house, Bonfils said: "You can't scare me."

O'Donnell replied: "You've not scared me so far."

Bonfils then said: "I'm unarmed, but I will go into a room with you and fight it out."

"All right," O'Donnell said. "Let's try to find a room here."

Witnesses agreed that Bonfils now lunged at O'Donnell, cursed him and dealt a full-arm swing to the attorney's left cheek-bone. O'Donnell swayed but kept to his feet, his brief case falling, his spectacles flying. He then prepared to defend himself, either with his cane, his fists or his gun—the witnesses being at variance on that score. Before he could go into action, O'Donnell felt his arms pinioned behind him. Body-guard Delaney, hiding in a doorway, had sprung upon the attorney's back.

With his foe in Delaney's embrace, Bonfils renewed the assault, striking O'Donnell perhaps three or four times. A body blow landed on the lawyer's watch-pocket, stopping the hands of his timepiece at five minutes to ten o'clock.

Although it was said that Delaney had drawn a revolver, it is probable it was O'Donnell's weapon, and that Delaney had relieved him of it during the wrestling-and-holding phase of the onslaught. Whatever the ownership of the revolver, court-house officials intervened. One of these, Assistant Attorney General Norton Montgomery, seized Bonfils and pushed him into a nearby office. The door closed and locked automatically. Bonfils crouched in a corner, holding his revolver and awaiting possible counter-attacks by O'Donnell's sympathizers.

Friends hammered at the door and at length persuaded Bonfils that his enemy had proceeded to the court room of Judge Teller on the fourth floor, where the injunction suit was to be argued. Although he was advised not to do so, Bonfils insisted on obtaining a warrant for O'Donnell's arrest. The warrant was issued under protest, charging O'Donnell with *assault with intent to kill.*

O'Donnell, his cheek cut, and blood coursing down his stout jowl, made an opening statement to the court:

"I appeared here last Saturday, Your Honor, as attorney for a large number of citizens with an interest in this suit separate from that of the Water Company. It is unnecessary to go into details of the case at this time, but it was put off until this morning, as Your Honor is aware, by the attorney for the plaintiff (Bonfils).

"On my appearance in the court house this morning, I was waylaid and brutally assaulted by the plaintiff in this case and a body-guard of thugs and assassins (sic), which is my apology for my present disheveled appearance. I have waited here since the assault and have not taken time to have the wounds dressed.

"I would ask Your Honor's permission for a few minutes' recess until I can have these ugly bruises attended to. I am not aware of what poisonous infection might have been injected into these wounds by the plaintiff's hands, and believe it better to have the wounds disinfected before entering into the work of this case."

O'Donnell left to receive first aid, refusing protection by attendants, although entitled to such, as an officer of the court. When asked if he intended sue Bonfils, O'Donnell glowered through his bandages:

"Men don't go to court over such matters."

Although the affair was dropped officially on both sides, Bonfils and O'Donnell maintained a fierce hatred for each other for many years, taking opposite sides on any and all questions that came before the people of Colorado.

Tammen occasionally was the victim of Bonfils' espionage system.

He once received a red-hot tip on a four-thousand-dollar bet at attractive odds on an election. The money had been put up at Sarconi's Billard Parlor and Bowling Alley, a place where Denver's important bets were booked. Tammen sent a friend to Sarconi's with instructions to take the wager. When the friend arrived, he found that Delaney had beaten him to the bargain, *covering the bet for Bonfils.*

Tammen called up Tony Sarconi, proprietor of the pool room. Sarconi had difficulty in digging up another bettor, but did so to keep face with Tammen.

Toward the end of the World War, a certain Denver youth received a commission as lieutenant. Bonfils ordered his editors to keep the new lieutenant's picture out of the *Post,* because he hated the young man's father. Tammen was in Kansas City at the time.

When Tammen had returned to Denver and Bonfils had gone on a vacation, the former saw the lieutenant's photograph on an editor's desk. He asked concerning it and was told of Bonfils' order.

"Now look here, son," Tammen said to the editor. "You know how this paper is owned, don't you? Fred owns one half, and I own the other. Well, you run that picture in *my* half!"

30

The Temple Is Sacked

SO PROFOUND was Bonfils' belief that he, the crusader, was the greatest friend "the people" had had since Lincoln that he never quite recovered from the shock of the Denver Street Car Strike in 1920. He could not explain satisfactorily the behavior of those whom he had championed for a quarter of a century. He made ticking noises with his tongue against his teeth.

Rich and powerful foes had abused him in public addresses and in print. Advertisers had tried boycotting his paper. Politicians had damned him. Rival journals had hung him in editorial effigy. Society had snubbed him, and ministers of God had grown hoarse in their pulpits from asking Jehovah why He, in His infinite wisdom, had allowed the baby Bon' to survive the whooping cough.

Such affronts, to be sure, did not please him, but he met them with fist and pen. So long as he could be the St. Bon' of everyday folk, he had a *mission.* He portrayed himself as a martyr, happy to go to the stake, "tranquil and unafraid." For he loved the common people, and they him—so he said. And now . . .

The motormen and conductors asked for an increase in pay on May 30, 1920. They didn't get it. More than a thousand trolley-car operators deserted their rolling stock at five o'clock the morning of August 1. Thousands of citizens walked. Tammen wrote a headline:

"KEEP THE DUST OFF YOUR AUTO SEAT
WITH A PASSENGER THESE MORNINGS"

Bonfils assailed the strikers in a "So The People May

Know" column, and Tammen put a sub-title beneath the stock head:

"WHAT FOOLS WE MORTALS BE!
HAVE WE ALL GONE CRAZY?"

The article referred to a certain agitator as the evil genius of the strike, and said:

"If a half-witted hobo, with a little too much hootch aboard, walked on our streets and let out a yell or so, he would immediately be arrested, carted off to jail and tried for disturbing the peace and dignity of the great city of Denver and her people; and he would promptly be convicted or kept in jail against any repetition of his offense against the peace and dignity of the people of Denver.

"But here we have the contemptible spectacle of a foreign agitator, a non-resident of Colorado, coming to Denver, setting the whole city into turmoil, violating the laws of the state and the ordinances of the city, fomenting and pulling off a strike among the employees of the Tramway and completely stopping operation of one of the most necessary and important utilities of the city—and he gets away with it and brags about it; all of which is done at a season of the year when the city is full of visitors, and the agitator goes so far as to even annul the constitutional rights and prerogatives of every citizen of the United States—and, as the German Kaiser did when he started the World War, this foreign agitator snaps his fingers at the Constitution of the United States and says, 'It is but a scrap of paper.'

"Every ordinary function of our city's government has been stopped or impaired or disarranged, and all of this is done to please this foreign agitator. In other words, the welfare, the comfort, the health, the business and the rights of three hundred thousand people are completely set aside and stopped by this one man, who represents one thousand, one hundred Tramway employees, or one-half of one per cent of our population, and democracy has been driven by the arrogance of this unimportant minority from our state; the constitutional rights of every citizen trampled upon and set aside, all to please the bigotry of this little band of selfish men.

"Certainly there is some force, somewhere, that can give the three hundred thousand people release from the domination of this little band of one thousand, one hundred willful and misguided men, or law is a farce and democracy a delirium."

This and other Bonfils polemics incurred the ire, not only of the striking crews but of other union workmen. The car windows were screened with dog-wire for protection of passengers. Strikebreakers, led by "Black Jack" Jerome of Los Angeles, began to operate the trams.

On August 5, mobs demonstrated by overturning three cars at the doorway of the Cathedral of the Immaculate Conception. Chief Hamilton Armstrong and nine members of his police force were injured by bricks and clubs, after an exploding firecracker had started the rioting. Two members of the mob were killed. Mayor Dewey C. Bailey asked that two thousand citizens volunteer as special officers to check the growing violence. From five-twenty o'clock in the afternoon until past midnight mobs worried the city. Every car barn was an Alamo, with beleaguered strikebreakers in the rôle of Davy Crockett.

After the rioters had upset the three trams at the cathedral door, breaking the windows of the cars, knocking out two strikebreakers and forcing the rest to flee into the church for sanctuary, someone voiced the gay idea:

"Let's get the *Post!*"

The cry "On to the *Post*" became the local *Marseillaise.* Several young Dantons in the garb of soldiers were in the van.

The marchers, growing more lusty and vehement every second, surged into Broadway. By this time there were a thousand singing, howling, threatening citizens—presumably not members of Bonfils' beloved "*Post* family."

As they turned into Fifteenth Street, the constantly increasing mob paused at the court-house plaza to refresh their spirits with that venerable hymn, "Hail, Hail, the Gang's All Here." Then there was a cheer and a round of catcalls theoretically intended for Bonfil's distant ears.

By the time the stimulated populace had reached Fifteenth and Champa Streets, leading to the block dominated by the *Post,* four thousand burlies were in the ranks. They began pushing and crowding into the thoroughfare on which the "Temple of Justice" stood, half darkened. A skeleton force of printers was at work on the third floor. On the second were a few editorial writers, worrying over nice phrases intended to make the bulldog edition of the following day as juicy as a stockyard steak. The first floor, housing the business office, was deserted, except for three city detectives and a Negro janitor, all of whom looked as though Gabriel had just sounded his long-awaited reveille. In the half-basement, and to be seen dimly through a heavy screen, were the four gigantic presses, and beyond them the store room of colored inks used to paint gaudy front-page pictures of murder, rape and arson, as well as to tint the many-hued comic supplements beloved by the kiddies. Also in the basement were stored thousands of dollars' worth of paper, in rolls as thick as the columns of an Egyptian temple. The press room and store rooms were deserted and sepulchral.

Although there was continuous movement in the street, the storm cloud seemed to hover in comparative silence as the great mob looked on the place. Then some boys set up a shout and began hurling stones through the large plate-glass windows of the business office. And now the hurricane descended. The young men in overseas uniform advanced to the front line and began to hand-grenade the building with paving blocks, bolts, nuts and the more rugged varieties of vegetables, such as the lowly rutabaga and the democratic potato.

All at once everyone seemed to be armed with something in the way of poles, bars of iron and here and there a grave-digging implement. It was hard to say how many spectators jammed the curb on the far side of the street, all cheering with boyish ardor as the stones flew and the glass fell.

The detectives and the janitor retreated after the windows began to decorate them with formidable crystal earrings. Upstairs, with a fusillade of bricks, iron bolts and other objects coming through the editorial-room windows, the writing force withdrew. The printers, too, turned down the gas beneath the melting pots of their linotype machines and left. As each group departed, the lights were snapped off and the harried workmen got away through rear doors and into an alley behind the *Post* building. They barely had made their escape, however, before the mob began filling the building. Several hundred rioters now learned of the alley exits and went thither to batter at doors, windows and walls. The din was of earthquake resonance.

The front windows having succumbed, a young soldier draped himself in an American flag and then grabbed a plank from a fellow-rioter. He advanced to the front door of the *Post* and began to ram it. The glass of that portal fell with a crash. Other men now assailed the door. Armed with pick handles, planks and iron bars, they savagely attacked the wood-work of the door.

While this was going on, a ladder appeared over the heads of the milling citizens. It was set against the iron balcony of the *Post* building, a kingly platform on which Bonfils was wont to show himself during *Post* celebrations, and to look down upon the faces of "his people." Three men now were ascending the ladder. Once on the balcony, they peered inside the Red Room, then kicked in the windows and entered that historic chamber. More invaders climbed the ladder.

The business office on the first floor now was filled to capacity. A drinking fountain was uprooted, the water gushing forth. This suggested other antics, and the vandals sought out every faucet, turning on the water to flood the building. The water dripped through the floor to the store room, ruining much of the newsprint paper in the basement.

Men were on top of the wire cages of the business office, tearing them down and twisting them with blows of sledges and iron bars. All drawers were pulled from the counters and the bookkeepers' accounts and ledgers torn and strewn about the floor. Someone started a fire after piling up torn books and files of documents, but the ankle-deep water prevented the blaze from spreading. An alarm box was pulled and the fire laddies responded. The fire fighters, however, were unable to lay a hose-line to attack the crowd, so thickly was the mob packed about the street hydrants.

The heavy grating was taken from the half-basement windows, the glass smashed, and the press room invaded. The light switch was found, and the mob in the basement began to attack the presses, using iron bars and throwing tools into the mechanism. The ink-rollers were slashed. Gasoline was poured on rags and waste and a fire started. An employee of the *Post,* loyal to the last, had wormed his way in, unnoticed by the wreckers, and as the fire spluttered, tossed a pail of water, quenching the flame. For his solicitude he was knocked down and might have suffered lasting injury, had not a few of the less-inflamed mobsters carried him to the street.

A half hour had gone by, and still the clamor rose as furniture was demolished, doors broken down and plumbing torn from the walls. Tammen's glass case, holding the stuffed body of the prized baby elephant, Prince Tambon, was given a terrific mauling. Pictures were jerked from the walls, their frames and glass smashed and the pictures themselves torn. Cash registers were put out of commission.

The iron stairway from the business office to the editorial and engraving floor was filled with rioters, all anxious to participate in the bedlam. There were many well-dressed men in the crowd.

When the engraving department was reached, one of the rioters shouted a warning and the wreckers steered clear of the great acid bottles, some of which might have spread death among the sacking men, had they been molested. A large bath of nitrate of silver in the etching room escaped injury. Eager to do something demonstrative, however, the invaders picked up a huge container of distilled water and threw it at a big and costly camera, badly damaging it. The lye vat was ruined, desks upset and more water faucets turned on to flood the second floor.

Due to their failure to locate the light switch in the composing room, the crowd spared the linotype machines. The hot lead in the pot was not a tempting thing in the dark. The editorial room was wrecked, but oddly enough, the typewriters were left unmolested.

From the press room came a great clangor. And now the men were hoisting newsprint paper to the street, where it was unrolled and torn into streamers. One roll was carried to the Tramway Building and placed there as a token of the crowd's displeasure.

And now, an hour having elapsed, the gendarmes, under command of Chief Hamilton Armstrong, arrived to charge the crowd. The chief's head was in bandages, a souvenir he had received at the cathedral when a brick struck him.

Warned of the arrival of the police—and Chief Armstrong was no weakling—the rioters began jumping from the iron balcony, scurrying out the alley-way exits and otherwise leaving the sacked building. But the Temple of Justice had been wrecked in a thorough manner, with many thousands of dollars' damage done.

Bonfils, furious at the demonstration, declared over and over that "the real people" hadn't been guilty of this; that it was the work of hot-headed young hoodlums. He issued Napoleonic orders.

"We'll not miss tomorrow's edition," he said. "The *Post* remains tranquil and unafraid. Get to work. Repair the damage."

Nor did the *Post* miss its next day's editions, although it emerged as if on crutches and with an emergency ward smell to its make-up.

Bonfils was "tranquil," but he sat down to write a "So The People May Know," as follows:

"If a mob can come into any man's place of business, and in violation of every law of the land wantonly destroy thousands of dollars' worth of valuable machinery, furniture and materials, without the slightest reason or justification, then indeed has every man's rights been assassinated, law made a sniveling mockery and life itself a ghoulish farce. And yet, all of this and more was done in the center of the city, and to the *Post* last evening.

"The *Post* for twenty-five years has fought to help labor secure every just demand it made, and, strange to say, the very union that is now striking was the first for which the *Post* drew the sword years ago, but because we differed from them in the present strike—unmindful of all past service—they or their associates and sympathizers unhesitatingly rushed to our building and tried to burn and destroy it, and their only reason for this outrageous and unjustifable act is that the *Post* does not agree with them.

"By the bludgeon and torch they seek now to control the world and force mental slavery upon all mankind.

"We do not ask you to feel sorry for the *Post,* nor for its partial destruction, for all material loss can soon be replaced, but we do ask you to feel sorry for such conditions as would make crimes like this possible anywhere in the United States.

"It means that Bolshevism, Sovietism and Anarchy, with gun and torch, have leaped from bloody and ravished Russia to our beloved land. It means that what has happened to the *Post* may happen to you. It means that Revolution and its red flag and bloodstained sword, is not a hideous nightmare, but is at the door of every man today.

"So do not feel sorry for the *Post*—feel sorry for yourself, for your children, for your country and for civilization. For no man may know who is next on the list.

"The *Post* remains tranquil and unafraid. It will continue to fearlessly expose wrong and evil by whomever practised. It will utterly refuse to be frightened or intimidated by any man or set of men, and if the laws of our country can't give us protection, we must and will protect ourselves."

Martial law was declared and three thousand soldiers patrolled the city. In a happier mood, and under a Tammen headline, Bonfils wrote:

A LITTLE DISFIGURED BUT STILL IN THE RING

"Thanks to our all-star staff and efficient and loyal mechanical employees of the *Post,* we are now back to a normal basis. . . . The courageous and loyal employees of this paper have worked day and night to get the paper in shape to serve, as usual, the interests and welfare of the *Post* family and the great public.

"Now let the community go about its business in the usual way; let sanity and tolerance prevail, but let us never forget our flag, our country, our duties, our rights and our blessed Constitution of the United States that guarantees equal rights to all mankind. And let us stand ready to enforce our laws, and let us all do as much as we can to remove the horrid blot and disgrace that this unfortunate and ill-advised strike brought upon the fair name of Denver and Colorado.

"Yesterday the total advertising in your *Denver Post* was three hundred and sixty-eight columns, and three thousand, six hundred and eighty-six thoughtful friends and members of the *Post* family brought their want ads to the *Post,* in spite of the turmoil, strife and general confusion.

"All of that is now happily over; our offices may not be quite as pretty as they once were, but are just as useful as ever.

"Come down and bring your neighbors and friends.

"The circulation of the *Denver Post* yesterday was 164,187, and this is 29,000 greater than the combined circulation of all the other Sunday papers printed in Colorado, Wyoming and New Mexico."

The street-car strike failed.

Bonfils continued to "fight for the people," but strive as he might to convince himself that the wrecking of the *Post* was the work of misguided *youths,* he occasionally became bitter about the "ingratitude of the masses." However, he was "tranquil and unafraid," continued year in and year out to emphasize the advantages afforded by our blessed Constitution and bragged about *Post* circulation and Colorado weather.

Bon's was not the only man to suffer impaired illusions because of the strike. There was John M. Mulvihill, manager of the world-famous Elitch's Gardens, a Denver summer resort which had an historic theater. On the first evening of the street-car walkout, Elitch's had scheduled a performance of Shakespeare's "Twelfth Night."

When the curtain rose, few patrons were there—the auditorium being graced mostly by ushers (one of whom was the now-celebrated Harold Lloyd of the movies). From that day to this there have been no Shakespearean performances at Elitch's, for, when an aide endeavored to solace Mr. Mulvihill for the week's all-time low gross of $1,200, saying it was because of the strike, Mr. Mulvihill replied:

"Strike, hell! The people don't like Shakespeare."

31

Anointed With Oil

BONFILS' CONNECTION with the Teapot Dome scandal of President Harding's cash-and-carry administration was incredibly weird. His was the first newspaper to break wide open the malodorous deal of the Salt Creek naval reserve—nevertheless, in the end, he profited handsomely.

In the summer of 1922, one of former Senator Fall's secretaries was in a sanatorium. Fall now was Secretary of the Interior. His ex-aide, a woman, had fallen ill while working night and day to further Fall's ambitions. And now she was experiencing that *ennui* common to many invalids, believing herself neglected and forgotten by those in whose ranks she had sacrificed her health.

The woman secretary decided to write confidentially to a girlhood friend, unburdening her mind. In a letter to a Denver woman of long acquaintance, the former Fall employee set forth that she had labored in the Secretary of the Interior's service, how illness had overtaken her, and how her former employer apparently had forgotten the tedious hours she had put in while he was busy distributing oil leases to friends. The simple letter, entirely without rancor, yet expressing wonderment at a world that passes one by, contained no hint that the woman knew Fall's acts to have been contrary to the public good. Her point was mainly that ambitious men might have silk hats, but there was wood underneath; they might have patent-leather boots, but there was clay inside. They forgot.

The recipient of this letter was touched. She consulted her husband, who happened to be a sub-editor of the *Post*. He saw in the letter a lead to a startlingly important story—the country was being stolen. He left his breakfast table on the run and soon was closeted in the Red Room with Bonfils.

After he had finished enlightening Bonfils as to the oil leases, which were being given away like kisses at a wedding, the sub-editor waited to be congratulated by an excited and crusading employer. He was amazed when Bonfils patted him on the shoulder in a matter-of-fact way and said quietly:

"Forget all about it, my boy. We'll just let this matter go by the boards. Don't say a word to a living soul."

"But, Mr. Bonfils . . ."

"Tut, my boy! Tut! Calm yourself, and remember—not a word to a living soul."

Mr. Bonfils watched the Washington news with a practiced eye, and waited. He also inquired as to what had been going on in the oil fields of Wyoming—the Government's naval reserves in particular—for the last several years. It developed that as far back as 1920 there had been underground activity on the part of oil men to obtain concessions from the Government.

Leo Stack, Western politician and oil man, had associated himself in 1920 with E. L. Doheny, the petroleum magnate, and Secretary Fall's life-long friend. That was during the Woodrow Wilson regime. Fall was then a Senator. The Secretary of the Navy was Josephus Daniels. Fall in those days had been active with Mexican investigations and peeping in on Wilson's sick room to learn if the President's mind was tottering.

Stack and Doheny had made an effort to obtain from Secretary Daniels a lease of a double row of offset wells along the boundary line, between the Naval Reserve No. 3 and the Salt Creek field of Wyoming. They failed in this.

Stack then became associated with the Pioneer Oil Company in some kind of agreement, under which Stack presumably was to assist, at Washington, in procuring a lease of the naval reserve, or some portion of it, for that company. In return for his services, Stack was said to have been promised an interest in any lease which the Pioneer Company might win.

The Harding administration now came into power, to return the nation to normalcy. Mr. Denby succeeded Mr. Daniels as Secretary of the Navy. While the great minds stood by, the oil-reserve land-leasing prerogatives were transferred from the Denby desk to that of the new Secretary of the Interior, Mr. Fall. In the light of history, there must have been a scene among the vying oil magnates resembling that in the paddock at Churchill Downs on Derby Day.

When the Pioneer Company decided not to utilize Stack's assistance in Washington, but entered an arrangement with Harry F. Sinclair, the Middle-Western oil czar, instead, Stack took his problem to Bonfils. He came to the right man.

Mr. Bonfils sat down to compose a "So The People May Know," denouncing the Teapot Dome lease as corrupt and contrary to public policy. The rights of the people must be defended. Bonfils was tranquil and unafraid as he wrote:

"Sinclair and his associates have received a gift of one of the richest proven, though undeveloped, oil fields of the United States through trickery that verges, if it does not encroach, on the bounds of crime."

Bonfils then dispatched one of his best reporters to New Mexico, there to examine into Fall's financial status. The *Post* investigator saw and learned much of interest. He found an oil magnate's private railroad car standing on a siding near the Fall ranch. He was told that the Secretary had come upon better days. When the reporter returned to Denver, he announced that Fall was "rolling in wealth." Bonfils did not publish this report.

Instead, Stack brought suit in the Colorado courts against the Pioneer Oil Company and Sinclair, alleging they had conspired against him. The complaint was never filed, according to a report by a Senate investigating committee, but a summons was issued pursuant to the procedure of Colorado courts.

Bonfils and Stack had entered an agreement: that in return for the *Post's* interest in the matter, and were Stack's activities fruitful, the *Post* would not go unrewarded.

Stack would take for himself the first fifty thousand dollars of damages. Thereafter the moneys or other valuable consideration would be split, fifty per cent to Stack and fifty per cent divided equally among Bonfils, his partner and attorney.

There were negotiations in New York and Kansas City, wherein Sinclair settled the suit with them by an agreement under which he paid two hundred and fifty thousand dollars in cash and promised to give nearly a million dollars more. The *Post's* attacks on Sinclair ceased. In fact, an article of laudatory nature appeared in the *Great Divide,* the *Post* weekly paper.

Bonfils now began to function in "big league" company. He was the guest on one of the Secretary of the Navy's warships in tropical waters. When Warren G. Harding went on his ill-fated journey to Alaska, Bonfils tagged along. Whether or not he was invited formally was a question for some debate in the state press of Colorado. It was pointed out that the White House had issued its usual routine notices, wherein the President's trip was announced and provision made for accredited correspondents to accompany the party.

Bonfils appointed himself as *correspondent* for the *Post!*

As a gesture of friendship, Bonfils took along a sealskin coat for Mrs. Harding.

Whether Bonfils brought up the subject of "oil" during his confabs with the frustrated and brooding President cannot be revealed.

All we do know is that Bonfils testified for several uncomfortable hours before the Senate Investigating Committee in regard to his part in the Teapot Dome matter, and that Senator Lenroot baited him until the publicist roared:

"You talk to me as though I was a common criminal."

When asked *why* he had not published the report of his investigator immediately on receiving it, Bonfils replied:

"A fear of committing libel."

Senator Lenroot—Is it not a fact that your contract with Mr. Sinclair was not based upon any legal rights of Mr. Stack?

Mr. Bonfils—That is not true.

Q—But that this whole deal was for the purpose of purchasing your silence in your newspapers?

A—That is absolutely false.

Q—Is it not true that you stopped your attack on Mr. Sinclair with the publication of that first article?

A—We printed the news every day.

Q—Any editorial comments concerning these transactions?

A—I do not think there were.

Q—Do you mean to testify that there was no change in the attitude of your newspaper concerning these transactions from the beginning?

A—I do.

Q—You had vigorously attacked Mr. Sinclair and these oil transactions up to a certain time?

A—We printed that it was a bad lease and I still think it is.

Q—You think it was a bad, corrupt deal, do you not?

A—We were not blaming Mr. Sinclair for it.

The Senator observed that Bonfils' attacks had ceased, once the million-dollar contract had been signed, sealed and delivered. In reply to which statement the publicist roared:

"They did not cease! They have not ceased! They never shall cease!"

The Senator then read a telegram from the *Post's* editor, in which it was said that no article reflecting on Mr. Sinclair or the oil leases had appeared for a year following the original "So The People May Know" blast on Sinclair.

Apropos of nothing, Bonfils suddenly leaned forward in his chair and announced:

"The *Denver Post* has the greatest circulation *per capita* of its publication city of any newspaper in the history of the world!"

Somewhat astounded by this grand trumpeting, Senator Lenroot replied: "I suppose Mr. Sinclair knew all that."

John C. Shaffer's *Rocky Mountain News* was a happy place the day that Bonfils was linked to Teapot Dome.

Mr. Shaffer was absent from the city and had not given any instructions as to the handling of the story, so the boys went the limit. Bonfils was portrayed as the Desperate Desmond of the West.

Then, testimony was given next day before the Senatorial Committee, tending to show that Mr. Shaffer, himself, had profited to the extent of ninety-two thousand, five hundred dollars from the Sinclair publicity funds!

The boys on the *News* were somewhat confused and nonplussed by this sudden awakening.

The American Society of Newspaper Editors thought they should do something about Bonfils' and Shaffer's behavior in regard to Sinclair and the oil barrels. Bonfils was a member of that association, which was regarded as a keeper of journalistic conscience. Shaffer was not a member.

The Society committee brought in a report, finding both publishers guilty of gross violation of the association's code of ethics. The report urged that Bonfils be thrown out of the organization and suggested a reprimand for Shaffer. Months went by, however, with no action, except an attempt to amend the constitution of the society. That amendment, it was felt, would provide a loophole for Bonfils to save his face. The amendment was blocked and Bonfils resigned.

Beset by criticism, assailed by libel suits, and harried by politicians, yet Bonfils was a tenaciously strong fighter. He had described one Ernest Morris as a "Hun", and in a cartoon by Wilbur Steele had depicted him in a German helmet, with the word "Skunk" attached to Morris' person. There was a suit for libel, and Morris received the offer of twenty-three attorneys, free of charge, to prosecute his libel action. Bonfils settled by publishing a complete retraction and by donating ten thousand dollars to charity.

Yet the bulk of Denver's citizens read the *Post,* and while many of them swore at it on occasion, they had to swear by it, insofar as news and features were concerned. Underneath all their circus-like activities, the *Post* owners were shrewd business men. They made sure of giving the sucker more news and more excitement and more features than it was possible for any competitor to do. That practice guaranteed circulation, and circulation guaranteed advertising.

32

The Welkin Ringers

THE *Denver Post* itself was a bigger and better show than the circus it owned. Bonfils might sit, "tranquil and unafraid," in the Red Room, mixing his daily metaphors and purpling those celebrated "So The People May Know" messages, but all about him rose a never-ending bedlam as the Big Brother of the West performed on a flying trapeze before a dumbfounded citizenry.

Tourists refused to believe their eyes when they encountered such surprises as a Houdini hanging by his heels from a rope attached to the *Post's* iron balcony and escaping from a straitjacket; or cages of roaring lions in the street at the *Post's* front door; or a tousle-haired Bonfils, a sudden wild gleam in his eye, prancing on the balcony, reaching into cloth sacks and pulling out fistfuls of new pennies, flinging them to fighting gamins below and shouting: "Lucky! Lucky! The *Post* brings you luck!"

So great was the hubbub, morning, noon and night in Champa Street that even the most hardened pioneer was apt to be startled, thinking for the moment that a hornet had become fouled in his ear-trumpet, or that the Indians were coming.

No other newspaper anywhere was as fertile in the science of ballyhoo, the art of self-aggrandizement, as was the Paper with a Heart and a Soul. Tammen saw to it that the promotional stunts were rip-roarers. Bonfils made sure that the expense thereof was low. The *Post* would give a picnic, but someone else had to provide the sandwiches and lemonade.

The *Post's* first noteworthy stunt was the sponsorship of a statewide beauty contest—the grandmother of the modern "Miss America" tournaments. Comely Miss Marguerite Frey captured the title. The *Post* then sent her to a national contest in Chicago, where other newspapers were submitting their respective Venuses. Miss Frey won that title, too. The *Post* was very happy, indeed, bragged of itself, of Miss Frey and the climate, but heartily regretted ever having hired a certain Baron E. von Pawel.

The Baron had worked for a semester or two, writing of Denver society from a nobleman's viewpoint. Then he clashed with Bonfils. Let us dip into an article published in the *News* on July 7, 1907, touching on this intramural conflict.

"This is an inside story of the Red Room, and also the story of the meanest graft of that past-master of mean grafters, Fred Bonfils. To say that this is the meanest graft Bonfils ever perpetrated is to say much, for his notorious connection with mining graft, stock graft, blackmail graft, lottery graft and circus graft is well known. But this is the meanest and smallest.

"And the best of it is that the story comes from one who recently was well known in and about the Red Room, and who was close to the chief grafter—so close that he was chosen to conduct the beauty contest. It is this contest that was the meanest graft.

"Baron E. von Pawel is the man who has turned state's evidence against the Red Room. He was close to the financial headquarters of the *Post*, but he says that when he found that the contest was nothing more nor less than a graft of Bonfils', he threw up his position rather than be the agent for working petty graft on a girl, and now he is out of a job.

"The women who were induced to enter the Bonfils' beauty contest are thanking their luck that they were not the winners.

"Baron von Pawel, who takes the witness stand, says that the girl who did win the prize, Marguerite Frey, has become the prey of Bonfils in a pecuniary sense. She was forced to sign a cast-iron contract agreeing to turn over all the money she might gain in the national contest to Bonfils, who in return, is to give her employment at ten dollars a week. Miss Frey, for this insignificant sum, is compelled by this contract to give all her time to the 'Paper with a Heart and a Soul,' although all the stories in the paper signed by her are written by Winifred Black, Fred Bonfils' sister-in-law.

"Baron von Pawel adds to these facts the following:

" 'A short time after Miss Frey was declared the winner of the Colorado contest, the trouble started. It became so aggravated that at last the *Chicago Tribune,* which conducted the contest for the most beautiful girl in Illinois, but did it on the square, announced that it would withdraw from the national contest, because, as a newspaper of high standing, it could not afford to be mixed up in such a shady transaction as that which Bonfils proposed, and of which the *Tribune* had heard.

" 'After she won, Bonfils began to tell Miss Frey, who is an innocent girl and not accustomed to men of his stripe, what he intended doing for her. I was then on the *Post,* and I saw what was coming. I told Miss Frey not to have anything to do with Bonfils and on no account to sign any paper he might give her.

" 'Shortly after this, Bonfils sent me from the office. When I returned I found Miss Frey sobbing bitterly. I asked her what the trouble was, and she said Bonfils had forced her to sign a contract.

" 'I at once went in to see Bonfils and denounced his action as a scoundrelly one. I resigned my position, although Bonfils, fearing I would tell the story, urged me not to do so. When he saw I was in earnest and would have nothing to do with his rascally scheme, he threatened me with bodily harm. I laughed at him and left the office.

" 'Miss Frey asked Bonfils for a copy of the contract, the nature of which she did not then know, and he refused to give it to her. Then she hired an attorney, and Bonfils was forced to give it. This copy, which I saw, makes this innocent girl give up all her rights in what she might get from the contest. As this amount is believed to be one thousand dollars or more, Bonfils got considerably the best of the bargain.'

"Asked last night about the Baron's statement, Bonfils, of course, denied it. He added that Miss Frey had been working in the art room of the *Post* for two years, which adds still another suspicious circumstance to the contest."

From beauty contests (the Old Faithful of journalistic circulation boosters) the *Post* went into all manner of stunts. It borrowed some and invented others. When the paper moved from Sixteenth and Curtis to Champa Street, it had open-air shows almost daily in front of its offices. The entire block between Fifteenth and Sixteenth on Champa was barred on these occasions to regular traffic, so that the thousands might witness *Post* antics.

Sometimes the entertainment would last for only a few seconds, such as the time when Gene Bedini, the vaudevillian, caught a turnip on a fork held in his mouth. The vegetable was thrown from the twelfth floor of the Foster Building across the street from the *Post*. The *Post* also sent Human Flies up the face of the Foster Building, and on occasion had circus ladies slide down a cable from that edifice to the street, their teeth clamped to pulleys.

The *Post* was one of the first newspapers to install an electric, play-by-play score board for world-series baseball games. The board was placed on the famous iron balcony from which Bon' looked down on the populace in an imperial manner.

There was a "time-ball," too, fixed to the flagstaff of the *Post's* roof. Precisely at noon each day, this ball would fall with a brave smash. Surprisingly enough, noon-time found hundreds of citizens standing at the *Post*, watches in hand, waiting anxiously for the flopping of the time-ball.

Of sirens there were a-plenty. And bombs and flares were commonplaces—although not of the strength and menace of those which rocked Kansas City the day its editor "displayed enterprise."

There were treasure hunts and housewives' fairs, prizes for the best crazy-quilts, costume fiestas, kids' band contests, cash premiums for the biggest trout caught in Colorado waters, awards for the best airplane models, bird houses, the best-shaped foot, the handsomest back, the oldest married couple, the best horseshoe pitcher and the finest lawns and gardens. The *Post* supervised mutt dog shows, with prizes for the ugliest dog, the loudest dog and the cur that scratched itself the most. There were rabbit shows, pigeon, chicken, fashion, automobile and athletic shows.

For its mid-summer automobile show, the *Post* used the civic center, a place intended by the late Mayor Speer for handsome buildings, Greek columns and bronze statues, and although the city charter specifically stated that this cultural area was to be free forever from commercial enterprises, the *Post* put up its signs regardless of the Athenian environment. If one grew weary of looking at the new cars, one could go across the lawn to the city's Greek Theater, there to see kootchie girls from Bonfils and Tammen's Empress vaudeville house writhing to the rhythm of saxophones.

The *Post* sponsored oratorical contests and spelling bees, swimming meets, kite-flying competitions, Christmas lighting contests, Easter egg rollings, sunset playlets by amateurs in a natural rock amphitheater at Morrison, parties at the Old Ladies' Home, baseball tourneys, cooking schools, dances, archery congresses, farmers' picnics, pioneer parties, juvenile rodeos, oldest sweetheart contests and a thousand other stunts. Merchants or other citizens who contributed prizes or refreshments to these affairs received honorable mention. The name of the man, the firm or organization which supplied the sinews for Bon's charitable engagements was honored by being hyphenated with that of the *Post*. The newspaper's title, of course, always came first: "The *Post*-Smith Roller-Skating Carnival", or "The *Post*-Motor Club Christmas Party."

Pilgrimages and junkets were a *Post* specialty. Tammen was sometimes to be seen in these dizzy caravans, but one might always expect to find Bonfils there, bubbling over with brotherly love and want ads.

The *Post's* annual special train to the Cheyenne Frontier Day celebration usually carried four hundred state and city officials, leading business and professional men from Denver to the land of whooping cowboys and bucking horses. It was a train of ten cars, the finest equipment of the Union Pacific's Columbine Limited to Chicago. The private car of the vice president of the road and that of the general manager were commandeered by Bon' for his

Cheyenne special. There were two dining cars and two big steel baggage cars, in which dance floors were placed, with an orchestra in each car, as well as bars at which free coffee, sandwiches, ice cream, cake, candy, doughnuts, cigars, cigarettes, punch—but no drop of liquor—were served all the way up and all the way back from the celebration. The guests were assured the best seats in the grandstand at Rodeo Park, not to mention free bus rides to and from the arena. There were no women on this train, except thirty or so pretty girls, who went along to dance with the guests. Guess who paid for all this? Bon'?

The *Post* fostered a yearly pilgrimage to the Mount of the Holy Cross in Colorado. This peak, containing a cruciform crevice filled with snow, is three hundred miles from Denver by rail and two hundred by automobile. It was almost inaccessible until the *Post* campaigned for a road and got in on the ground floor with a summer camp at the base of the sacred hill. Bon' grew very pious about the camp, ballyhooed its beauties and chronicled many miraculous cures that occurred when pilgrims gazed on the natural cross. He said it was "a non-profit affair." There were cooks, saddle horses, guides, an orchestra, post-office, lecturers and trails where the young could vary their religious routine with moments of love. The annual pilgrimage in July was absolutely non-denominational.

In emulation of James Gordon Bennett, Bonfils was keen for exploration. In his employ was perhaps the greatest and nimblest of all promotion geniuses, Al Birch. He was so brilliant as a publicity man that New York tried several times to procure his services. But Birch preferred the West and its excitements.

Bon' sent Birch and a party to explore the Yampa River cañon in Northwestern Colorado. Government survey officials had reported this cañon to be about sixty miles long and from two thousand five hundred to four thousand feet deep. No white man ever had lived through its rapids, although several had attempted the feat. It was said that old-time bandits of Wyoming used to rob Union Pacific trains and then flee into a "pocket" of the cañon and "hole up." One man could withstand a posse here, where only a narrow trail extended across the face of a sheer cliff.

Birch built two stout boats, and with three other men tackled the job. They almost perished. One boat was smashed to bits; the other was damaged badly. The party was in the cañon for three weeks, with no way to get up the steep walls. They found the remains of the old dugout where the train robbers had lived, although the trail long since had crumbled. When the party finally shot the rapids for the glory of the *Post,* they were so badly stove in that they were laid up for several months.

Birch afterward explored the Marble Caves in the Sangre de Cristo Mountains for the *Post.*

The *Post* sponsored three rabbit hunts each winter on the plains of Eastern Colorado. From three hundred to a thousand hunters would respond to help the farmer get rid of these pests. The hunters would go to Hugo, Colorado, form companies of one hundred and proceed on military lines. A captain and two lieutenants commanded each group. It was surprisingly seldom that the hunters shot each other.

After the hunt, the grateful farmers muttered prayers for the *Post* and gathered the dead rabbits, loading them onto trucks and transporting them to the railroad station. There the rabbits were spread on the ground overnight to freeze, with sentinels guarding against marauding coyotes. In the morning the whole male population of Hugo would turn out to load the rabbits onto freight cars (donated). In these hunts there never were bagged fewer than eight thousand rabbits, and on one occasion there were one hundred and ten thousand rabbits that had died for the *Post.*

In Denver the rabbits were placed in wagons. The wagons drew up in front of the *Post,* and, while the band played, the frozen hares were passed out to the poor by sixty patrolmen.

Bonfils would look down on "his people" from the iron balcony that fronted the Red Room. He seemed very happy at such times.

The *Post*-Fire Department Christmas Toy Shop was a novel enterprise. Each November, the *Post* began urging the public to give its old, broken or discarded toys to the Denver Fire Department. Such donors would telephone the Department and the fire laddies would call for the toys. Business firms of the city lent tools and machinery to the Department, donated paint, lumber, hardware and other materials for repairs. All the city firemen, in their off-hours, devoted their time to mending and repainting toys.

Children would write to the *Post*, giving their addresses, their ages and stating what kind of toys they wanted. Firemen then would investigate, and make up packages suitable for each child. These bundles, carefully numbered, were laid aside. A child was given a tag corresponding to the number of his package. Then there was a tremendous "morning before Christmas" at the *Post*, with thousands of children standing in line to receive their gifts. Girl employees of the *Post* would make five thousand or more sandwiches from bread, meat, butter and other foostuffs donated by restaurants, so that the children might eat while waiting for their gifts.

Bon' was strong for the Yuletide spirit. Although the orchestra at his Empress Theater was non-union and had been known to fight the musicians' association bitterly, the *Post* succeeded in getting the Denver Musicians' Union to donate a big band that visited hospitals, sanatoriums and asylums to play carols on Christmas Eve.

He also procured for the city an electric cross on an airplane each Christmas Eve. The *Post* induced the Neon sign people to donate a huge cross of red tubing. This was put on the underside of the largest National Guard airplane. The emblem-bearing plane would fly for an hour and a half over Denver and suburbs, while Bon' stood in the back yard of his great home and was "touched" by the beauty of his flying crucifix.

The *Post* promoted a Business Men's Motor Cavalcade from Denver to the Black Hills of South Dakota one summer. Five hundred gentlemen paid their own expenses, the object being to "cement good will between Denver and that good business region."

When the International Advertising Clubs of the World held their convention in Denver in 1928, the *Post* presided over a spectacular entertainment. The Moffat Tunnel was not yet completed. The *Post* promoted a special train and took seven hundred and fifty delegates to the east portal of the bore. Bon' had a table *one third of a mile long* set up in the tunnel.

This table was covered with brown paper. Rough board benches flanked it. Miners' utensils were laid. The little rock-dumping cars used in tunnel excavation were employed to haul food up and down the "dining-room." Beans, bacon, boiled potatoes, sour-dough biscuits, coffee and pie—the miners' diet—was the bill-of-fare.

This gargantuan gesture earned the *Post* immense publicity. It worked slightly to the disadvantage of the persons who footed the bills, however, for at the eleventh hour it was found that the tunnel seepage was not conducive to the comfort of diners. A costly roof and floor were installed.

The *Post's* relationship with the local theaters, motion-picture houses and other amusement enterprises was impressive.

Bonfils became intensely interested when, in 1917, the motion picture had definitely outdistanced the legitimate theater in the amount of advertising placed in newspapers. The cinema heads were pouring fortunes into their promotional and advertising budgets. Bonfils' own paper showed a three hundred per cent increase in amusement advertising in one year. He decided to take personal charge of the situation, much to the dismay of press agents and business managers of playhouses. Their day of free publicity was definitely done. If they wanted photographs of lovely actresses in the *Post*, they had to pay according to Bonfils' standards.

At Bonfils' behest, special checks were made as to each theater's newspaper space. The *Post's* drama columns were measured, as were those of opposition papers, and the expenditures of amusement companies in each paper compared carefully.

To his horror, Bon' found that certain theaters and parks were using as much lineage in other papers as they were in the *Post*. To his way of thinking, this constituted some form of disloyalty. When he discovered that several playhouses were not advertising in other sheets as copiously as they were in the *Post*, he formulated a code, with a sliding scale. First, he ruled that there must be an immediate and substantial increase in advertising rates for all Denver's amusement places, but that those theaters which elected to advertise exclusively in the *Post* would be given a "bonus" of a *decreased* rate. Next, he announced that those houses which advertised seventy-five per cent in the *Post*, and twenty-five per cent in all other papers, would be given an advantage also—but not *quite* as generous a "bonus." For example, any theater spending *all* its appropriation with the Post ("And we don't care how little it is, son") would be billed at thirty cents a line. Those that used seventy-five per cent of their total space in the *Post* would be given a thirty-five cent rate, and those that used one-half their space in the *Post* would be given a forty-cent rate.

If, however, any theater used *more* space in other papers than in the *Post*—treason!

It soon developed that theaters which used the *Post* exclusively for advertising not only received columns of free publicity, but were assisted by the wildest stunts imaginable.

A show called "The Under Dog" came to the Broadway Theater. It was a political drama. The press agent said he would spend an extra pile of money with the *Post*, provided the paper would concoct some effective publicity stunt.

This order came on short notice, but Mr. Birch was ready—as ever. He visited the dog pound and induced the pound master to "lend" him all the curs incarcerated there. The pound master also assigned two expert dog catchers, and gave Birch himself a loop with which to snare absent-minded mastiffs. After a few hours' work, Birch and his aides had collected more than a hundred dogs of plebeian ancestry but of splendid voice. He hired a rather rickety wagon, put the dogs in a makeshift pen fashioned of chicken wire and bed slats nailed to the side-boards of the vehicle. Then he painted a red sign on the aged van, inviting the public to "Come see 'The Under Dog.' "

Everything went reasonably well until the buckboard of shanghaied curs approached City Hall. The heat of the July day made these crammed-in beasts restless and they were sending up a genuinely deafening clamor; but so far there had been no death-dealing bites—just a few hot-weather nips and three or four lovers' quarrels.

The City Council was in session this warm day. The Supervisors were so loudly criticizing the Mayor, or *vice versa*, that none of them paid much attention to the on-coming wagon with its cargo of throaty mongrels. Supervisor Kindell claimed afterward that he had heard the din, but had mistaken it for applause.

In turning a corner, the ancient wagon shed a rear wheel. The front wheel on the same side collapsed out of sympathy. This accident gave the vehicle a sudden list. With an almost human groan, the van lay on its side, as though dying. The makeshift coop bellied with dogs, then gave way. And now the one hundred and more beasts of all sizes and breeds were scrambling, snapping, howling, baying, flea-jostling and singing something that sounded suspiciously like a hymn to sex. They charged on City Hall.

Some distrait public servant had left the basement door ajar, the one leading to the Police Department. Twenty of the stampeding dogs, with a homing-pigeon instinct, made for a barred door giving on the jail. That door frustrated them momentarily. The leader of the detachment, a Dachs-hund-Terrier-Dalmatian, sniffed at the bewildered trusty's leg, then bit him and led the way to the office of the Chief of Police. Chief Felix O'Neill was that moment comparing some finger prints—and thinking. The score of dogs charged in, upset a brass cuspidor and began to play a desperate kind of leap frog. A fat sergeant and several patrolmen arrived, drawing their night sticks.

A second group of dogs made for the stairs leading to

the Health Department, where they interrupted one vaccination and two Wassermann tests. Still another and larger pack of yammering mutts raced into the council chamber. The issue of Mayor *vs.* Supervisors became confused as the city fathers struggled, knee-deep in dogs.

Nor was this all. One sly and rascally cur, a renegade with an Airedale body and a Sealyham chassis, loped into the Mayor's office. There he studiously outraged a clerk's desk, a mahogany relic of pioneer days—the clerk himself barely escaping a consequence as impertinently foul as any criticism ever leveled at a conscientious administration.

Birch scored so many sensational successes in his line that it seems hardly fair to refer to his defeats. However, the case of Violet Mann, the beauteous blonde appearing at the Orpheum, was a temporary setback for this genius.

One summer night Birch hired a stage hand, Bo Brown, to "kidnap" Violet as she left the stage door with her sister and brother-in-law. She was "lifted" at a street corner and put into an automobile. Violet was wearing a white dress. There was a grip on the floor of the car, containing a black dress, black hat and black shoes. Violet made the change while police cars pursued.

Detectives at the Union Depot were on the lookout for "a woman in white." Birch managed to slip Violet past the gate, boarding a train to Colorado Springs. When they had reached that town, Birch took Violet to a secluded place, beat himself on the nose to supply blood, then smeared the actress' hands and cheeks. Next he gave her some medicine. After she had downed it, there was no need to tell her it was ipecac.

Violet staggered into the police station to inform the night captain that she had been abducted, attacked and compelled to fight for her honor, then hurled from an automobile. She actually was so ill by now that the captain ("I have a sweet little daughter myself, girlie") was ready to lynch any suspect. He called the Denver police by telephone, and was enlightened.

"Just kick that gal out! It's all a fake. We nabbed the guy that rented the car for this press-agent gag and made him come through."

The story of Mr. Carter's monkeys is a part of the history of Denver ballyhoo. Mr. Carter was a jeweler, but his real love was not for diamond brooches and wristwatches—but for monkeys. He had a lot of 'em, all trained and full of tricks. The more he saw of his monkeys, the less he cared for jewels. He yearned to display the talents of his charges in a big, public way. So Mr. Carter petitioned Elitch's Gardens to book his monkey act, which the management did with a proviso that some impressive publicity stunt be devised.

Mr. Birch thought that an "escape" of the animals would attract attention. Mr. Carter agreed to this, once he was assured that it was merely a fake. The town was only fairly excited when it heard of the roaming monks. One of them had been "found" in the apartment of a society lady, and another in a South Denver grocery store. The general mood of the city on learning of these captures was "What of it?" So Mr. Birch set out to make the "escape" more startling.

He arranged with the night watchman of the State Capitol to keep eight monks overnight, and in the morning to release them on the coping of the state-house dome at an hour when the Governor and other tax-eaters were due at their offices. Birch then proposed to have the police and fire departments capture the monkeys, using great extension ladders while the populace gaped.

Everything looked propitious until an early-bird delegation called at the Capitol to visit Governor Ammons. There was quite a party when the mild-mannered Governor came out on the Capitol steps to be photographed while shaking hands with the voters. The cameras were lined up and his excellency was grasping a delegate's right hand in fellowship when there was a sudden and loud explosion, and then another, and another. A spirit of panic prevailed. It appeared as though some assassin were bombing his excellency.

One of the Governor's staff officers glanced up and saw monkeys unscrewing electric light bulbs from the dome's ribs. The animals were chattering and prancing and bombing away while the delegation scattered. Mr. Birch found it necessary to appear before the board of Capitol managers to explain.

To glorify the beauties of Estes National Park—Colorado's Yosemite—Mr. Birch publicized it as a Garden of Eden. This phrase suggested a need for an Eve, so Mr. Birch obtained one. He prevailed on Hazel Eighmy, a clerk in a photography studio, to pose as Eve, principally because she had handsome legs.

The *Post* one afternoon announced that Miss Eighmy, bored by the mockeries of civilization, was about to retire to the wilderness of Estes Park, wearing only a leopard skin and taking with her nothing else but her own fair self. She would outdo Robinson Crusoe, subsist on roots, berries and herbs until she could fashion a spear, a bow and some arrows for game. She would make her own fires, by rubbing sticks together. She would vanish from human ken for a month.

On the appointed day, Miss Eighmy arrived at the cabin of Enos Mills, the renowned naturalist, where she changed from civilized garb to the hide of a defunct leopard. She certainly looked fine as she let down her corn-colored hair and stood before the cameras. It seemed a shame to allow her to go alone into the woods, but a group of forest rangers restrained several sympathetic fellows who sought to accompany her. One ranger led her to the wilderness, so as to make it official.

That ranger guided Miss Eighmy to a back trail, where Mr. Birch and his automobile were waiting. Miss Eighmy's mother was there, too, as chaperon. They drove several miles farther on to Birch's cabin. The Modern Eve was to hide there until time came to emerge from her tussle with nature.

While the host, Mr. Birch, was opening a can of beans for his fair guest, she suffered a nostalgia for the city's pavements and bright lights. Nor could Mr. Birch persuade her that the scent of pines and murmur of brooks was preferable to the tootlings of a nine-piece dance orchestra. The ranger had committed a terrible *faux pas,* revealing to Miss Eighmy that a hoe-down was scheduled the following evening at a nearby village.

"I'm going to the dance," Miss Eighmy said. "What's the point of staying cooped up here like a rube?"

Mr. Birch spoke of the glory that was to be hers on emerging from the wilderness, famous and the recipient of vaudeville offers. Already, he explained, she was participating in the profitable business of supplying the Hearst newspaper syndicate with articles depicting the struggle with nature in the raw.

"I'm going to the dance," Miss Eighmy said.

"If you've simply *got* to dance," said the host, "I think there are some records here for my Victrola."

"Yeah?" said Miss Eighmy. "Well, I saw them, and they go back to the time the phonograph was invented, almost."

Miss Eighmy went to the dance. Birch accompanied her, fought off four romantic tourists and finally dragged her away when he overheard some woman or other say she "thought the girl's face looked familiar."

Birch had to return to Denver, to resume his duties as city editor of the *Post*. Hazel's mother promised she would keep Eve incommunicado.

Again at his desk, Birch found the entire nation interested in the Modern Eve's brave conflict with the primitive. Queries arrived hourly from newspapers, magazines and even book-publishing houses, asking for photographs and special stories. Things such as these excited Mr. Birch no little.

He decided there should be a cave man angle to the story. He assigned a photographer and reporter to find an Adam.

"I don't want any goof actually to go up there," he said. "Just look around Berkeley Bathing Beach and pick up some stray tourist, a raw-looking geezer. Take a bearskin along and pose him. Make him a present of the bearskin if he'll stand for a pipe story."

The results were magnificent. The newspaper men had found a telegrapher on vacation from his native Omaha. He was big and fierce looking, and an addict of rare beefsteak. Birch ran his picture—with the bearskin aboard—and wrote the story himself, a chronicle of a man's brave heart that beat for a lonely and imperilled Eve.

Thirty-six hours later Birch was horrified to hear from Captain Claud L. Way, supervisor of the national park at Estes, in a long-distance telephone call.

"There's a nut up here," said the puzzled Captain, "who claims you appointed him as Adam to find your Eve. I tried to argue with him and he put up a fight."

"Where is he now?" asked Birch.

"Well, he was roaring and going around naked, almost, with a bear rug over himself. He had a club and was beating the bushes. So I just kicked him out and told him I'd throw him in the can if he ever showed up again."

Birch wrote a story about the ejection of Adam by an angel of the law. It created much excitement. There was talk that Adam was about to sue Captain Way, manhandle Birch, bring action against state and nation. To add to Birch's discomfiture, Captain Way protested against the *Post* revealing that he had kicked a taxpayer out of a national playground. And finally an Estes Park banker came to town, angry because the Park was receiving such undue publicity.

Birch brought Hazel back to civilization ahead of schedule, booked her in vaudeville, where she earned some seven hundred dollars and received scores of proposals of marriage. Birch was to be seen near the stage door, armed with a club in case Adam ever showed up—which he never did.

The *Post* coöperated to the fullest with its motion-picture clients, provided they gave Bonfils' paper the bulk of their local advertising. One movie concern, resenting "high-handed methods" withdrew advertising of its Denver houses and tried getting out a dodger to be distributed from door to door. It was a short-lived fight. The public always read the *Post* for theatrical information.

The *Post* gave Leo, the Metro-Goldwyn-Mayer lion, a great reception. Leo, as may be unnecessary to state, is the melancholy fellow who is seen at the beginning of all MGM films. As a promotion enterprise, Leo made a personal tour of the key cities of the country one year. A report that there were sixteen of these original Leos simultaneously on tour is beside the point. The one that came to Denver seemed authentic enough. He was even more melancholy in life than the screen portrays him and was inclined to sleep overmuch.

A "birthday party" was on Leo's agenda in each town visited by him—or them. The *Post* decided to accord this Rip Van Winkle of the zoos a party such as no king of beasts ever had enjoyed.

Champa Street was roped off and a cage set up on a platform which once had been the prize ring wherein Jack Dempsey and other pugilistic stars had risked their chins. Due to some oversight, the people who had been using this platform each noonday for a week were not notified. These folk were delegates to a Presbyterian convention. They had received glad permission from the *Post* to preach every day while the presses were rolling and the newsboys howling. Any group might gather and cavort in front of the *Post*, provided they were certified patriots and promised enough noise and human interest. It was a Roman forum—with hippodrome trimmings.

The *Post* had advertised the importance of Leo's birthday anniversary fête. Bauer's Confectionery and Catering Company had baked a whopping big cake with sixteen candles (Leo was nineteen that same year in Seattle). This was to be lighted and shown to Leo—provided he could be roused from his sleeping sickness—and then cut up and distributed among the kiddies.

Shortly before the ceremony, the Presbyterians brought their Bibles into Champa Street. Presumably they had not been reading the newspaper, for they mistook the huge throng for sinners waiting to be saved. The presence of the cage puzzled them somewhat, so they asked the *Post* janitor: "Where do we perform?"

"Perform hell!" said the janitor (a non-conformist, no doubt). "It's Leo the Lion's birthday party. You'd better get going."

The delegates, however, stood by, while flunkeys wheeled forth the great cake, gleaming with white icing and containing *Post*-MGM greetings done in pink on the frosting. Leo woke up once and yawned, then relapsed into a jungle dream. A keeper lighted the sixteen tapers.

The ignited cake was very close to Leo now, almost singeing his mane.

Leo sniffed, then awakened, growled and swatted the cake, snuffing the candles and smashing the great pastry into bits. Then he lay down again, licked his paw and went back to sleep. There was no party that day.

If there was a national candy week, the children knew that the *Post* would distribute two tons of sweetmeats, donated by manufacturers. Apple week brought tons of that fruit to Champa Street. The *Post*—when there was a shortage of "weeks"—formulated its own campaigns. One of these was "Shop in Denver Week."

The railroads reduced fares and merchants put on bargain sales for the benefit of outdwellers. To enliven the evenings, the *Post* gave a big free show in front of the Temple of Justice. There were athletic events, boxing and wrestling. The *Post*, obviously, never besmirched an athlete's amateur standing by forcing money on him.

One wrestler carried a grudge against the *Post*, and against Bonfils in particular, for injury to his pride. He had been wheedled into appearing on the platform during "Shop in Denver Week," and told that it would mean much publicity. He got in the ring, but almost immediately was butted by his opponent. His bridgework caved in, and he resigned. When the wrestler failed to receive money to repair his damaged molars, he went up and down the curb, pointing to his vacant gums and roaring: "Where's this here Bonfils guy? Let me at him! He's *got* to pay for my teeth."

One day at City Park, a boy and a girl found a nickel. These children were about six years old and truly appreciative of the things a nickel could buy. They held an excited consultation beneath a maple tree. The girl suggested they invest in an ice-cream cone. The boy agreed that ice cream was a remarkably fine thing, but pointed out that a division of a single cone might lead to innumerable arguments. The girl proposed a rather forward-looking technique. They would take alternate licks at the cone. Of course, in carrying out this equitable plan, each would be placed on his or her honor, not to take porcine licks.

"We'll get peanuts," said the boy.

The girl asked him to reconsider. Ice cream, after all, was sweeter, cooler, more exciting in every way than peanuts. No shells to crack. No thirsty after-feeling.

"I say we'll get peanuts."

The girl tried to effect a compromise on soda water. Then candy.

"We'll get peanuts."

Of course peanuts were much better than nothing at all, so the young lady gave in with the melancholy resignation of a wife.

A swarthy alien's peanut stand was to be seen near the zoo, and the youngsters went thither with their nickel. The boy asked for and received a sack of peanuts. The children then returned to their maple tree, to divide the goobers. There was some argument as to the size of the nuts that the boy kept for himself. Also there was an odd nut, which he took as a commission for doing the supervisory work.

The children then bit into the nuts. The girl made a wry face and spat. The nut had been bitter. The boy tried several nuts. He found them withered and bitter. Then the boy took what remained of the nuts and returned them to the wrinkled sack. He and his girl went to the vendor's wagon, made a complaint, placed the sack of nuts on the sill of the stand and demanded their nickel.

The swarthy vendor looked at them with a ferocious air of reproval. He refused to return the nickel. Then, with more tears and a few shouts from the boy, the vendor told the children to get the hell away from the peanut stand.

The boy and girl once again retreated to their maple tree and sobbed bitterly. At this point a stalwart citizen, who had been taking the air, came upon Hänsel and Gretel. He made inquiries. As the story of the withered and bitter peanuts was unfolded, he became irate.

He visited the alien vendor and made a demand for the nickel. The vendor told the irate citizen to go to hell. At first the irate citizen was going to beat up the vendor, but a better plan dawned on him. He told the little girl to go on home. To the boy, he said:

"Son, you come with me. I know a place where we'll get even with all the peanut sellers in the world."

The irate citizen and the buncoed boy went downtown. They arrived at Bonfils' Red Room and were given immediate hearing. Bonfils listened to the sad story, then pressed buttons and began roaring for his troops.

He assigned three reporters to look into the peanut situation. They were to find out all the facts, prices, conditions and licensing requirements for peanut vendors.

For days there were tremendous stories on the peanut situation in Denver. There were attacks on defective peanuts and on their handlers. It got so that merchants were afraid to sell peanuts at all. When they did sell them, it was with great foreboding, and the price of peanuts dropped. One could buy a huge sack for a nickel. Denver never before had seen such finely roasted and healthy peanuts as now graced a trembling market.

The city fathers made the licensing of peanut vendors a truly complex ritual. Many permits were revoked. And it is said that the man who sold the withered and bitter peanuts to the Hänsel and Gretel of City Park not only lost his license and was kicked bodily from the people's playground—he actually was *deported* from the country!

Bonfils never let the people lack for a champion.

Pioneers always appealed to Bonfils' sympathy. He liked them so much that he simply couldn't think up enough contests to entertain them. A great chance came one year—I think it was during "The *Post*-Jantzen Learn-to-Swim Week," or possibly "The *Post*-Rocky Ford Watermelon Week"—when a reporter stumbled upon a rare case. An aged pioneer was found in a house which he himself had built shortly after Civil War days. He had lived there peacefully, and without modern conveniences, such as electric lights, telephones, running water, gas stove or phonograph. But he was content.

The dear old fellow drew his water from an ancient well, of which he was quietly proud. He had a small pension and wanted nothing more of the world. Nor did he bother anyone. He loved solitude.

Bonfils saw this as a "pitiful case" and got very busy about it. To think that a pioneer, "one of the bulwarks of our society," should spend his old age in such primitive fashion!

It was an indictment of modern civilization.

The old fellow didn't appear to know what all the fuss was about, but looked on mildly as the *Post* campaigned, editorialized and crusaded to give this pioneer his due.

A music store donated a phonograph. The Public Service company provided a year's supply of gas and electricity. Artisans gave their services, and dealers supplied the materials to equip the pioneer's house with gas, water and electricity. The Water Company offered a year's supply of that fluid free to this mild old pioneer. He had a telephone, without cost for a whole year, unless he got in the habit of calling Chicago. Stores gave groceries, and meat was promised for twelve months. Civilization literally went hysterical in giving of itself to this patriarchal gentleman.

He didn't appear to know why so many bundles were coming into his house, where he had been content to inch along in a poor but tranquil fashion for so many years. Nor did he grasp the reason for the presence of workmen with their hammer-throwings and pipe-wrestlings, although he got to know them from their comings and goings and once offered to tell a story about General Phil Sheridan to a plumber who was putting in the bath fixtures and kitchen sink.

As a generous climax to this civilized activity, Bonfils personally decided to do something real and fine for the blessed old pioneer. Others had donated this and that for a year. He would top it. He ordered that a free subscription to the *Denver Post,* daily and Sunday, be given him *for life.*

It is said that the old man didn't feel so comfortable amid his new splendor. The telephone buzzed constantly, cranks, people of whom he never before had heard—and of whom he didn't want to hear—gabbing with him. Then, when the year of free things was up, a lot of correspondence began to come his way, envelopes with tissue windows in them. And now, if he cared to continue the blessings of civilization, he must pay. The year was up.

The old fellow, robbed of his peace and former resignation, run ragged by knocks at his door, and beset by bills, did the sensible thing. He died.

33

A String Of Pearls

THE SMALLEST words of criticism were enough to earn Bonfils' life-long enmity. Yet Tammen was the one man who could, and did, twit him constantly—especially with regard to his parsimony. Occasionally Bonfils would do some sudden and showy deed, as though to refute the claim that he loved his nickels.

One day he issued an excited order: "Get all the freckled boys you can find and have them here within an hour."

No assignment by Bonfils ever was questioned. Reporters went out to hunt speckled lads for delivery at the Bonfils door. It was summer, and the supply of freckled young men was plentiful. About a hundred were brought to the *Post* and there lined up to await whatever the Bonfils whim dictated.

Bon' came from his office, seemed somewhat disappointed at seeing *so many* boys, then regained his composure. He asked a lieutenant to "get me a great big bag of silver at once."

While the aide was getting the great big bag of silver, Bonfils addressed the mystified lads on the glories of America, the blessed Constitution and recited circulation and advertising figures as applied to the *Post*. Then the bag of silver arrived.

Bonfils frowned slightly when he found that *dollars,* not *dimes,* were in the bag. Nevertheless, he was exceptionally game. He dipped into the bag, and passed hurriedly down the line of boys, pressing a dollar into each outstretched palm. He paused, wordlessly, to reprove with an eagle-stare one ambitious fellow who had put forth *both hands.* Then he announced suddenly:

"No one ever goes away from the *Post* empty-handed."

He sighed and disappeared.

The Christmas following this eposide, Bonfils was in the East, and Tammen had put gold pieces in each pay envelope of *Post* employees. When Bonfils returned, he was as nearly angry with Tammen as anyone ever had known him to be.

"We pay our people what they're worth, and more. This sort of thing only spoils them."

On one occasion, Senator Ham Lewis of Illinois was visiting Bon' in the Red Room. Tammen entered and at once began to harangue the pink-whiskered solon on Bonfils' worship of the golden calf.

"Senator," said Tammen, "you may not know it, but you're sitting across the table from the smartest and ablest money-maker in America."

"I well believe that," the Senator agreed.

"But," Tammen added, "he's also the tightest. Why, he's so damned tight that he's been using the same automobile, an aged Franklin, for the last six years. It's a disgrace to me and to this newspaper. Can't you use your influence to get him at least to repaint it?"

Senator Lewis tugged at his cerise beard. "Fred, don't you think you put too much emphasis on money?"

Mr. Bonfils was ill at ease. "It's not true that I think money is everything. But it means power. And with power, one can then do good."

"What did I tell you?" Tammen said. "Money is his life."

"Fred," said the Senator, "two prominent men died recently in Chicago—died within the same hour. A Chicago newspaper combined their obituaries in a single story, and I'll never forget one paragraph that said: 'Smith died leaving three million friends. Jones died leaving three million dollars.' "

Mr. Bonfils turned as red as the Senator's beard. He didn't like the story, and he never seemed thereafter to be interested in Senator Lewis, personally or politically.

In defense of his nickel-nursing policies, Bonfils once told M. Koenigsberg, the famous Hearst executive:

"Certain people say I'm stingy. Well, I cannot forget my boyhood back in Troy, Missouri. We were then a large family and very poor. We seldom had meat. My good father once took a steak and cut it up so that each child would have a piece no larger than another child received. I remember him giving my mother *his* share. Then he kept the bone for himself, pretending it was by far the best and largest piece of all. I wasn't fooled, and then and there I decided that I would get a fortune from the world, that my family never would suffer, and that I would *keep* what was mine and let others waste their substance."

Although it was Bonfils' rule to make his traveling companions share all expenses, his familiar, Volney Hoggatt, was an exception. He would pay Vol's way, but did not allow for fancy food or drinks. In fact, they once put up at an inn where the coffee was good, but scanty. Volney wanted an extra cup, but Bonfils wouldn't permit it.

"Too much coffee ruins the digestion," he said.

Volney discovered that the one pot which was served for Bonfils and himself actually contained two and one-half cups. Bon' usually finished his coffee in time to drain the extra half cup. To checkmate his host, Volney armed himself with a small hot-water bottle. During the meal he surreptitiously would pour the extra half cup into the rubber container, thereby gaining the added refreshment for himself.

A clergyman once indicated the difference between Bonfils' monetary viewpoint and Tammen's by citing an incident wherein the partners gave him a reward for performing a certain valuable service. Tammen wrote the cleric a check for a thousand dollars, with a notation, "For services," scrawled across the margin. Bonfils handed over a check for five hundred dollars, with the word, "Charity," written along the side.

Tammen's closest friend was Ogden Armour, Chicago meat packer. These comrades communicated daily, no matter how widely separated. Armour's private car often was part of the Sells-Floto Circus train, for Tammen was fond of traveling with the show.

The Tammens frequently visited the Armour country place, and it was there that Mrs. Tammen one evening was admiring Mrs. Armour's string of perfectly matched pearls. Tammen, also, became interested in the pearls.

"Say, sister," he said, "take those things off and let me have a squint at 'em."

He held the necklace for a while, then asked Armour: "How much did these baubles set you back, Ogden?"

"I forget off-hand," Armour said. "Half a million I think."

Tammen handed them back to their owner. "I'm no tightwad, God knows, but if I was a billionaire, instead of just a bum multi-millionaire, I wouldn't pay half a million for such junk!"

Tammen had a long talk with his wife in their Humboldt Street home one evening. They were discussing the making of their wills. They had no children to inherit their fortune. They once had planned on adopting a child, but had decided they were too old to act as foster parents.

The Tammens thought it would be a fine thing to leave their money, all of it, to charity. But which charity? There were three groups which appealed to their sympathies. One was the destitute tubercular patients. A second was the Old Ladies' Home. A third was the Orphans' Home.

Beyond agreeing that one or all these groups would share in their riches, the Tammens postponed the actual drawing up of their wills until another day. Tammen then went to the office, where he found Bonfils in conference with a group of club women.

Bonfils explained to Tammen: "These good ladies, Harry, have started a drive to build a wing to the Children's Hospital. They need fifty thousand dollars for that purpose. They have asked that we each contribute a thousand dollars."

"What did you tell 'em?" Tammen asked.

"I have pointed out to them," Bonfils said, "that we can

give something much more valuable than money—publicity. Thousands of dollars in publicity."

"But they still want the money. Is that it?" Tammen asked archly.

Bonfils cleared his throat. "Mmmm. Yes. They want us to contribute. What do you think, Harry?"

Harry turned to the expectant ladies. "Could you come back tomorrow afternoon at this time?"

The chairman of the committee said they would be back.

After the ladies had gone, Bonfils handed Tammen a check for a hundred thousand dollars. It was a *Post* dividend. Little more was said about the Children's Hospital, and Tammen left for his home.

He greeted Mrs. Tammen by giving her the check for a hundred thousand dollars. "There you are," he said. "There's your string of pearls. I was going to order 'em myself, but I'll be damned if I want it on my conscience, paying out all that money for something that diseased oysters produce. So you take the sin on your own shoulders."

Mrs. Tammen's secretary said someone was on the telephone. Mrs. Tammen was gone a long time. When she came back, she said:

"Harry, I've an idea that may appeal to you."

"You never had one that didn't," he said, "except wanting pearls, perhaps. And that's all right, too."

"The pearls have something to do with it," she said. "Listen. The Children's Hospital committee is asking me to give a thousand dollars to their fund. A new wing is needed."

"I know," Tammen said, "they were down to see Fred and me today. Well, what do you want to do?"

"Harry, I think pearls are beautiful things, but there's something more beautiful. And I want to give this check to the Children's Hospital."

"All of it?"

"All of it."

He looked at her for a while. Then he put his arms about her and kissed her. "I think you're the most beautiful thing there is in life," he said. "I think you're swell."

Tammen was very happy about this decision. The Children's Hospital was asking for fifty thousand dollars. He would give a hundred thousand in his wife's name. Double what was wanted. That sort of gesture was peculiarly dear to Tammen's spirit. He liked to spring surprises—no matter what sort.

"We'll play a joke on Bon'," he said to his wife. "Call your friend, the committee head, and swear her to secrecy. Tell her under no circumstance to tell the other ladies or Bon'. Tomorrow I'll spring it when they call."

The next day, the club women gathered in the Red Room. Tammen purposely was late. Bonfils was sitting there, twiddling his thumbs, a bit nervous over a decision he himself had reached—to contribute a thousand dollars!

When Tammen came in, Bonfils indicated by his manner more than by words that the delay was unwarranted. "Harry, these good women have been waiting an hour."

"That's too bad," said Tammen. "I should have telephoned."

"What do you mean?" Bonfils asked.

"Simply this, Fred. I can't give a thousand dollars, any way you want to look at it."

"Harry!"

"I can't do it. Sorry."

The women (all except the committee head) were rather deflated. Bonfils then decided to deliver a moral lecture.

"I am surprised at you, Harry," he said. "We have been the closest of friends for many, many years. And I can truthfully say I never knew you to be this way before. You always were generous and sensible. Now you're just being stubborn. Why?"

"Ladies," said Tammen, "I won't give a thousand dollars, and that's final. However, I'll tell you what I *will* do. I'll give you a hundred thousand dollars!"

Bonfils looked stupefied. "Harry! Harry! What is this you're saying?"

"I'm saying I'll build that wing, not for fifty thousand dollars, but a better wing for a hundred thousand. But I won't give just a measly one thousand."

Tammen was laughing. The women were almost hysterical. Bonfils thought his partner suddenly daft. In fact, he inferred as much when he said:

"Harry, I hope you *know* what you're doing."

"You bet your life I know what I'm doing," Tammen said. "I'm giving these women a string of pearls."

That cryptic remark led the ladies to think Mr. Tammen a bit eccentric, but they were at pains not to be too inquisitive.

And now Tammen made out his will, leaving one half his entire fortune outright to the Children's Hospital, and the remainder in trust for Mrs. Tammen. At her death that half also would go to the Children's Hospital.

When the hospital wing had been built, Tammen (under protest) was called on to be present at the dedication. He stood at the rear of the throng of notables. When one of the speakers urged that he "come forward to say a few words," Tammen said:

"Not in a hundred years."

"But why won't you stand up front, so the people can see you, the one who made possible this great wing?"

"Because," said Tammen, "I've done so little."

The hospital people now needed more funds to provide up-to-date equipment for the fine wing. A committee called on Tammen, the chairman saying:

"You have done so much that we hesitate to ask more."

"Hesitate, hell!" Tammen said. "Did you think I was going to let the place go unfurnished?"

He gave another fifty thousand dollars.

Of the millions he left, Tammen had only one ironbound proviso in his benefaction: that poor patients, unable to pay, would not be singled out as "charity" cases. Not even the nurses were to know which was a "charity" case and which a paying patient. Each child was to be treated like any other, rich or poor.

The work of Mrs. Tammen, then and thereafter, in taking care of what she and her husband called their "string of pearls," made her one of the best-beloved women in Colorado's history. The Children's Hospital became one of the leading institutions of its kind.

Tammen always had looked healthy. His skin was

smooth and fair, his eyes bright. Yet in fact he now began having trouble with his stomach, and grew anemic. He traveled more and more, visiting doctors at Johns Hopkins Hospital and elsewhere. He spent much time writing letters and giving out interviews.

He liked to tell how he "had captured a sucker," Bonfils, adding: "And I've still got him." He once said to Ashton Stevens, veteran drama critic of Chicago:

"Every man is a genius, if you can only place him. But we don't always know genius, not even when it is laid out before us. Bon' and I don't. Once we paid Charlie Chaplin thirty dollars a week [at the Empress Theater, in a sketch: "A Night in a London Music Hall"] and thought he was worth it. Later we paid him a hundred thousand [with Broncho Billy Anderson] and weren't sure. We didn't know Charlie was a genius till somebody offered him a million. That's the first time I ever laughed at Chaplin—when they offered him a million."

Jack Dempsey gave Tammen a valuable police dog, "Mox." Tammen was overfond of this animal. It lay stretched at his feet whenever he was in his office. It rode in Tammen's automobile. He would permit no one to touch Mox, nor, if he could prevent it, to speak to the dog.

"He's a one-man dog and no mistake," Tammen would say.

Unknown to Tammen, Mox secretly was everyone's friend. The moment Tammen would go from the room, Mox would visit in democratic fashion, suddenly showing a frisky and affable nature.

One day Mox was riding proudly beside Tammen in the publicist's automobile. Charlie Jacobs, a theatrical man, said "hello" to Tammen, then called to "Mox." The dog astounded his master by leaping from the car and running to Jacobs, who said:

"A one-man dog, eh?"

"*That's* something I can't understand," Tammen said, almost sulkily.

The Sells-Floto Circus was sold to a syndicate in

November, 1920. Jerry Mugivan headed the group of buyers, which also owned the Hagenbeck-Wallace Circus, the John Robinson Shows, the Howes-London and the Yankee-Robinson enterprises.

Tammen's health really began to decline, once his beloved circus had left his care, although he was not dangerously ill until 1924.

In May of that year he began growing tired and weak. He had blood transfusions. He wrote to a friend:

"I have had three blood transfusions, among other things. The first one was a pint of blood given by a red-headed American girl—the second from a black-eyed, black-haired Irish girl, but who supplied the blood for the third transfusion, I don't know. I will probably have another transfusion about next Saturday, May 31. Taking it all in all, I am on the road to recovery and am down at the office for a little while, twice each day, but don't do much work, go out riding, eat and sleep well, and think I am going to be better than I have been for a long time."

But Tammen's improvement was of short duration. There was an operation performed at Johns Hopkins Hospital in Baltimore. Ogden Armour took Tammen thither in his private car. He stood with Mrs. Tammen beside the bed as Tammen roused from the anesthesia.

"Well, doctor," said Tammen, "what is it, and how long have I got to go?"

"It's as I expected," the doctor said.

"Name it."

"*Carcinoma.* Cancer."

"That's all right, too. How long have I got to go?"

"One cannot say for certain. A year at most. I would think six months the probable limit."

"That's great! Think what a man can do with six months of life, knowing almost for sure that he has to go. Well, let's start off by giving a party to these fine nurses. They shall have champagne, and I'll sip just a little, whether it is good for me or not."

When Armour had taken his friend to Denver in his private car, Tammen gave orders that no one should know that he was dangerously ill. "I'm just indisposed," he would say to callers. Trained nurses were in attendance, and Mrs. Tammen, too, stayed with him constantly.

On the night of July 18, Tammen said to his wife:

"Agnes, send the nurse away. I want to talk to you."

After the nurse had gone, Tammen asked his wife:

"Would you mind talking to me all night? Or are you too tired?"

"I'm not tired, but you mustn't drain your energy by talking all night."

"Honey," he said, "there is no more energy to drain. The fact is I'm dying . . ."

"Please don't say that, Harry."

"Let's be calm. What a wonderful life we have had together! And how wonderful you have been. Really, it isn't so hard to go away, knowing what fun we've had. Please talk all night with me."

Mrs. Tammen was brave. "Very well, Harry. . . . We'll talk . . . all night."

He was silent for a while. Then he said: "Now look here. Let's get business over with first. My will is made. But there's another thing. First, you know I wouldn't fool you?"

"Of course not."

"I wouldn't give you the wrong steer on anything."

"No."

"All right. A lot of people hate Fred (Bonfils). They think he is crooked. They think he would trim Santa Claus while he was busy putting presents on a Christmas tree. Now get this straight, Agnes. I *trust Fred Bonfils.* He is not a crook. A hard dealer, and close-fisted. But we've never had a scratch of a pen between us in all these thirty years. Not one scratch saying that I own this or he owns that. We simply were partners. I trust him implicitly. Do you get that clearly?"

"I do."

"Very well. Our agreement is that there is one share of stock in the *Post* held in escrow. That share is to go to the survivor, principally because we don't want to have any women, however good and fine you all are, to be in a position to hamper the judgment of the partner that survives. Now it looks certain that Fred is to be the surviving partner. He will get the extra and controlling share. I am willing that this is to be so, for you never will suffer at Fred's hands, because you will be Harry Tammen's widow. Trust Fred in everything. Never question him. I am dying. And I would not tell the woman I love to trust a man who would be unfaithful to such a trust."

"I believe everything you say."

"Now let's talk about ourselves, Agnes. It has been such a happy road to travel with you. The two greatest things in life are thoughtfulness and gratitude. You have been thoughtful, and I have been grateful. That's why we're so happy."

He then talked on through the night, recalling the time they first had met, of the rides near the old cemetery to see the sun set—ground on which their home now stood.

At five o'clock in the morning, Tammen said: "Be sure, Agnes, to see that the ones I have named in my will, the fellows and girls down at the office, get their money as soon as possible." Then he clutched at his side and said: "I have a pain."

After the doctor and nurses had worked over him for some minutes, Tammen said to his wife: "Agnes, let me hold your hand a little while."

She put her hand in his. Then he asked her: "What is this thing that is coming to me?"

And then he died.

34
The Lone Hetman

BONFILS NOW was sixty-three years old, an age when most generals either have retired or are expecting to do so. Yet this hetman worked and schemed even more energetically than ever before. He and Tammen had drawn a nominal salary of one thousand dollars a week each. Bonfils now "raised" himself, first to fifteen hundred and then two thousand dollars a week.

He told friends that he was worth sixty millions of dollars. Tammen's will showed an estate "in excess of two million dollars." He probably had been worth ten. An offer for the purchase of the *Post,* in which Tammen had held a full partnership, was fifteen million two hundred thousand dollars, possibly the largest bid ever made for a newspaper property outside the cities of New York and Chicago.

An observer, conversant with the inside facts regarding these two men, sees Bonfils at this stage of his career as trying to emerge tardily from a long and scowling isolation. True, he had on occasion shown himself to the populace, loudly proclaiming himself as humanity's benefactor; but by temperament and through circumstance he was a lonely man. Even in his home, which he dominated with a patriarchal discipline, stern and unyielding, he did not bare his secrets. Only to Tammen had he revealed his soul. After Tammen's death, Bonfils at first seemed a defiant, yet frightened, man. He had won notoriety and fortune, but didn't appear to know just what to do with either. There was an air of frustration about him as he set about the job of increasing the tumult that he and Tammen had inspired. He inaugurated bigger and better contests and sponsored gaudier and greater stunts than ever before.

It was as though he were endeavoring to add Tammen's characteristics to his own. He tried his hand at composing headlines in the Tammen manner. For example, when the trans-Pacific fliers had been heard from, after many hours of doubt as to their fate, Bonfils wrote:

"BLESS GOD—THEY'RE SAFE!"

He engaged in philosophical letter-writing. In one missive he advised a former *Post* reporter:

"Now, my boy, you know how pleased I am to see you progress and develop and grow. Be careful of yourself; be careful of your habits. Remember that in renunciation there is the finest growth and development—in gratification, paralysis and death of the best things of life."

During Tammen's lifetime, there had been an unwritten rule that pictures of *Post* owners must never appear in that newspaper. A new editor one day ran a large portrait of Bonfils. The staff trembled, fearing that Bonfils would be enraged at this departure from policy, but no reprimand came from the Red Room. From that time on the Bonfils face and form—still very handsome—graced the *Post* pictorial columns, and with astounding frequency.

Bonfils seemed thirsty for honors. He campaigned in his newspaper for the bestowal of the state's No. 1 automobile license plate on "Colorado's worthiest citizen." He held that some favored politician or person of wealth usually got the No. 1 license, whereas it should be given annually to that citizen acclaimed as the one who had proven during the year that he or she had most benefitted the community.

"When you see Automobile License No. 1," he wrote, "you will know that Colorado's foremost citizen is faring forth."

Miss Emily Griffith of the Opportunity School won the automobile-license honor the first year. Fred G. Bonfils won it the second!

He beamed when he was made a member of the Burnt Thigh Tribe of Sioux Indians. This ceremony was performed in the Red Room. Chief Iron Shell inducted Bon' into membership as "Big Chief White Eagle," a blood brother, with an honorary post-office address at Rosebud Reservation, Parmalee, South Dakota.

The officiating sachem displayed an historic war bonnet, said to have been worn by the great warrior, Chief Telequah. Cameras clicked as Iron Shell pressed the headdress of eagle feathers and heron plumes on Bon's appreciative brow. Chief Little Thunder, tribal spokesman, orated and grunted profoundly in praise of the new blood brother, and then the Redmen adjourned to the street in front of the *Post* where they kicked up their moccasins in a noble shindig, out of respect for Bonfils and the midday sun.

That Bonfils felt a proprietary interest in the sun was indicated by his "supervision" of a solar eclipse. For days the *Post* screamed with astronomical data, and there was a grand invitation extended to every man, woman and child of Denver to come to the *Post,* there to witness the antics of the heavenly bodies. All persons, however, were advised to bring *their own* smoked glasses.

The rival *Rocky Mountain News* annoyed Bonfils no little when it stole a march on his eclipse. While a crowd stood in front of the *Post,* and during the dark seconds, an electric sign suddenly gleamed from a hotel roof down the street. That sign had been installed secretly by the *News.* It read:

"The Sun Is Our Only Rival—and He Is Off the Job."

Although the *Post* spent more and more money to obtain the products of syndicates and wire services, Bonfils never once let down in effecting small enonomies. He announced one day that all middle names of citizens either should be omitted entirely from news stories and feature columns, or, at most, the initials only should be used. He said he had had an exhaustive survey made, and had found the setting up and printing of middle names cost the *Post* more than a thousand dollars a year.

He took pleasure in applying this rule to such foes as Charles MacAllister Wilcox, a leading merchant and society man. To Mr. Wilcox's chagrin, he found himself referred to frequently as "Charlie Wilcox."

One day in 1928 Bonfils astounded his city and state with what he believed to be his greatest deed. He announced that he was putting his entire fortune into "The Frederick G. Bonfils Foundation for the Betterment of Mankind." The news stories published in the *Post* dealt with this act in thundering syllables, but try as one might, nobody could determine exactly what it meant.

One story said:

"Every dollar of Mr. Bonfils' fortune goes into the Frederick G. Bonfils Foundation for the Betterment of Mankind. But more and greater and bigger than all his millions, is the character, the soul of the man, which will be behind the gift, pervading its spirit and directing its work for ages to come. For into this, Mr. Bonfils is putting every atom of wisdom, culture, kindness, love and intelligence that he has gleaned in his long and wonderfully vivid career. Into it goes the very heart and brain and soul of the man who is and has been better loved and more misunderstood than possibly any public man in the history of this great Western country."

The Foundation terms—when they were explicit enough for the bewildered layman to grasp—spoke of prizes for scientific feats. A permanent cure for cancer and for tuberculosis would bring a quarter of a million dollars each to the scientists who achieved such boons. An influenza cure would be rewarded with twenty-five thousand dollars. The other awards were hidden behind a screen of mystery, with rumors that substantial premiums would go for a non-stop flight around the world—the start and finish to be in front of the *Post* building—and for the establishment of communication with the planet Mars.

Bonfils himself never grew publicly articulate as to the exact terms of his Foundation, beyond saying:

"For the betterment of mankind—I wish those five words to be the ruling and governing spirit of the Foundation. The idea is as broad as the world, as the earth, and yet I have in my mind a special love for our people of this great inter-mountain territory, particularly the people of the states of Colorado and Wyoming. I want the Foundation to be so administered that it will result in better homes, better schools, better and more intelligent people, healthier and happier conditions of life, greater morality, and more widespread regard for the love of God and the gospel of Christ. Whatever may achieve this end will have the support of the Foundation. I do not favor hobbies and 'isms' and strange, unproven cults. Science and intellect and morality and true wisdom are the agencies that will help mankind, and these are the agencies that I have most in mind."

The mystified citizens wondered much about this vague Foundation. Was it a subterfuge designed to simplify Bonfils' income-tax worries? Or did he intend to hold in death the fortune which he had guarded so ferociously in life?

And what of his heirs?

The Bonfils will seemed to ignore the Colorado law which gives a widow a dower interest of one half an estate. The testament provided that the widow, Mrs. Belle Bonfils, was to receive a life-income of fifty thousand dollars a year. At her death, this income would pass to the unmarried daughter, Helen. The latter would be given an income for life of twenty-five thousand dollars, provided she remained single. If she married, the grant would be cut in half.

There was a provision which certain observers regarded as an indication that Bonfils' other daughter, May, had married against the publicist's wishes. To her he left a life income of half that given to Helen Bonfils. If, however, she left her husband, Mr. Berryman, May would be given an income equal to that of Helen. This provision, it is said, may afford grounds for an attack on the will, as a matter contrary to public policy.

A petition for the court to change the will is now believed imminent. Inasmuch as members of the family are the executors of the estate, and therefore the only parties to enter a court contest on the petition, it appears likely that the Bonfils Foundation will be materially affected.

Bonfils, in addition to making his will, bought a room in the new Fairmount Cemetery Mausoleum. It was one of the two most expensive vaults, costing thirty thousand dollars. Visitors entering the granite pile would see it first, to the left of a stained-glass window depicting the Mount of the Holy Cross.

The mausoleum is on a hill, looking toward the western range, and also high above many of the graves of Bonfils' ancient enemies, including that of Senator Thomas M. Patterson. Tammen's grave, with a monument of three Grecian columns upholding a frieze, is at the foot of the hill.

Aside from some stirring libel suits and political fights, the outstanding episode of Bonfils' career during the year, 1929, was his offer to Calvin Coolidge. Bonfils wrote to the White House as follows:

"The *Denver Post,* the largest paper in the United States between the Missouri River and the Pacific coast, wants you as its editor-in-chief, and as the press states that you are considering newspaper work, the *Post* will pay you a salary of seventy-five thousand dollars a year to start. Your policies and those of the *Denver Post* are so entirely in harmony with each other that you would feel at home on this paper. This offer is made in the utmost good faith, and a guarantee indorsed by every national bank in Denver will assure you of the earnestness of this offer. Denver is the most beautiful and delightful city in the world in which to live, and we want you to seriously consider this offer.

(Signed) "Frederick G. Bonfils,
Owner and Publisher of
the *Denver Post.*"

Mr. Coolidge declined the honor.

35

The Donnybrook Fair

THIRTEEN YEARS of editorial in-fighting had set John C. Shaffer of Chicago on his journalistic heels. He wobbled out of Denver with a pair of editorial tin ears and chronic pains in his financial solar plexus. All his prayers, hymns and timid decencies had failed to stop the *Post's* bull-like rushes. He got up from the canvas, shook his head to clear it, rubbed the resin from his eyes, and then sold his *Rocky Mountain News* and *Denver Times*—oh, so gladly—to the Scripps-Howard chain.

The city once again rose to greet a new champion, hoping this one would not spurt vanilla when wounded. Roy Howard, chairman of the Scripps-Howard board, was a young and resourceful newspaper man, and a fighter. He advanced, in November, 1926, with pepper in his sling-shot to meet Goliath Bonfils.

The purchase price was said to have been "around one million dollars," The Scripps-Howard chain already owned the *Denver Express*, an evening paper of much courage but little prosperity or prestige.

Mr. Howard's opening speech, delivered before the Denver Chamber of Commerce, was as sweet an uppercut as anyone had cheered since the days of Bob Fitzsimmons. He stood up and, without polite preamble, said:

"We are coming to Denver neither with a tin cup nor a lead pipe. We will live with and in this community, and not on or off it. We are nobody's big brother, wayward sister or poor relation.

"We come here simply as news merchants. We are here to sell advertising, and sell it at a rate profitable to those who buy it. But first we must produce a newspaper with news appeal that will result in a circulation that will make that advertising effective. We will run no lottery.

"We have sense of humor enough to know that a challenger never looks as good as a champion—before the fight."

It was the signal for a battle—an expensive one. The *Times* was consolidated with the *Express,* and became the *Denver Evening News* on November 23. The following statement by Scripps-Howard executives was carried in the first consolidated edition:

"In a newspaper sense, Denver is unique. For years the city has been marked journalistically by one large and three comparatively small publications. The trend has been more and more toward a monopoly by the largest, the *Denver Post,* published by Fred G. Bonfils.

"That trend has threatened Denver for some time with a newspaper dictatorship. Such a situation is attributable to a number of causes.

"First, the very agile and adroit publishing ability of F. G. Bonfils, who by one means or another has been able to force his newspaper ahead, against a divided field.

"Second, an over-crowded competitive condition that has made possible the growth of one at the expense of the others. That condition could be corrected by only one process—a merger such as that which has now been brought about.

"We believe that a dictatorship of Denver's newspaper field by the *Denver Post* would be nothing less than a blight, and we believe, furthermore, that because of recent developments the time is ripe for challenging that dictatorship. Hence the merger and the pledge that the resources of the Scripps-Howard organization are behind this move to correct what we consider a sinister journalistic situation."

The *Post* met the Howard lead with its long-threatening ace—a morning *Post.* That had been the menacing card which Tammen had played in procuring from Shaffer the *Associated Press* franchise for his Sunday paper.

Howard had been competing for some time with the *Associated Press,* building up the Scripps-Howard *United*

Press, using the latter service in Scripps-Howard papers and discarding the *A.P.* He now announced he was ready to throw the *A.P.* out his Denver windows.

The *Post* stood by, hoping to catch that valuable asset.

The *Morning Post* arrived in Champa Street with all the excitement attendant on the birth of a baby brother. The first edition was published the morning of January 3, 1927. It was a bouncing baby, indeed, *containing fifty-two pages!* Howard was not the sort to nod over his journalistic porridge. Word had come to him concerning the *Post's* elephantine intentions.

On that same day, Howard's *News* appeared with *sixty-eight* pages from stem to stern. It looked as though the paper-pulp forests were in for a terrific beating.

There were raids by both offices, with the rival papers bidding for employees. Salaries went up. The *Post* began importing printers and editorial men. It was called "the newspaper battle of the century."

Floyd Clymer, veteran motor-race driver, and the man who gave to the nation its first automobile spotlight, broke all speed records between Cheyenne and Denver, driving the one hundred and ten miles in one hundred and nine minutes, carrying exclusive pictures of the Stanford-Alabama football game for the *Sunday News*.

That same day, the *Evening News* displayed enterprise by sponsoring a parade of girls, whom the caption-writers called pretty, to celebrate the start of "Joy," a serial story. The paper gave away thirty thousand copies of its Peach Edition the day this serial commenced.

In January next, the *News* scored a victory that may seem tame to the uninitiated, but which was a right-hand chop to Bonfils' heart. That paper took the lead in classified advertising—the want ads—for the first time in twenty years. The circulation of the *Rocky Mountain News* was 64,987, as against 54,666 the previous year, and that of the *Evening News* was 39,041, as against 23,-524 for the old *Times*. During the month of January, the morning and evening *News* printed 63,583 classified ads against 50,844 for the morning and evening *Post*.

The invaders presented localized serial stories, together

with elaborate and costly exploitation. The *Post* countered with similar enterprise in launching the story, "Chickie."

The *News* tried to sink the *Post's* "Chickie" by giving away ten thousand copies of the book with Sunday want ads. The *News* met the *Post's* stunts with showy affairs, including limerick contests, bowling tourneys, cooking schools, and opened a free travel-service bureau with Fay Lamphier, national beauty-contest winner, as hostess. The furniture and decorations for the bureau cost forty thousand dollars and included a mural panel of the Scripps-Howard lighthouse, for which Allen True, Denver artist, received fifteen hundred dollars.

And now started the great "gasoline war." On February 4, 1927, the Sunday *News* came out with a one hundred and twelve page paper, containing twenty-six pages of classified ads, against fourteen pages of such want ads in the ninety-six page *Post*. The *Post*, through a dealer, offered two gallons of gasoline as a premium for each want ad inserted in its forthcoming Sunday paper. The arrangement was that the *Post* pay *nothing* for the gasoline; the dealer was to be honored for his petrol by the *Post's* fanfare, which would lure new customers to his pumps.

The *News* boosted the ante to three gallons for each want ad placed in its Sunday edition. The *Post* saw the bet and raised it to four. The *News* then tilted its offer to five gallons per twenty-cent ad. The dealer, who first had rallied to the *Post* cause, began to squirm. After all, his coöperation had not anticipated a Niagara of gasoline in return for publicity. After some wrangling, the *Post* stuck to its offer of four gallons, agreeing to *pay* the dealer for two gallons per person.

The mob of motorists became so huge at the *News* that the Howard troops moved desks to the sidewalk, the full length of the building. It looked like a Parisian street café with ink bottles instead of decanters on the commercial tables. A squad of policemen was required to keep want-ad customers in line. The clerks worked fourteen hours Friday and twelve on Saturday, without leaving their desks and counters. Food was brought to them as they wrote, so they could dine and work at the same time.

So great was the burden of composition that the *News*

had to farm out fifty-nine columns of classified advertising to job printers. The defunct *Express* plant was brought, Lazarus like, from the tomb and operated during the gasoline war. The circulation of the Sunday *News* went beyond 76,000, all that the presses could handle of the one-hundred-and-twelve-page paper.

This gasoline war continued for one week, then both sides retired to lick their wounds and to contemplate the inroads on their cash boxes. The *Morning Post* was eating into the profits of the *Evening Post*. But Bonfils said he would rule or ruin. The *News* suspended its ornate and expensive service bureau four months after it had opened. Circulation of the *News* (Sunday) reached 110,000 but advertising lagged. After the Scripps-Howard organization had spent *three million dollars* on the invasion, Roy Howard negotiated peace terms with Bonfils. A truce was declared on November 5, 1928. Bon's losses were believed to have been at least two millions.

The city was wearying of the spectacle of two organizations, with their four papers, battling for leadership. Scripps-Howard was tired of losing money at the rate of more than two thousand dollars a day. Two years of this desperate struggle had passed, and the only ones who had benefited were gas-hungry motorists and winners of contests. The terms of the peace were far from unfavorable to Bonfils.

Bonfils (joyously) suspended his till-draining *Morning Post,* the biggest elephant he ever had owned. The Scripps-Howard people scrapped their *Evening News.* Denver was an "afternoon paper town," both by training and by preference, and the balance here was in favor of the *Post*. Denver now had but two newspapers. A rise in circulation price to three cents for the dailies and ten cents for the Sunday papers was announced, together with an agreement covering hours of publication, the elimination of bonuses for street hustlers, and barring all circulation contests involving premiums, prizes, insurance or coupons.

In the peace-pipe puffing exercises, Howard said:

"This development in Denver represents nothing more or less than a successful attempt to bring common sense and a rule of reason to bear on a newspaper situation which for two years past has been singularly devoid of either. . . . Scripps-Howard has no apology to offer for the *Denver Evening News.* For two years that paper has been one of the very best and one of the most complete produced anywhere in our organization. But it would be poor sportsmanship for us to fail to admit that despite the loyalty of the *Evening News* readers, F. G. Bonfils more successfully met the evening newspaper tastes of this community as a whole than we did.

"For a generation Denver has been accustomed to a jazzy, colorful type of journalism featuring big type, colored paper and all that goes with newspapers of that type. In the *Denver Post* it has had all of this, plus a most thorough news and feature coverage.

"The *Evening News* was a distinct departure from this style. It was more subdued in both text and typography, in conformity with Scripps-Howard style. Denver preferred the old order. We preferred not to compromise in the matter of a technique which is serving us well in twenty-four other cities."

In September of 1929, Eugene Greenhut, capitalist, was organizing a newspaper merger. The finances were provided by Hambleton & Company of Baltimore. M. Koenigsberg, for many years an outstanding executive for William Randolph Hearst, and formerly president of Hearst's *King Feature Syndicate, Universal Service* and *International News Service,* was to be the head of the Greenhut group.

Koenigsberg, partly because he was one of the finest organizers in the newspaper world, and partly on account of his acquaintance with Bonfils since 1910, was chosen to make an offer for the *Post's* purchase by the banking group.

There were twenty long-distance telephone calls and Koenigsberg made several trips to Denver to consult Bonfils. The bankers were deferring other acquisitions, believing the *Denver Post* would serve as a good bell-wether for a chain of newspapers in the Central, Middle-Western and Western states.

Bonfils finally visited New York. There was a meeting in Koenigsberg's office, at which Greenhut formally offered Bonfils fifteen million two hundred thousand dollars for the *Post*. Bonfils temporized for a day or two, and finally left for Denver, ostensibly to get the consent of the minority stockholders (who were Mrs. Tammen and the trustees of the Children's Hospital). The banking group felt that this was a "stall," as Mrs. Tammen would have followed Bonfils' lead in the deal without question.

A few weeks afterward, the stock market crashed. The debacle, naturally, hog-tied all mergers in process of formation, including the Greenhut proposition. Nevertheless, as the time came for the picking up of the financially dead and wounded, Greenhut found a banker willing to pay twelve million seven hundred and fifty thousand dollars for the *Post*. Koenigsberg, also, in the meantime had enlisted the interest of Eastman, Dillon & Company in the deal. In January of 1930, Koenigsberg telephoned Bonfils for an appointment.

The chief obstacle was Bonfils' reluctance to permit a certified accountant *to audit the books* of his company. A definite offer for the newspaper could not be made unless it were based on authentic audit of the books. Bonfils submitted some figures, scrawling these on a piece of note paper, but would not guarantee them. Throughout these negotiations, Bonfils insisted that he would *denounce* anyone who intimated that the *Denver Post* was for sale.

In this connection, he said: "The financial standing of a newspaper must be safeguarded with all the care with which we surround a woman's reputation for chastity. Both are equally delicate."

Finally it was agreed that Koenigsberg should go to Denver, ostensibly as general manager of the *Post,* and, while there, he might bring in Eastman, Dillon & Company's expert, who would go through the books, apparently as Koenigsberg's assistant—his identity to be hidden from everyone, except Bonfils.

Frank Gannett, the publisher, meanwhile had been invited into the situation by Eastman, Dillon & Company. Herbert Cruickshank, general auditor of the Gannett

Newspapers, accompanied Koenigsberg to Denver. Finally, in the middle of May, Cruickshank completed his audit of the *Denver Post.* Melville Dickenson, new business manager of Eastman, Dillon & Company, went to Denver to conclude the negotiations.

Cruickshank and Koenigsberg had been living at the Park Lane Hotel. Dickenson joined them there. On May 23, 1930, Dickenson shook hands with Bonfils—in Koenigsberg's presence—on an oral agreement to purchase the *Denver Post* for fifteen million two hundred thousand dollars. Bonfils left the two men shortly thereafter, saying he would return within the hour, accompanied by his lawyer.

Fifteen minutes after he had left, Bonfils got Koenigsberg on the telephone. He said:

"You must understand, Koenigsberg, these fellows get nothing but the plant and the property. They can't have any of my newsprint. I wouldn't give them enough to wipe their hands on."

Mr. Koenigsberg didn't quite understand this attitude. He did not then know of the time when Bonfils and Tammen had worked the same trick on Mr. Dickey, during the sale of the *Kansas City Post.* But, as Bonfils continued speaking, his meaning became clear to Koenigsberg, a veteran in major-league newspaper negotiations.

Bonfils intended to withhold from the sale *all the paper's accumulated profits.* In other words, the surplus on hand would be declared out in dividends before the sale was concluded. This meant, in exact figures, that $1,634,976.46, as shown in Cruickshank's audit, would be withheld from the buyers, or, to put it another way—Bonfils was tacking this sum on the sales price!

Bonfils' greed caused him to overplay his hand. The deal fell through. More, he took occasion to place a sinister construction on Koenigsberg's commission in the deal. It was known to all parties that Koenigsberg was to receive three hundred thousand dollars for his work in effecting the sale. Bonfils had written the figures down on a piece of his favorite scrap paper. He later told Koenigsberg that he understood the commission to be two hundred thousand dollars.

Koenigsberg is a large and tremendously strong man. He had gained many a physical decision in his younger days, when he was an ace editor. And when Bonfils implied that Koenigsberg was not telling the truth, the former Hearst executive rose, and very firmly put his hands at Bonfils' throat. Bonfils turned white; then Koenigsberg thought it beneath him to throttle a man older than himself, no matter what the pretext.

He walked from the Bonfils office, never to speak to him again.

The following table may be skipped by those who do not care for statistics; the importance of it lies in the remarkable profits shown for this newspaper over a period of its ten best years. These closely guarded figures never before have been available to anyone other than Bonfils' circle and are the result of the Cruickshank audit of the *Post* Printing and Publishing Company.

Year	*Gross*	*Income Tax*	*Net*
1920	$1,207,110.97	$357,371.50	$ 849,739.47
1921	1,223,754.33	243,790.33	979,964.00
1922	1,474,067.77	171,717.63	1,302,350.14
1923	1,539,732.55	176,129.37	1,363,603.18
1924	1,418,617.36	171,958.87	1,246,658.49
1925	1,444,606.91	154,851.41	1,289,755.50
1926	1,354,019.65	164,480.17	1,189,539.48
1927	779,358.37*	61,299.27	1,045,758.86†
1928	981,716.47*	91,980.00	1,203,408.57†
1929	2,094,818.86	229,903.08	1,864,915.78

* Showing shrinkage of profits in 1927 and 1928, occasioned by expenses incurred in the operation of the *Denver Morning Post*, which was launched only to fight the Scripps-Howard group.

† These figures are attained by restoring as profit what the auditors described as "non-recurrent expenses"—the cost of publishing the *Morning Post*.

When Herbert Hoover was campaigning for reëlection as President, Bonfils visited Washington. On his return he went hook, line and sinker for Mr. Hoover. On the election of Mr. Roosevelt, Bonfils was very sad. One of the things that galled him, also, was the fact that he had to cough up some heavy income taxes, his arrears in that respect having been said to have reached almost a million dollars.

36

The Deluge

THERE WAS a lively meeting of the Jane Jefferson Club at the Brown Palace Hotel on the eve of the primary campaign in August of 1932. Walter Walker, retiring state chairman of the Democratic party, and Edwin C. Johnson, one of the Jeffersonian designees for Governor, threw a few fire balls at each other.

They suddenly dropped their personal row when the name of Fred G. Bonfils arose in connection with a keynote speech. Both men began to flay the *Post* and its proprietor. Chairman Walker said, among other things:

"Despite the boast that the *Post* is advertised by its enemies, we know that the vulture gnashes its teeth and shakes with rage when attacked.

"A resolution condemning the *Post* should have been passed by the Democratic assembly, and it would have been, save for those who truckle and fawn at the feet of the *Post*, in order to get their pictures in the news columns and that of the wife in the society sheet.

"The day will come when some persecuted man will treat that rattlesnake as a rattlesnake should be treated, and there will be general rejoicing. . . . Bonfils is a public enemy and has left the trail of a slimy serpent across Colorado for thirty years. . . . Frederick G. Bonfils is the only one of his kind in the world, and the mold was broken after him. . . . Colorado has stood too much from the contemptible dog of Champa Street."

When a reporter for the *Rocky Mountain News* asked for a stenographic transcript of this castigation, it was given with the remark that the *News* wouldn't dare print it. The *News* nevertheless published the Walker blast. A few days afterward Denver read that Bonfils had sued that newspaper for libel, claiming two hundred thousand dollars for damages.

Mr. Bonfils' counsel was the facile Philip Hornbein. Colonel Philip Van Cise, former district attorney and a leader of the Western bar, represented the *News*. As the *News*' ownership was nonresident, testimony was taken by deposition. There was much thumbing of law books and not a little squabbling in the presence of learned judges ere Colonel Van Cise gained the right to question Bonfils before a notary public.

The publicist appeared for thirty minutes, more or less, answered a few preliminary questions, then suddenly and flatly refused further replies. Colonel Van Cise's examination indicated clearly that the Bonfils past was to be resurrected in full detail, from corn-field lottery to Wyoming oil-money reward. Attorney Hornbein advised his client to retire from the hearing room. Bonfils, thoroughly enraged, strode from that chamber and to the street, his black derby jammed low on his large head, his tan overcoat flapping in the November breeze.

He had answered questions pertinent to his fifty-one per cent ownership in the *Post*, but when Colonel Van Cise had veered to the personal category, the Bonfils jaw muscles stood out and the Bonfils moustache quivered. He refused to tell if he were married or not, if he would be seventy-two years old, come December, or where he went to school and what marks he had received in various class rooms.

Attorney Hornbein objected to these questions, pleading irrelevancy, and presumably seeking a ruling on trivial inquiries that he might set a precedent for subsequent rulings on personal matters not quite so savory. Bonfils' counsel resisted a claim that the issue involved his client's entire life history.

A few hours after Bonfils had huffed out of the hearing room, Colonel Van Cise appeared in district court to file a petition asking that Bonfils be cited for contempt of court. The plea was granted, the citation made returnable the following Friday, at which time Bonfils was ordered to show cause why he shouldn't be punished for contempt.

Colonel Van Cise's petition declared that if Bonfils had answered questions put by *News'* counsel, the defendant intended to prove by Bonfils himself the authenticity of forty-one salient points in justification of the purported libel.

The petition was a narration of alleged sinister antics by the *Post's* czar. It began with bunco charges, then proceeded to describe alleged fist fights in the streets of Guthrie, during land-rush days, an extortion suit, town-site promotions of a flimflam quality, spoke of aliases and lottery projects, swindles and reputed short-weight dealings in the coal yards, plus an accusation that Bonfils had been implicated in attempting to bribe a jury in a profoundly revengeful move to convict Lawyer W. W. Anderson, the man who had shot him and his partner during the famous Packer, man-eater affair, and that Bonfils had provided money for a defense witness in that case to leave the country for several months.

The petition claimed that in 1904, in Alva, Oklahoma, Bonfils secretly had put through a fraudulent contract for the erection of a court house at a big profit to himself. The courts had charged fraud, according to the *News* petition, and had held the contract null and void. The alleged bribery of a commissioner to facilitate that contract was cited in the *News* petition. Nor, the *News* added, was that the only court-house transaction which had enriched Bonfils.

The defendants charged that between 1905 and 1909 Bonfils and his partner had received more than forty thousand dollars from the Rock Island Railroad for advertisements in the editorial and news columns of the *Post* in violation of newspaper ethics and contrary to postal rules and regulations.

Fraudulent stock transactions in connection with the mining excitement in Goldfield, Nevada, in 1905, were advertised in the *Post,* the petition said, with the consent and participation of the plaintiff, and with great profit to himself.

The maintenance of a newspaper "blacklist," attacks on merchants who refused to advertise in the *Post,* the browbeating of laundrymen who would not buy *Post* coal, spurious colonization schemes, the "hijacking" of Sinclair in the Teapot Dome matter, the claim that he had incorporated the Bonfils Foundation for the Benefit of Mankind, but had failed to give any sum to that Foundation for prizes which he had advertised, attacks on the establishing of a municipal airport, because the founders would not purchase lots in which he was interested—all these things and more were laid at Bonfils' door by the petition.

A rather fantastic paragraph of that document follows:

"That because of the many and varied transactions in which he has been involved since 1881, he is subject to violent nightmares, and fears that in one of them he may reveal some of the shady transactions of his past, and requires a constant companion when asleep, so that he may be instantly awakened when seized by such spasms."

Another paragraph spoke of Bon's capture of a "tame fish" as evidence of his sporting prowess, and added:

"That when the plaintiff takes parties fishing on the North Platte, he hogs the river, and insists on his boat being first, so that he can get the best fishing."

There were further legal duels concerning Bonfils' suit, involving not only his appearance in court to face the contempt charge, but in regard to his answering further questions before the notary. A judge ruled Bonfils must answer such questions in January of 1933.

He was a sorely troubled man. For the first time in his seventy-two years of struggling, fighting, yearning, brooding and working, he became slightly stooped. He was still handsome and dapper, but the lines deepened in his cheeks and brow. An unfamiliar look of defeat was on his face.

One of his intimates said: "If Harry Tammen had been alive, you can bet Fred never would have brought *this* suit."

37

Crucifix

AND NOW the brooding Corsican sat in the Red Room with his gods, Money and Power. They did not seem to give him the solace he needed. The burden was great—harsh questions, exhumed by the spade of the relentless Colonel Van Cise, a man not only learned in the law but a cross-examiner trained by years as a district attorney, rose to haunt him.

It had been Bonfils' habit to take his poodle, "Dixie"—successor to "Trixie"—for sunset walks in Cheesman Park, a pleasant common that adjoined his own garden. Now he did not choose to go for those walks, and his failure to wander in the park perhaps saved him from a kidnapers' plot.

Charles Boettcher, II, son of a wealthy family, had been abducted as he stood with his wife at their garage door. The kidnapers held him for sixty thousand dollars ransom. One of the leaders was quoted as having said to the victim:

"It's just your hard luck that *you're* here. We'd planned to snatch Bonfils while he was walking with his poodle, and hold him for a hundred thousand."

Bonfils, the veteran of so many wars of abuse, grew fidgety and was subject to fits of annoyance. Reports that the *Post's* circulation on a Sunday was 299,000, nearly two hundred thousand more than the combined paid circulation of all other Colorado papers, brought a momentary gleam to Bonfils' eyes. Then he lapsed again into a sullen mood. There were questions to be answered—at least forty-one galling questions.

He kept at his work with the dogged gameness that had

marked his long career. When his managing editor, W. C. Shepherd, put his head through the doorway and said: "Coolidge is dead," Bonfils stared incredulously.

He always had been a loud partisan of the silent New Englander. Had he not offered him the high post of editor of the *Post?* When he heard this news, it was as though he had lost his last friend on earth.

When two of his intimates visited him some weeks later, he emerged from his almost stoical mood to talk of life and death. He evaluated his gods—at last.

"Coolidge has gone," he said, looking about the room, as though the great Calvin himself had been accustomed to sit there, and had left a vacant chair. "And one wonders why we keep up the struggle, and what it all means."

He went on—a Napoleon looking at a sunset from St. Helena and communing with himself. "Life is a constant struggle. The battle is only worth while if the thing for which we fight concerns the common good, or if it gives us another chance to help those who are weak or dependent."

He waved toward a door leading to the reception room. "By stepping through the doorway called death, we could find peace and enter upon the larger experience."

He kept looking at that door. "It all seems so simple. Why should a man depend solely on material things, on wealth?"

A lieutenant entered to confer on a winter sports carnival at Genesee Park. Bonfils gave directions with something like his old-time peremptory voice. The carnival was to be free, for the people, with ski-jumping and toboggan rides. The lieutenant left the room and Bonfils watched the door through which he had gone.

"Just like going from one room to another," he repeated. "And that is what we call death. Rather simple."

One of his listeners asked: "Don't you think you might concentrate on religion?"

His eye flashed. "No!"

The woman who had put the question was a long-time associate. She persisted with her comment. "Many persons find something in religion to buoy them in a troublesome time. Many have found peace in the Catholic faith. Why don't you seek admission to the Catholic Church?"

Bonfils stared hard. Then he drew a deep breath and roared: "Never! Never!"

He prepared to go home. He thought he had a slight cold. He was sniffling. He squared his shoulders as he passed through the doorway as though to shake off evidence of physical weakness. He had a grand pride of body.

He did not come to the Red Room next day. It was said that he had influenza, with a threat of pneumonia, and an infection of the middle ear. A minor operation was performed to afford him relief. He was in great pain.

He grew worse the next day. He was suffering from toxic encephalitis, an acute inflammation of the brain. His family gathered beside him at his home. An oxygen tent was ordered by attending physicians. Still the world at large didn't know the seriousness of Bonfils' illness.

Then, with his attorneys explaining why their client could not appear in court to answer those hated questions, Bonfils had a glimpse of what lay beyond. He lately had roared down a suggestion that he turn to the Church for comfort. Now, with his eyes fixed on something other than life, he spoke weakly:

"Send for a priest."

The Reverend Father Hugh L. McMenamin, rector of the Cathedral of the Immaculate Conception, arrived at the home of the dying publisher. He baptized him in the Catholic faith.

Bonfils looked at the crucifix, and then lapsed into a long sleep, the morning of February 2, 1933—his last sleep.

THE END